CELLULAR PROTEOLYTIC SYSTEMS

MODERN CELL BIOLOGY

SERIES EDITOR

Joe B. Harford
RiboGene, Inc.
Hayward, California

ADVISORY BOARD

Richard G. W. Anderson
University of Texas
Health Sciences Center
Dallas, Texas

Bill R. Brinkley
Baylor College of Medicine
Houston, Texas

Don W. Cleveland
Johns Hopkins University
School of Medicine
Baltimore, Maryland

James R. Feramisco
University of California, San Diego
La Jolla, California

Elaine Fuchs
The University of Chicago
Chicago, Illinois

Volker Herzog
Universität Bonn
Bonn, Germany

Ann L. Hubbard
Johns Hopkins University
School of Medicine
Baltimore, Maryland

Richard D. Klausner
National Institutes of Health
Bethesda, Maryland

William J. Lennarz
State University of New York
at Stony Brook
Stony Brook, New York

Daniel Louvard
Institut Pasteur
Paris, France

Mary Lou Pardue
Massachusetts Institute of Technology
Cambridge, Massachusetts

Alan L. Schwartz
Washington University
School of Medicine
St. Louis, Missouri

RECENT VOLUMES PUBLISHED IN THE SERIES

Volume 11 — Antisense RNA and DNA
James A.H. Murray, Editor

Volume 12 — Genes in Mammalian Reproduction
Ralph B.L. Gwatkin, Editor

Volume 13 — Microtubules
Jeremy S. Hyams and Clive W. Lloyd, Editors

Volume 14 — Growth Factors and Signal Transduction in Development
Marit Nilsen-Hamilton, Editor

Volume 15 — Cellular Proteolytic Systems
Aaron J. Ciechanover and Alan L. Schwartz, Editors

CELLULAR PROTEOLYTIC SYSTEMS

Editors

Aaron J. Ciechanover
Technion
Israel Institute of Technology
Haifa, Israel

Alan L. Schwartz
School of Medicine
Washington University
St. Louis, Missouri

A JOHN WILEY & SONS, INC., PUBLICATION
NEW YORK • CHICHESTER • BRISBANE • TORONTO • SINGAPORE

**Address All Inquiries to the Publisher
Wiley-Liss, Inc., 605 Third Avenue, New York, NY 10158-0012**

Copyright © 1994 Wiley-Liss, Inc.

Printed in the United States of America

Under the conditions stated below the owner of copyright for this book hereby grants permission to users to make photocopy reproductions of any part or all of its contents for personal or internal organizational use, or for personal or internal use of specific clients. This consent is given on the condition that the copier pay the stated per-copy fee through the Copyright Clearance Center, Incorporated, 27 Congress Street, Salem, MA 01970, as listed in the most current issue of "Permissions to Photocopy" (Publisher's Fee List, distributed by CCC, Inc.), for copying beyond that permitted by sections 107 or 108 of the US Copyright Law. This consent does not extend to other kinds of copying, such as copying for general distribution, for advertising or promotional purposes, for creating new collective works, or for resale.

Library of Congress Cataloging-in-Publication Data

Cellular proteolytic systems / editors, Aaron J. Ciechanover, Alan L. Schwartz
    p.   cm. — (Modern cell biology ; v. 15)
   Includes index.
   ISBN 0-471-02189-X
   1. Proteins—Metabolism. 2. Proteolytic enzymes. 3. Lysosomes.
4. Ubiquitin. I. Ciechanover, Aaron J. II. Schwartz, Alan L.
III. Series.
   [DNLM: 1. Proteins—metabolism. 2. Lysosomes—physiology.
3. Ubiquitin—physiology. W1 MO124T v. 15 1994 / qU 55 C3929 1994]
QH573.M63 vol. 15
[QP551]
574.87 s—dc20
[574.19´245]
DNLM/DLC
for Library of Congress   94-19184

The text of this book is printed on acid-free paper.

Contents

Contributors vii

Preface
Aaron J. Ciechanover and Alan L. Schwartz ix

UBIQUITIN SYSTEM

The Ubiquitin-Mediated Proteolytic Pathway: Mechanisms of Recognition of the Proteolytic Substrate and Involvement in the Degradation of Native Cellular Proteins
Aaron J. Ciechanover and Alan L. Schwartz 3

Molecular Genetics of the Ubiquitin System
David Keith Gonda 23

LYSOSOMAL/VACUOLAR SYSTEMS

Selective Degradation of Cytosolic Proteins by Lysosomes
J. Fred Dice and Stanley R. Terlecky 55

Autophagy: Its Mechanism and Regulation
Glenn E. Mortimore and Motoni Kadowaki 65

Hepatic Endosomes Are a Major Physiological Locus of Insulin and Glucagon Degradation In Vivo
François Authier, Barry I. Posner, and John J.M. Bergeron 89

Proteolysis in the Yeast Vacuole
Elizabeth W. Jones and Deborah G. Murdock 115

PROTEOLYSIS IN THE ENDOPLASMIC RETICULUM

Degradation of Proteins Retained in the Endoplasmic Reticulum
Juan S. Bonifacino and Richard D. Klausner 137

PROTEOLYSIS RELATED TO ANTIGEN PRESENTATION

Protein Catabolism and Antigen Processing
Clifford V. Harding 163

VIRAL-RELATED/HIV PROTEASE

Viral Proteases: Structure and Function
Tudor I. Oprea, Chris L. Waller, and Garland R. Marshall 183

Index 223

Contributors

François Authier, Department of Anatomy and Cell Biology, McGill University, Montreal, Quebec H3A 2B2, Canada [89]

John J.M. Bergeron, Department of Anatomy and Cell Biology, McGill University, Montreal, Quebec H3A 2B2, Canada [89]

Juan S. Bonifacino, Cell Biology and Metabolism Branch, NICHHD, National Institutes of Health, Bethesda, MD 20892 [137]

Aaron J. Ciechanover, Department of Biochemistry, Faculty of Medicine, Technion-Israel Institute of Technology, Haifa 31096, Israel [3]

J. Fred Dice, Department of Physiology, Tufts University School of Medicine, Boston, MA 02111 [55]

David Keith Gonda, Department of Molecular Biophysics and Biochemistry, Yale University School of Medicine, New Haven, CT 06520 [23]

Clifford V. Harding, Institute of Pathology, Case Western Reserve University, Cleveland, OH 44106 [163]

Elizabeth W. Jones, Department of Biological Sciences, Carnegie Mellon University, Pittsburgh, PA 15213 [115]

Motoni Kadowaki, Department of Applied Biochemistry, Niigata University, Niigata, Japan [65]

Richard D. Klausner, Cell Biology and Metabolism Branch, NICHHD, National Institutes of Health, Bethesda, MD 20892 [137]

Garland R. Marshall, Department of Molecular Biology and Pharmacology, Washington University School of Medicine, St. Louis, MO 63110 [183]

Glenn E. Mortimore, Department of Cellular and Molecular Physiology, The Pennsylvania State University, Hershey, PA 17033 [65]

Deborah G. Murdock, Department of Biological Sciences, Carnegie Mellon University, Pittsburgh, PA 15213 [115]

Tudor I. Oprea, Center for Molecular Design, Washington University, St. Louis, MO 63130 [183]

Barry I. Posner, Department of Medicine, McGill University, Montreal, Quebec H3A 2B2, Canada [89]

Alan L. Schwartz, Departments of Pediatrics, Molecular Biology and Pharmacology, Washington University School of Medicine, St. Louis, MO 63110 [3]

Stanley R. Terlecky, Department of Biology, University of California at San Diego, La Jolla, CA 92093 [55]

Chris L. Waller, Environmental Toxicologic Division, United States Environmental Protection Agency, Research Triangle Park, NC 27711 [183]

The number in brackets is the opening page number of the contributor's article.

Preface

Cellular proteins exist in a dynamic state. Their steady-state levels are maintained by a delicate balance of synthesis and degradation. Investigations within cellular biochemistry during the past several decades have provided a refined understanding of the nature of protein synthesis, its constituents, enzymatic reactions, and regulatory mechanisms. Indeed, in the current textbooks of cellular biology, the subject of protein synthesis occupies substantial space. Not so for the subject of protein degradation. In fact, the biology of protein degradation is only recently coming to light. Little more than a decade ago, the lysosomes, with their substantial complement of broad-specificity proteases, were considered the locus of cellular protein degradation. The past decade has seen enormous progress in understanding the complexities of how a cell is able to distinguish between two of its constituent proteins: to allow one to remain undegraded for months while the other is degraded within minutes.

At present, there are two predominant cellular systems for the degradation of cellular proteins—the vacuolar pathway (including lysosomes, endosomes, endoplasmic reticulum, and so forth) and the cytoplasmic ubiquitin-mediated pathway. Nearly 100 years ago, Metchnikoff and his followers began studies of the uptake and digestion of particles in acidified, enzyme-containing intracellular structures. However, it was not until the 1950s that deDuve and colleagues were able to fractionate cells and identify the "lysosome" as the organelle within which the wide array of proteolytic enzymes function. More recently, the biology of endocytosis has broadened the view that intracytoplasmic degradative vacuoles include early and late endosomes as well as, recently, elements of the endoplasmic reticulum.

Unfortunately, the existence of the vacuolar proteolytic systems did not provide a suitable explanation for the specificity of degradation of cytoplasmic proteins. In the late 1970s, Hershko and Ciechanover observed that abnormal cytoplasmic proteins were degraded at neutral pH in a process that required ATP. Additional studies revealed that multiple factors were required to support this property. The first such factor isolated was the 76aa polypeptide ubiquitin.

At present, there is sufficient understanding of the overall biology of the vacuolar systems (lysosomes, endosomes, endoplasmic reticulum and so forth) as well as the cytoplasmic ubiquitin-mediated pathway to permit a concise overview of the field of cellular proteolytic systems. The present volume attempts to integrate several fundamental areas—basic cellular biology of protein turnover, the molecular machinery involved, the model systems and strategies upon which our current understanding is based, and several major pathobiological consequences of ordered and disordered protein degradation.

The two chapters on the ubiquitin system review the organization of the system, including the identification of its components and the current understanding of the mechanisms responsible for the recognition of physiological cellular proteins. Ciechanover and Schwartz focus on the cellular physiology in higher eukaryotes, while Gonda defines the components of the system and their functions using molecular approaches in yeast.

The lysosomal/vacuolar systems are addressed in four complementary chapters. The first by Dice and Terlecky examines the selective degradation of cytoplasmic proteins by lysosomes and provides one bridge to the cytoplasmic ubiquitin system. Mortimore and Kadowaki dissect the cellular process of regulated vacuolar degradation, autophagy, its mechanisms and regulation. The proteolytic systems within endosomes (and lysosomes) are the focus of the chapter by Authier, Posner, and Bergeron with much focus on the liver. All of these vacuolar systems of higher eukaryotes have complementary components in yeast, the subject of Jones and Murdock.

Bonifacino and Klausner address the recently identified system of cellular proteolysis within the endoplasmic reticulum. Finally, two critically important general pathophysiological systems of cellular proteolysis are the focus of the last two chapters. Harding dissects protein catabolism and antigen presentation, which encompasses both the vacuolar and cytoplasmic systems discussed above. Oprea, Waller, and Marshall focus on the viral-related proteases, especially the HIV protease, in terms of biology, pharmacology, and mechanisms of action.

We have tried to emphasize basic concepts with generalized features. Each section contains both a broad overview as well as detailed experimental systems and results. We have tried to select themes that transcend specific proteins and unique control mechanisms or are restricted in their significance. We are now emerging from a decade of initial exploration in the area of cellular systems of protein degradation. The coming years will no doubt reveal the details of mechanisms and regulation.

Aaron J. Ciechanover
Alan L. Schwartz

UBIQUITIN SYSTEM

The Ubiquitin-Mediated Proteolytic Pathway: Mechanisms of Recognition of the Proteolytic Substrate and Involvement in the Degradation of Native Cellular Proteins

Aaron J. Ciechanover and Alan L. Schwartz

OVERVIEW

Ubiquitin modification of a variety of protein targets within the cell plays important roles in many cellular processes. Among these are regulation of gene expression, regulation of cell cycle and division, involvement in the cellular stress response, modification of cell surface receptors, DNA repair, import of proteins into mitochondria, uptake of precursors into neurons, and biogenesis of mitochondria, ribosomes, and peroxisomes. The best studied modification occurs in the ubiquitin-mediated proteolytic pathway. Degradation of a protein via the ubiquitin system involves two discrete steps. Initially, multiple ubiquitin molecules are covalently linked in an ATP-dependent mode to the protein substrate. The targeted protein is then degraded by a specific and energy-dependent high molecular mass protease into free amino acids, and free and reutilizable ubiquitin is released. In addition, stable mono-ubiquitin adducts are also found in the cell, for example, those involving nucleosomal histones. Despite the considerable progress that has been made in elucidating the mode of action and roles of the ubiquitin system, many problems remain unsolved. For example, little is known of the signals that target proteins for degradation. While mechanistic aspects of recognition via the N-terminal residue have been studied thoroughly, it is clear that the vast majority of cellular proteins are targeted by other signals. The identity of the native cellular substrates of the system is another important, yet unresolved, problem: only a few proteins have been recognized thus far as substrates of the system in vivo. The scope of this review is to discuss the mechanisms involved in ubiquitin activation, selection of substrates for conjugation, and degradation of ubiquitin-conjugated proteins in the cell-free system. In addition, we summarize what is currently known of the physiological roles of ubiquitin-mediated proteolysis in vivo.

INTRODUCTION

Cellular proteins are in a dynamic state of synthesis and degradation. The process is extensive and highly selective: specific proteins are degraded within cells at widely different rates. Protein turnover is involved in basic cellular functions such as the regulation of the levels of regulatory proteins, cyclins, for example; adjustment to stress such as starvation and heat; and preferential removal of defective proteins. Intracellular protein breakdown in all organisms has an absolute requirement for metabolic energy. This, apparently paradoxical, energy requirement probably reflects mechanisms that

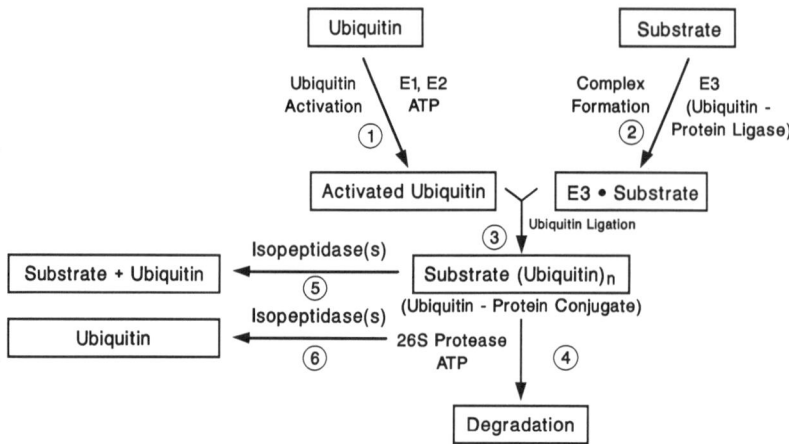

Fig. 1. *Overview of ubiquitin conjugation and degradation pathway. 1: Activation of ubiquitin by E1 and E2. 2: Complex formation of the protein substrate and E3. 3: Conjugation of ubiquitin(s) to the protein substrate. 4: ATP-dependent degradation of ubiquitin-protein conjugates to peptides and free amino acids (mediated by the 26S protease complex). 5: Release of reutilizable ubiquitin and protein substrate via isopeptidase(s). 6: Release of reutilizable ubiquitin via isopeptidase(s) during proteolysis of the protein substrate to peptides and free amino acids.*

endow the proteolytic system(s) with a high degree of specificity toward their protein substrates. In mammalian cells there are separate lysosomal and nonlysosomal mechanisms that are involved in different aspects of protein degradation. Proteins that enter the cell from the extracellular milieu (such as receptor-mediated endocytosed and pinocytosed proteins) are degraded in lysosomes. Lysosomal degradation of intracellular proteins occurs mostly under stressed conditions such as starvation. Nonlysosomal mechanisms are probably responsible for the highly selective turnover of intracellular proteins that occurs under basal metabolic conditions [for a monograph on general aspects of proteolysis, see Glaumann and Ballard, 1987].

A nonlysosomal ATP-dependent proteolytic system has been characterized and partially purified. The system consists of several essentially required components. One of these has been identified as ubiquitin, an abundant, 76-residue polypeptide that is found in all eukaryotic cells. The C-terminal glycine of ubiquitin is activated by ATP to a high-energy thiol-ester intermediate in a reaction catalyzed by the ubiquitin-activating enzyme E1. Following activation, E2 (ubiquitin-carrier protein) transfers ubiquitin from E1 to a ubiquitin–protein ligase, E3. E3 catalyzes the last step in the conjugation process, isopeptide bond formation between the activated C-terminal of ubiquitin and ε-amino groups of lysine residues of the protein substrate. This reaction involves specific binding of the substrate to E3 prior to the reaction with the activated ubiquitin. E3, therefore, appears to play a major role in selection of proteins for conjugation and degradation. Following formation of a conjugate between ubiquitin and the protein substrate, the protein moiety of the adduct is degraded by a specific ATP-dependent protease into free amino acids and free and reutilizable ubiquitin. Figure 1 is a model describing our current understanding of the events leading to degradation of a protein via the ubiquitin pathway [for recent reviews on the ubiquitin system, see Finley and Chau, 1991; Rechsteiner, 1991; Varshavsky, 1992; Hershko and Ciechanover, 1992; Jentsch, 1992].

It should be noted that modification of proteins by ubiquitin may also serve nonproteolytic functions. As described above, modification by ubiquitin plays a role in a variety of cellular processes. While the underlying mechanisms involved in all these processes are still obscure, the genetic basis and the enzymatic reactions leading to ubiquitin modification in some of these processes have been revealed. In some of these processes, ubiquitin is transferred directly to the protein target by E2 in a reaction that does not require E3 (Fig. 1). Most of these conjugates are of the mono-ubiquitin type, and the proteins thus targeted are not destined for degradation.

Despite the considerable progress that has been made in isolating the enzymes and dissecting the mechanisms involved in ubiquitin-mediated proteolysis, many problems remain unsolved. Such problems involve, for example, the identification of the structural signals that render proteins susceptible to ubiquitin ligation and subsequent degradation, and the characterization of the native substrates of the system. In particular, the primary and secondary (post-translational) signals involved in programmed and cell cycle–controlled degradation of certain proteins, cyclins, for example, are of obvious biological interest. In this review we shall discuss mostly these two aspects of the ubiquitin pathway.

EXPERIMENTAL FINDINGS
Activation of Ubiquitin

"Covalent affinity chromatography" over immobilized ubiquitin [Hershko et al., 1983] played a crucial role in elucidating the three-step mechanism involved in activation of ubiquitin and conjugate formation. In this process, the C-terminal glycine of ubiquitin is first transformed into a high-energy thiol ester in a reaction that is catalyzed by the ubiquitin-activating enzyme E1 (Fig. 2). This activation of ubiquitin is the initial step for the participation of ubiquitin in any biological function. Using the ubiquitin-affinity column, Hershko and colleagues have identified two additional fractions, E2 (ubiquitin-carrier protein) and E3 (ubiquitin-protein ligase), that were required along with

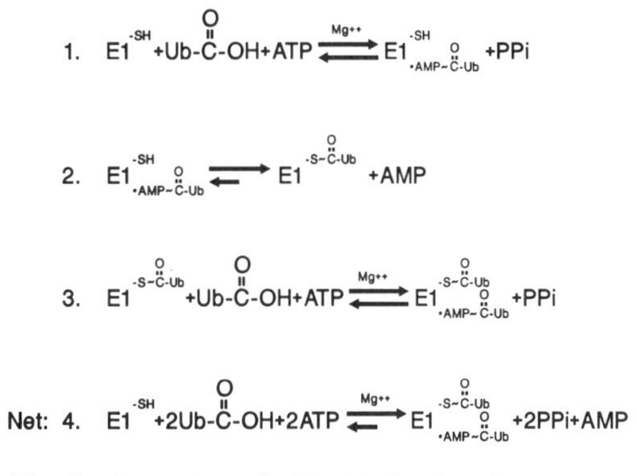

1. $E1^{-SH} + Ub\text{-}\overset{O}{\underset{\|}{C}}\text{-}OH + ATP \underset{}{\overset{Mg^{++}}{\rightleftharpoons}} E1^{-SH}_{\cdot AMP-\overset{O}{\underset{\|}{C}}-Ub} + PPi$

2. $E1^{-SH}_{\cdot AMP-\overset{O}{\underset{\|}{C}}-Ub} \rightleftharpoons E1^{-S-\overset{O}{\underset{\|}{C}}-Ub} + AMP$

3. $E1^{-S-\overset{O}{\underset{\|}{C}}-Ub} + Ub\text{-}\overset{O}{\underset{\|}{C}}\text{-}OH + ATP \overset{Mg^{++}}{\rightleftharpoons} E1^{-S-\overset{O}{\underset{\|}{C}}-Ub}_{\cdot AMP-\overset{O}{\underset{\|}{C}}-Ub} + PPi$

Net: 4. $E1^{-SH} + 2Ub\text{-}\overset{O}{\underset{\|}{C}}\text{-}OH + 2ATP \overset{Mg^{++}}{\rightleftharpoons} E1^{-S-\overset{O}{\underset{\|}{C}}-Ub}_{\cdot AMP-\overset{O}{\underset{\|}{C}}-Ub} + 2PPi + AMP$

Fig. 2. *Activation of ubiquitin by the ubiquitin-activating enzyme E1. E1 catalyzes the activation of ubiquitin, which is essential for any subsequent step in the ubiquitin pathway. The overall reaction consumes two ubiquitin moieties and two ATP molecules, ultimately binding ubiquitin in a high-energy thiol ester. AMP and pyrophosphate are released, and an E1–ubiquitin complex is formed that contains ubiquitin and AMP in a 2 to 1 ratio.*

E1 to reconstitute conjugate formation and subsequent degradation in the cell-free system from reticulocytes. When E1 was incubated in the presence of affinity-purified E2 fraction, ATP, and radiolabeled ubiquitin, multiple low-molecular-mass E2–ubiquitin adducts were formed. Addition of E3 and a protein substrate was necessary to catalyze the last step in the conjugation reaction, transfer of activated ubiquitin from one or more of the E2 adducts to the substrate's $\varepsilon\text{-}NH_2$ groups [Hershko et al., 1983]. Therefore, proteins in the E2 fraction acted as intermediate carriers of activated ubiquitin. Further studies have shown that, like E1-S–ubiquitin, the E2–ubiquitin intermediate is also a high-energy thiol ester between the C-terminal residue of ubiquitin and an –SH group of a cysteine residue of the enzyme.

Several major species of E2 enzymes have been identified by covalent affinity chromatography of various extracts [see, for example, Pickart and Rose, 1985], whereas others have been characterized via genetic studies. Over 30 genes encoding E2 enzymes have been isolated from various organisms [Jentsch, 1992] (Table I). Structurally, all known E2s share a conserved domain of approximately 16 kDa. This domain

TABLE I. Ubiquitin-conjugating Enzymes, E2s*

Yeast gene	Functions (yeast)	Protein size (kDa)
UBC1	Sporulation	24
UBC2/RAD6	DNA repair, protein degradation	20
UBC3/CDC34	G1–S cell cycle	34
UBC4	Protein degradation	16
UBC5	Protein degradation, stress	16
UBC6	Protein secretion	28
UBC7	(Stress)	18
UBC8	(?)	23
UBC9	(Viability)	—
UBC10/PAS2	(Peroxisome function)	—

*Summary of yeast family of ubiquitin-conjugating enzymes (UB) (E2s).
(Adapted from Jentsch [1992] and Jentsch et al. [1990].

contains the Cys residue required for the formation of ubiquitin–E2 thiol ester. Certain E2 enzymes contain additional typical domains such as a highly acidic C-terminal domain that promotes interaction with basic substrates, histones, for example. Functional studies of E2 enzymes have shown that they can be clustered in two major groups. Most E2 enzymes catalyze transfer of ubiquitin to small amines or small basic proteins in a reaction that does not require E3. The reactions catalyzed by these enzymes result in mono-ubiquitinated derivatives that do not serve as proteolysis intermediates. The cellular roles of these enzymes are not known, though genetic studies indicate that some of these E2s play important roles in a variety of basic cellular processes. For example, UBC3 (ubiquitin-conjugating enzyme, a term assigned to the E2 enzymes by Jentsch [1992] and colleagues), the product of the yeast gene CDC34, is essential for viability [Goebl et al., 1988]. Mutations in this E2 affect G1–S cell cycle progression, DNA replication, and spindle pole body separation. The enzyme interacts with CDC4 and CDC53 and appears to exert some of its effects via this interaction. UBC10 has been identified as the product of PAS2 and is essential for peroxisome biogenesis [Jentsch, 1992]. The roles of other species of mono-ubiquitinating E2s are not known. For example, disruption of the UBC8 does not yield any detectable mutant phenotype [Jentsch, 1992]. The second group of E2 enzymes is involved in multiple ubiquitinations that target the protein substrate for degradation. The mechanism of this reaction is quite different from that involved in mono-ubiquitination: here, the first activated ubiquitin moiety is transferred from E2 to a specific Lys residue of the substrate that is bound to E3. In successive reactions, a polyubiquitin chain is synthesized by processive transfer of activated ubiquitin moieties to Lys^{48} of the previous (and already conjugated) ubiquitin molecule [Chau et al., 1989] (Fig. 3). In some other, though infrequent, cases, E2 and E3 mediate conjugation of single ubiquitin moieties to multiple Lys residues of the protein substrate [Hershko and Ciechanover, 1992]. During the entire reaction, the substrate remains bound to E3. The binding of the substrate to E3 probably facilitates the synthesis of the poly-ubiquitin chain. This branched structure serves, most likely, as a recognition marker for the 26S protease complex that degrades the protein moiety and releases free ubiquitin: mono-ubiquitinated proteins are not recognized by the protease. It is assumed that E2 enzymes involved in proteolysis recognize distinct species of E3s: they do not appear to interact directly with the proteolytic substrates. This task is accomplished, most probably, by the different ligases.

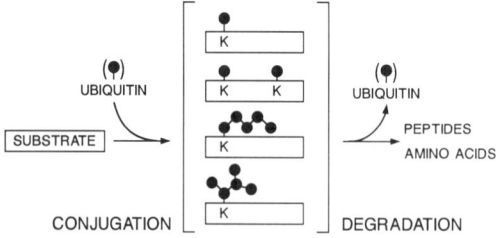

Fig. 3. *Protein substrates (boxes) are ubiquitinated to yield different ubiquitin–protein conjugates (mono-ubiquitin, multi-ubiquitin chain, and multi-ubiquitin tree). Lollipop symbol, ubiquitin; K, lysine residue.*

Recognition of Substrates by Ubiquitin–Protein Ligases

Role of the α-amino group of the protein substrate. Several studies suggested a role for the N-terminal residue of the protein substrate as a conjugation/proteolysis recognition marker. Hershko and colleagues found that selective modification of the α-NH$_2$ group of lysozyme greatly decreased the ability of the protein to be conjugated to ubiquitin and degraded in a cell-free reconstituted system (see, however, Role of Non-NH$_2$-Terminal Signals in Ubiquitin-Mediated Degradation, below). Addition of polyalanine side chains (which add free α-NH$_2$ groups) to the modified protein restored its susceptibility to ubiquitination and degradation [Hershko et al., 1984a,b]. The researchers suggested that the α-NH$_2$ group of a protein plays an important role in its recognition by the ligase and may even serve as a target for ubiquitin conjugation. In contrast, the ε-NH$_2$ groups are not obligatory for degradation and most probably serve as modulators of the rate of this process. Chin and colleagues [1986] reported that lysozyme in which the ε-NH$_2$, but not the α-NH$_2$, groups were selectively blocked was degraded in reticulocyte extracts in an energy- and ubiquitin-dependent mode. Bachmair and colleagues [1986] found that the identity of the N-terminal residue of a protein is an important determinant of its metabolic stability in vivo. These researchers engineered 20 different species of *Escherichia coli* β-galactosidase that differed from each other *solely* in the identity of their N-terminal residue. Analysis of the half-lives of the mutated proteins following their expression in yeast revealed that they varied from less than 2 minutes to more than 20 hours. Degradation of the short-lived species was accompanied by the formation of ubiquitin–protein conjugates, suggesting that the process is mediated by the ubiquitin system. It was concluded that the 20 amino acids at the N-terminal position can be divided into "stabilizing"" and "destabilizing" residues with respect to the half-lives they confer on the different mutated proteins ("N-end rule") [Varshavsky, 1992; Bachmair et al., 1986].

The role of E3 in recognition of protein substrates were examined by Hershko and colleagues [e.g., Heller and Hershko, 1990]. These researchers purified an E3 protein from rabbit reticulocytes and showed that the binding site for the proteolytic substrate resides in this enzyme. They were able to cross-link labeled proteolytic substrates to the ligase in a specific manner and to identify two groups of protein substrates that are recognized by two distinct binding sites. Type I substrates are proteins that have a basic NH$_2$-terminal residue (Arg, His, Lys), whereas type II substrates are proteins with bulky-hydrophobic N termini (Leu, Trp, Phe, and Tyr) (Fig. 4). The ligase binding, ubiquitin conjugation, and degradation of these proteins is specifically inhibited by N-terminal derivatives of the three amino acids (such as His-Ala but not Ala-His, or Phe-Ala but not Ala-Phe). Successive studies by Baker and Varshavsky [1991] have demonstrated that similar derivatives can also inhibit E3α in vivo. The two binding sites reside on the same enzyme molecule; however, they are independent: Inhibition of one site does not interfere with the activity of the other [Hershko and Ciechanover, 1992]. The ligase was designated E3α. More recently, type III substrates have been identified [Reiss et al., 1988; Gonda et al., 1989]. These proteins have a Ser, Ala, or a Thr residue in the N-terminal position and are recognized by a distinct species of E3 designated E3β [Heller and Hershko, 1990].

How are substrates with acidic NH$_2$ termini recognized and degraded? Ciechanover and colleagues have shown that ribonucleases inhibit strongly and specifically the degradation of proteins with either glutamic or aspartic acid at the N-terminal position. The inhibitory effect could be abolished by inhibiting the nucleases before adding them to the proteolytic system [Ferber and Ciechanover, 1987]. Addition of individual species of RNAs to RNA-depleted systems demonstrated that only tRNA could restore the proteolytic activity. While studying the underlying mechanisms responsible for the selective effect of the nucleases on the degradation of proteins with acidic N

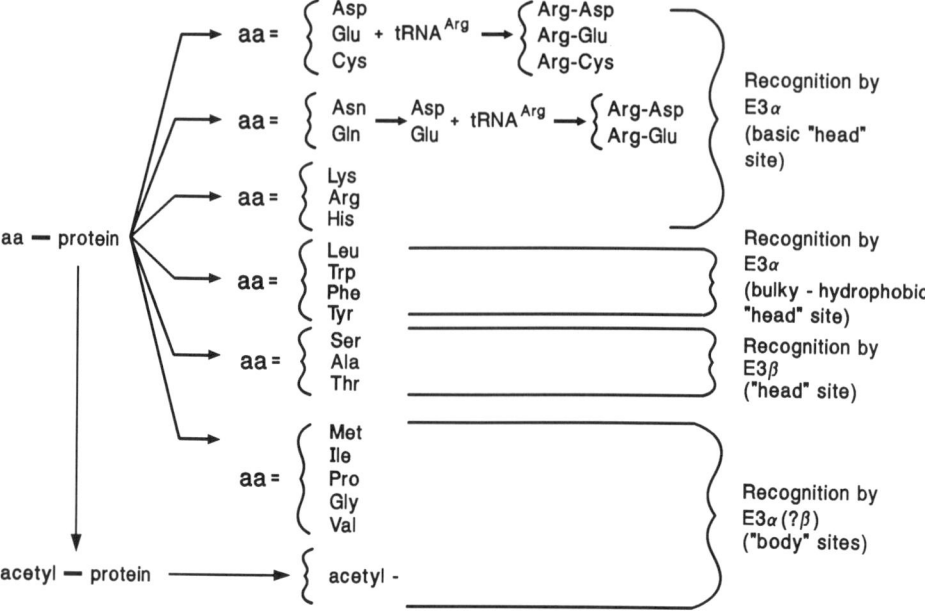

Fig. 4. *Recognition of the proteolytic substrates by ubiquitin–protein ligases (E3s). Distinct sites of E3s recognize a specific group of substrates that are characterized by the nature of their N-terminal amino acid residue as described in the text. Some of these residues require post-translational modification prior to recognition. Other proteins are probably recognized by signals that reside downstream from the N-terminal residue ("body" sites).*

termini, the investigators noted that eukaryotic cells contain arginyl-tRNA-protein transferase. This enzyme catalyzes the transfer of arginyl residue from charged arginyl-tRNA to acidic N termini [Soffer, 1973]. Further studies have shown that reticulocyte lysates also contain arginyltransferase and that the modification of acidic N-termini substrates by this enzyme is required for their degradation by the ubiquitin system [Ferber and Ciechanover, 1987]. While ribonucleases inhibited the degradation of ^{125}I-labeled acidic amino termini substrates, the degradation of the same substrates in which the N terminus was modified by the addition of arginine was insensitive to the nucleolytic activity. Proteins with NH_2-terminal Asn, Gln, or Cys residues are also post-translationally modified by the addition of an arginine moiety prior to ubiquitin-mediated degradation. For these proteins, the Asn and the Gln residues must first be converted to Asp and Glu prior to the modification by the arginyltransferase [Gonda et al., 1989]. Although the enzymatic steps involved in this post-translational modification were delineated, the mechanistic explanation for this process was still missing. It is possible that the N-terminal binding site of E3 is negatively charged. Therefore, proteins with negatively charged carboxyl groups at their N termini (acidic amino acids) cannot bind to this site. Conversion of the amino acid residue at the N-terminal position into a positively charged residue (basic amino acid) allows the binding between the substrate and the ligase to occur.

Bachmair and colleagues have shown that, in addition to a free N-terminal residue, the recognition of β-galactosidase involves also one or two lysine residues that are spatially proximal to this residue [reviewed by Varshavsky, 1992]. Chau and colleagues [1989] studied the role of this second recognition marker and showed that one of the lysine residues is specifically tagged by multiple ubiquitin residues that create ubiquitin–ubiquitin branched structures (the C-terminal glycine of one ubiquitin molecule is conjugated to the lysine residue at position 48 of the neighboring ubiquitin molecule; see above). These ubiquitin chains are probably recognized by the 26S high-molecular-mass protease complex that degrades ubiquitin–protein conjugates (see below) and thus contribute to labilization of these adducts.

A more detailed analysis of the mechanism of action of E3α revealed that in addition to the specific binding sites for ubiquitin and the substrate, it also interacts with one (or more) of the E2 enzymes with which it acts in concert [reviewed by Hershko and Ciechanover, 1992]. The formation of an E2•E3 complex probably facilitates the transfer of activated ubiquitin from E2 to the protein substrate bound to the ligase. Sung and colleagues [1991] have shown that the ability of UBC2 (RAD6) to promote poly-ubiquitination and degradation in vitro is dependent on the presence of E3α. A similar study carried out in yeast demonstrated that UBC2 is physically associated with UBR1, the yeast homolog of E3α: immunoprecipitation of UBR1 also precipitated UBC2 [Dohmen et al., 1991]. In addition, certain UBR1 substrates ("N-end rule" substrates) such as Arg-βgal and Leu-βgal were dramatically stabilized in *UBC*2 null mutants.

Role of non-NH$_2$-terminal signals in ubiquitin-mediated degradation. Whereas many mechanistic aspects of the N-end rule have been elucidated, a compelling body of evidence indicates that its physiological functions are rather limited and N-terminal recognition is not essential. 1) Approximately 80% of the cellular proteins are N-α-acetylated [Brown and Roberts, 1976]. As for the remaining free N-termini proteins, the rules that govern removal of the initiator Met by methionine aminopeptidase suggest that this residue will be cleaved only if the penultimate residue is a "stabilizing" amino acid [Hershko and Ciechanover, 1992]. Thus, proteins with "destabilizing" N-terminal residues appear to be sparse. 2) Mayer and colleagues [1989] have shown that N-α-acetylated proteins are degraded via the ubiquitin system. Removal of the modifying group is *not* necessary for recognition and degradation to occur. 3) Reiss and colleagues [1988] found that the recognition of certain proteins with free N termini (either "destabilizing" or "stabilizing" according to the N-end rule) is not dependent on the identity of their N-terminal residue (type IV or "body" type substrates). 4) Most convincing, mutational inactivation (deletion of the gene that encodes the yeast E3α) of the N-end rule in the yeast *Saccharomyces cerevisiae* is not lethal and does not result in a characteristic phenotype [Varshavsky, 1992; Bartel et al., 1990]. The data suggest that recognition of type IV substrates (N-α-acetylated and non N-end rule free N-termini proteins) is mediated via ubiquitin–protein ligases that are distinct from E3α or E3β and are recognized by motifs that reside, most probably, downstream from the N-terminal residue. Unfortunately, the nature of these signals has not been revealed yet. Interestingly, Gonen and colleagues [1991] have recently purified a novel factor that is required specifically for the *degradation* of ubiquitin *conjugates* of N-α-acetylated proteins. It seems, therefore, that the selective recognition of these proteins occurs during both ligation and degradation and involves complex signals.

Degradation of Ubiquitin–Protein Conjugates

Following marking of a protein with ubiquitin, the protein moiety of the conjugate is selectively degraded with the release of free and reutilizable ubiquitin. Hershko and colleagues have shown that the process requires ATP [Hershko et al., 1984a,b]. A protease that specifically degrades proteins conjugated to ubiquitin in an ATP-dependent mode was described by Waxman and colleagues [1987] and Hough and colleagues [1987]. Purification of the protease from rabbit reticulocytes revealed that it has a native molecular mass of approximately 1,300 kDa and an apparent S value of 26. A similar protease was also described by other researchers [reviewed by Varshavsky, 1992; Rechsteiner et al., 1993]. During the purification, the researchers also purified an ATP-independent protease with a broad substrate specificity. This protease has a native molecular mass of approximately 700 kDa and a sedimentation coefficient of 20S. The 20S enzyme has been previously described by a number of investigators and is commonly known as *macropain,* the multicatalytic proteinase complex (MCP; this term will be used throughout the text), or the proteasome [reviewed by Rechsteiner et al., 1993].

Structural analysis of the 26S protease complex revealed that it is a multisubunit enzyme. It contains two large subunits with molecular weights of 100 and 110 kDa, about 10 subunits of 40 and 62 kDa, and several subunits of 21 and 34 kDa [Hough et al., 1987; Rechsteiner et al., 1993]. Two-dimensional PAGE of the purified MCP complex revealed 15–20 different proteins with molecular weights ranging from 21 to 32 kDa [Hough et al., 1987]. Although the MCP and the 26S protease complexes have many distinguishing characteristics, they share some similar subunits as judged by comparison of SDS-denaturing gels. This structural resemblance together with similarities in substrate specificity, optimal pH for activity, and sensitivity to common inhibitors, led Rechsteiner and colleagues to propose that the 26S and the MCP protease complexes may be related to each other [reviewed by Rechsteiner et al., 1993] and that the MCP is part of the 26S. Indeed, recent experimental evidence supports the hypothesis that the two proteases are indeed related to each other. Three distinct factors, CF-1, CF-2, and CF-3 (for conjugate-degrading factors) all of which are required for ATP-dependent breakdown of ^{125}I-lysozyme–ubiquitin conjugates, have been resolved from reticulocyte extracts [Ganoth et al., 1988]. A detailed study of the role of ATP in the degradation process has shown that, following incubation of the three factors in the presence of the nucleotide, they associate to generate a high-molecular-mass (>1,000 kDa) active complex. The researchers suggested that one role of ATP in conjugate breakdown is the formation of an active multienzyme complex. They did not, however, rule out additional roles for ATP hydrolysis, such as peptide bond cleavage or movement of the substrate along the enzyme and release of the cleavage products in this process. Recent studies by Eytan and colleagues [1989] and Driscoll and Goldberg [1990] have demonstrated that MCP is indeed part of the 26S protease complex. MCP (which was identified as CF-3) is incorporated in an ATP-dependent manner into the 26S protease complex [Eytan et al., 1989]. Western blot analysis of the 26S complex that was assembled from its various components demonstrates that is contains subunits of MCP [Driscoll and Goldberg, 1990]. A study by DeMartino and colleagues [1991] has shown that a serum raised against one of the subunits of MCP precipitates the 26S complex and thus inhibits ubiquitin-mediated degradation of labeled substrates in BHK extracts [DeMartino et al., 1991]. It appears that MCP constitutes the "catalytic core" of the 26S protease complex and contains all its proteolytic activities [reviewed by Hershko and Ciechanover, 1992; Rechsteiner et al., 1993]. It is possible that the ATP-independent proteolytic activity of the MCP is due to "uncoupling" of the "catalytic core" (CF-3) from the putative "regulatory" subunits, CF-1 and CF-2. The presence of these "regulatory" subunits is most probably required in order to confer specificity and energy dependence to the 26S protease complex, thus ensuring that it will degrade only substrates destined for degradation. It is possible that CF-1 and CF-2 are involved in binding of the ubiquitin moiety or the poly-ubiquitin chain of the conjugate, thus bringing the substrate within the desired proximity to the active site of the protease. Electron microscopic studies have shown that the 26S protease complex composed of two large terminal cap-like structures attached to a thinner central core having four protein layers [Ikai et al., 1991]. The central structure is the MCP, whereas the cap structures are most probably the regulatory subunits [Ikai et al., 1991; Dahlmann et al., 1991].

The 26S protease complex contains also a variety of ubiquitin C-terminal hydrolytic activities that are probably involved in the cleavage of the poly-ubiquitin chains and the release of the terminal ubiquitin moiety from the substrate's Lys residue(s) [reviewed by Hershko and Ciechanover, 1992; Rechsteiner et al., 1993]. A recent study by Murakami and colleagues [1992] has demonstrated a role for the 26S protease complex in the ubiquitin-*independent* degradation of ornithine decarboxylase [Murakami et al., 1992]. Thus, it appears that the protease can recognize and degrade in a specific manner nonubiquitinated proteins as well, and the "regulatory" subunits CF-1 and CF-2 may have additional functions beyond recognition of ubiquitin. Little is known of the regulation of the 26S protease complex. Depending on purification procedures, the MCP can be purified in either a latent or active form. This variable activity probably reflects the presence or absence of copurified regulatory factors. Indeed, recent studies from several laboratories have reported the purification and characterization of both inhibitors and activators of the protease [reviewed by Rechsteiner et al., 1993]. Whether these modulators are also active within the context of the 26S protease complex enzyme is not yet clear.

Several recent studies have demonstrated that the MCP is found in the yeast *S. cervisiae* [see, for example, Hilt et al., 1993]. Cloning and sequencing of several subunits of the yeast enzyme have revealed a substantial homology to similar subunits isolated from MCP complexes of other eukaryotic cells. Tetrad analysis revealed that disruption of the structural genes encoding most of the studied subunits is lethal to the haploid cell. Nonlethal mutations lead to a phenotype that is defective in degradation of abnormal proteins and in the ability of the cell to withstand stress. Also, the mutants accumulate ubiquitin-protein conjugates that cannot be degraded. Thus, genetic analysis further corroborates the notion that the MCP is indeed an integral part of the ubiquitin conjugate proteolytic machinery [for recent reviews, see Hershko and Ciechanover, 1992; Rechsteiner et al., 1993].

Ubiquitin–C-Terminal Hydrolases

It is essential for the operation of the ubiquitin system that ubiquitin recycle. Several

TABLE II. Various Functions of Ubiquitin–C Terminal Hydrolases

Intracellular protein degradation
   Release of ubiquitin from Lys residues of end proteolytic products
   Disassembly of poly-ubiquitin chains
   Release of "mistakenly" ubiquitinated proteins ("proofreading")
   "Trimming" of abnormal poly-ubiquitin structures
Processing of precursors in ubiquitin biosynthesis
   Cleavage of ubiquitin from "tail" to "head" poly-ubiquitin precursors
   Release of ubiquitin from "tail" extensions of ribosomal proteins
   Removal of biosynthetic extra C-terminal amino acid residues
Release of ubiquitin from small, nonprotein adducts such as amines and thiols
Reversal of nonproteolytic modifications

ubiquitin–C-terminal hydrolases (i.e., enzymes that hydrolyze the linkage between the C-terminal glycine residue of ubiquitin and various adducts) have been described, but in most cases their functions have not been delineated. Table II lists some of the possible functions of such enzymes. In protein degradation, ubiquitin–C-terminal hydrolase(s) is required to release ubiquitin from isopeptide linkage with Lys residues of the protein substrate at the final stage of the proteolytic process. In addition, a C-terminal isopeptidase activity is required in order to disassemble poly-ubiquitin chains linked to the protein substrate, following or during the degradative process. A "proofreading" function has been proposed by Rose [1988] for hydrolases to release free protein from "incorrectly" ubiquitinated proteins. While there is no direct evidence for such a function, many highly specific biochemical pathways are accompanied by suitable editing systems. Another possibility is that ubiquitin–C-terminal isopeptidases are required for "trimming" abnormal polyubiquitin chains, so that the conjugates will now be recognized and bind to the 26S protease complex.

In addition, some ubiquitin–C-terminal hydrolases are probably required for the processing of biosynthetic precursors of ubiquitin, since most ubiquitin genes are either arranged in linear polyubiquitin arrays or are fused to ribosomal proteins (which are encoded as C-terminal extensions in certain ubiquitin genes; reviewed by Finley and Chau [1991]. Thus, maturation of functional ubiquitin molecules involves release of ubiquitin from linear polyubiquitin precursors, cleavage of ubiquitin from ribosomal proteins, and removal of extra amino acid residues that are encoded at the C termini of some poly-ubiquitin molecules.

Ubiquitin–C-terminal hydrolases may have other functions as well. Rose [1988] pointed out that products of high-energy E1-ubiquitin and E2-ubiquitin thiol esters may react with intracellular nucleophiles (such as glutathione or polyamines). Such reactions may lead to rapid depletion of free ubiquitin unless such side products are rapidly cleaved by appropriate ubiquitin–C-terminal hydrolases. In addition, in cases when modification by ubiquitin serves nonproteolytic functions, de-ubiquitination by an appropriate hydrolase will reverse the modification. For example, histones H2A and H2B are rapidly de-ubiquitinated during mitosis and re-ubiquitinated shortly afterwards [Mueller et al., 1985]. Similarly, the ubiquitination catalyzed by UBC3/CDC34 is probably cell cycle dependent and thus reversible [Goebl et al., 1988].

In view of the numerous and important functions of ubiquitin–C-terminal hydrolases, it is not surprising that a great number of such enzymes exist. Information available prior to 1988 was reviewed by Rose [1988]. One of the most characterized enzymes is a hydrolase from reticulocytes and erythrocytes, discovered by Rose and Warms [1983] and Rose [1986]. It was initially detected as a thiolesterase that cleaves ubiquitin–DTT thiol ester that was formed by the reaction of E1 in the presence of DTT. Subsequently, it was found to act on a variety of C-terminal amide derivatives of ubiquitin with small compounds. The enzyme was purified to near homogeneity and was found to be a monomer with an apparent subunit molecular mass of 30 kDa. Studies on the mechanism of inhibition of this enzyme by borohydride led to the discovery of ubiquitin–C-terminal aldehyde, a powerful inhibitor of this [Hershko and Rose, 1987] and some other ubiquitin–C-terminal hydrolases.

More recently, Mayer and Wilkinson [1989] have resolved from calf thymus four hydrolases that act on ubiquitin ethyl ester. Three of these appear to be isoenzymes of ~30 kDa. The fourth is larger, estimated at ~100 kDa. The gene that encodes the major 30-kDa enzyme was cloned and sequenced and found to be 54% homologous with PGP 9.5, a major (1%–5%) protein in neuronal cells [Wilkinson et al., 1989]. It is not known why this isoenzyme is expressed at such high levels in neurons. The role of the 30-kDa enzymes is not known. While some of them act on ubiquitin thiol/amide derivatives, the 30-kDa hydrolase from erythrocytes also acts on ubiquitin–protein fusions such as ubiquitin–metallothionein or the natural fusion protein of ubiquitin to the 52-amino acid human ribosomal extension protein [Hershko and Ciechanover, 1992]. Thus, the 30-kDa class of hydrolases may also have roles in the processing of at least some biosynthetic precursors of ubiquitin. However, Tobias and Varshavsky [1991] have recently reported that the yeast 30-kDa hydrolase does not act on engineered fusions of ubiquitin, such as ubiquitin-β-galactosidase. The latter fusions are efficiently cleaved by another, 90-kDa yeast hydrolase, the product of the UBP1 gene(45). The 90kDa enzyme also processes efficiently natural ubiquitin–protein fusions, such as ubiquitin linked to ribosomal proteins or linear polyubiquitin. Null mutants of the 30- and 90-kDa hydrolases are phenotypically normal [Tobias and Varshavsky, 1991]. This is presumably due to the presence of several enzymes with similar activities in this organism.

Another ubiquitin hydrolase of apparently different functions is a 100-kDa enzyme from erythrocytes, characterized recently by Hadari and colleagues [1992]. The enzyme is an abundant erythrocyte protein that was purified by affinity chromatography over immobilized ubiquitin. It acts preferentially on ubiquitin–Lys^{48}–ubiquitin linkages in branched polyubiquitin chains. The enzyme stimulates the breakdown of poly-ubiquitin chain(s)–conjugated proteins by the 26S protease complex. It acts probably by disassembling protein-free, polyubiquitin chains that were released from conjugates and remained bound to the 26S protease complex. Thus, it prevents competition between these chains and chains bound to substrates that have not been degraded yet. More recently, Eytan and colleagues [1993] reported the characterization of an ATP-dependent hydrolytic activity that is an integral part of the 26S protease complex. This enzyme cleaves both isopeptide and linear bonds and appears to release ubiquitin in the terminal stages of degradation.

Involvement of the Ubiquitin System in Degradation of Cellular Proteins In Vivo

The role of ubiquitin conjugation in marking proteins for degradation was discovered while studying ATP-dependent degradation in rabbit reticulocyte lysates. The hypothesis that ubiquitin marking of a protein is an obligatory intermediate step in its degradation received considerable experimental support from various laboratories using diverse genetic, biochemical, and cell biological approaches. Experiments in cell free systems have shown that the three enzymatic components of the ubiquitin ligase system, E1, E2, and E3, as well as the 26S protease complex, are required for ligation of ubiquitin to the substrate and for the subsequent degradation of the marked protein. Experiments in intact cells have lead to similar conclusions. Hershko and colleagues found a strong correlation between the rate of degradation of amino acid analog-containing proteins and the formation of ubiquitin conjugates of these proteins. Similarly, Chin and colleagues have shown a correlation between the rate of degradation of oxidized hemoglobin and the formation of ubiquitin conjugates of this protein, though in this case Fagan and colleagues have demonstrated that oxidant-damaged hemoglobin can also be degraded in ubiquitin- and ATP-independent process [reviewed by Hershko and Ciechanover, 1992]. Bachmair and colleagues found a strong correlation between the degradation of short-lived species of β-galactosidase and the formation of ubiquitin conjugates of the unstable proteins [Varshavsky, 1992; Bachmair et al., 1986; see also above]. Experiments using a mouse cell cycle arrest mutant with a thermolabile E1 have shown that inactivation of E1 strongly inhibited the degradation of short-lived and abnormal proteins [Ciechanover et al., 1984]. Analogous results were obtained using other cell lines that harbor similar mutations [reviewed by Hershko and Ciechanover, 1992]. Recently, Seufert and Jentsch cloned two genes that encode similar, however distinct, species of ubiquitin-carrier proteins (E2s) [see Jentsch et al., 1990]. The enzymes encoded by the two genes *UBC*4 and *UBC*5 generate high-molecular-mass ubiquitin–protein conjugates in vivo, consistent with the notion that attachment of multiple ubiquitin molecules to proteolytic substrates is required for their selective degradation. Turnover of short-lived proteins and canavanyl-peptides, but not of long-lived proteins, is markedly reduced in mutant cells in which the *ubc*4 and *ubc*5 genes were disrupted. Furthermore, loss of UBC4 and UBC5 activities impairs cell growth and the ability of cells to withstand stress such as heat or incubation in the presence of amino acid analogs [reviewed by Finley and Chau, 1991; Varshavsky, 1992; Jentsch, 1992]. Taken together, experiments carried out both in a cell-free system and in intact cells strongly support the hypothesis that conjugation of ubiquitin to at least some proteins renders them susceptible to proteolysis.

To gain a better understanding of the mechanisms of selectivity and regulation of the ubiquitin system, studies on rapidly degrading specific cellular proteins are obviously required. To date, the possible involvement of the ubiquitin system in the degradation of specific cellular proteins has been suggested in a few cases reviewed recently [Rechsteiner, 1991] and discussed briefly below.

Phytochrome. Phytochrome is a plant regulatory photoreceptor that consists of a linear tetrapyrrole linked to a polypeptide of ~120 kDa. The protein exists in two interconvertible forms. The Pr form, which absorbs red light at 670 nm (P_{670}), is converted to a new spectral state, Pfr or P_{730}, following exposure to far red light. This photoconversion initiates, in yet unknown mechanisms, a series of developmental and morphogenetic responses that facilitate the adaptation of plants to photosynthetic light. Interestingly, it is accompanied by a 100-fold increase in the degradation rate of the protein ($t_{1/2}$ of ~100 hours →$t_{1/2}$ of ~1 hour) and a concomitant increase in ubiquitin–phytochrome conjugates [Shanklin et al., 1987]. The significance of the proteolytic burst is not understood: it may serve to downregulate the active form of the receptor. An important, yet unresolved, problem relates to the biochemical alterations and signals that lead to the recognition, conjugation, and degradation of the Pfr form of the photoreceptor. Following irradiation, the protein aggregates and becomes associated with undefined subcellular fractions. It may well be that Pfr first aggregates and the aggregated form is recognized by the ubiquitin ligase. Also, the possible light activation of a specific phytochrome–ubiquitin ligase cannot be ruled out.

Oncoproteins. Many nuclear oncoproteins have rapid turnover rates [Rogers et al., 1986], but little is known about the mechanisms involved in their degradation. The degradation in cultured cells of the tumor suppressor protein p53 [Gronostajski et al., 1984] and of c-myc and c-myb [Luscher and Eisenman, 1988] were shown to be nonlysosomal and ATP dependent. Studies in reticulocyte lysates have demonstrated a role for the ubiquitin system in the degradation of the tumor suppressor protein p53 and the nuclear oncoproteins N- and c-myc, c-fos, and E1A [Ciechanover et al., 1991]. Immunodepletion of the ubiquitin-activating enzyme E1 resulted in complete inhibition of the degradation of these proteins. Degradation resumed following addition of purified E1.

Scheffner and colleagues noted exceptionally low levels of p53 in human cervical carcinoma cell lines transformed by the "high-risk" human papilloma viruses HPV-16 and -18. They tested the hypothesis that HPV E6 proteins target the tumor suppressor protein for degradation. Such a mechanism could explain, at least in part, the tumorigenicity of these oncoproteins. Indeed, E6 proteins encoded by the high-risk HPV-16 and -18 stimulate ATP-dependent conjugation and degradation of p53 in crude reticulocyte lysates [Scheffner et al., 1990]. Several lines of evidence indicate that the in vitro system reproduces faithfully the mechanisms involved in the processing of p53 in vivo, at least in HPV-transformed cells: 1) Degradation of p53 in the cell-free extracts and in intact cells requires metabolic energy [Gronostajski et al., 1984; Ciechanover et al., 1991; Scheffner et al., 1990]. 2) In both systems the degradation occurs at neutral pH and is not inhibited by lysosomotropic agents (51, 53, 54). 3) The ability of E6 to direct degradation of mutant p53s was related to the ability of the p53 proteins to bind to the HPV oncoprotein [Scheffner et al., 1990]. It should be noted that while E6 associates with p53 and thus stimulates its degradation, it remains stable throughout the proteolytic cycle. The mechanism of targeting of p53 by E6 is not known. It is possible that E6 is recognized by the ubiquitin–protein ligase but cannot be targeted as it does not have accessible ubiquitination sites. In contrast, p53 cannot bind to the ligase, but it harbors Lys residues that can be readily conjugated. It should be emphasized that E6 stimulates the degradation of p53, but is not essential for the process to occur. Significant degradation in vitro occurs in the absence of the HPV protein [Ciechanover et al., 1991]. Also, p53 is extremely short-lived in vivo in many cell lines that are not transformed by E6 [Rogers et al., 1986]. Similar E6/E7-dependent transrecognition and targeting has been demonstrated in the case of the retinoblastoma protein [Scheffner et al., 1992].

This mode of transrecognition and targeting is not an exception that is unique to oncoproteins. In a well-studied case, Johnson and colleagues have shown that in a heterodimeric protein the N-terminal residue that serves as the ligase-binding site and the Lys residue that is targeted by ubiquitin can reside in different subunits of the protein [Johnson et al., 1990]. In this case, only the ubiquitinable subunit is degraded. The recognition of p53 seems to be somewhat more complicated, as the association between p53 and E6 requires an additional protein, E6-associated protein (E6-AP) [Huibregtse et al., 1993]. E6-AP is a 100-kDa protein that contains a novel reading frame encoding 865 amino acids. Possible trans recognition may occur also during the degradation of c-Fos. In this case, c-Jun serves as the targeting protein.

Interestingly, this process is regulated by signal-dependent phosphorylation and dephosphorylation of the Jun protein [Papavassiliou et al., 1992]. A different case of trans recognition was reported in the degradation of ornithine decarboxylase (ODC). Here, the protein is targeted for degradation both in vitro and in vivo by antizyme in a ubiquitin-independent process that is mediated by the 26S protease [Murakami et al., 1992; see also above]. As described above, this protease also degrades ubiquitin-conjugated proteins. Thus, the mechanism of targeting of ODC appears to be different from that described for E6- and N-terminal–dependent mediated ubiquitination. The mechanisms of degradation of all these proteins appear to share, however, a common characteristic: One subunit of the complex probably provides the recognition site, whereas the other is targeted for degradation. In all these cases, the targeting protein appears to function as a chaperon: the substrate binds to it in a manner that probably exposes ubiquitination sites that are otherwise buried in the three-dimensional structure of the protein.

MATα2 repressor. The MATα2 repressor is a transcriptional regulator of mating type switching in the yeast *S. cerevisiae*. Hochstrasser and Varshavsky [1990] have shown that this regulatory protein turns over rapidly. Deletion analysis revealed that the protein has two degradation signals, one requiring residues 53–67 in the N-terminal region (Deg 1), while the other resides in residues 136–140 in the C-terminal domain (Deg 2). Initial studies revealed that the two signals are recognized by different mechanisms. More recently, Hochstrasser and colleagues [1991] have shown that the degradation of the repressor is carried out, at least in part, by the ubiquitin system. They were able to demonstrate ubiquitin adducts of MATα2. However, since the protein is degraded rapidly throughout the cell cycle, it was not possible to learn from this finding whether the adducts are indeed proteolysis intermediates. Two experiments indicated that the ubiquitin system is involved in the degradation of MATα2. 1) Overexpression in yeast of ubiquitin^{Arg-48} (which cannot generate polyubiquitin chains) slowed down twofold the rate of degradation of the protein. 2) The rate of degradation was significantly lower in the $\Delta UBC4/UBC5$ yeast mutant, indicating that the two ubiquitin-carrier proteins are involved in the degradation of the repressor molecule [Hochstrasser et al., 1991]. In a more recent study, the researchers have demonstrated that multiple UBCs are involved in the degradation of the repressor. UBC6 and UBC7 target protein for degradation via the Deg 1 signal, whereas UBC4 and UBC5 degraded the protein following recognition of a yet unidentified signal [Chen et al., 1993]. Interestingly, UBC6 and UBC7 form a complex that is necessary for their function (Fig. 5).

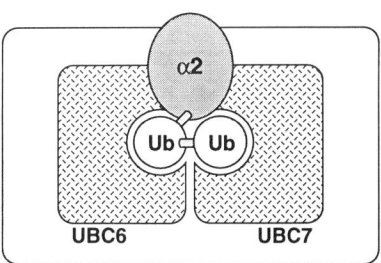

Fig. 5. *Ubiquitin and degradation of the α2 repressor. Model for α2 multi-ubiquitination by a UBC6/UBC7-containing ubiquitin–protein ligase complex. The UBC6 and UBC7 proteins may interact with each other directly, indirectly, or both through binding sites on an E3 ubiquitin–protein ligase (large white rectangle). [Adapted from Chen et al., 1993, with permission of the publisher.]*

Cyclins. Cyclins are proteins involved in cell cycle control in eukaryotes. Several types of cyclins exist, including A-type, B-type, and G1 cyclins. Cyclin B was first discovered in fertilized sea urchin oocytes as a protein that is synthesized during the interphase and then *rapidly* degraded at the end of the metaphase [Evans et al., 1983]. All cyclins appear to be associated with p34^{cdc2} protein kinase or related kinases and presumably act by regulating the activity of the kinases. The programmed degradation of cyclins at specific stages of the cell cycle is a dramatic example of regulated protein degradation and raises interesting questions about the mechanisms involved. The elucidation of such mechanisms became possible with the establishment of cell free systems that carry out regulated cyclin degradation. Using fertilized oocytes from different sources, Luca and Ruderman [1989] and Murray and Kirschner [1989] prepared extracts at the onset of mitosis and showed that, following a short incubation, there is a specific degradation of mitotic cyclins.

Characterization of the process revealed that it reproduces faithfully the in vivo events. Glotzer and colleagues [1991] studied the proteolysis targeting signals and the possible involvement of the ubiquitin system in the cell cycle–dependent degradation of cyclins. They showed that the N-terminal region of the cyclin molecule is necessary for recognition and degradation: Incubation of interphase extracts with truncated and stable cyclin B that lacks the N-terminal 90 amino acid domain led to arrest at a stable metaphase state. Mutational analysis demonstrated that the segment of amino acids 42–50 in the sea urchin cyclin B molecule, RAALGNISN, is required for degradation. When the invariant Arg residue in this sequence was replaced by Cys, the resulting derivative was stable. As degradation is cell-cycle dependent, it is clear that primary signals as well as post-translational modifications must be involved in this complex process. The researchers examined the possible involvement of the ubiquitin system in the process. They showed that incubation of *degradable* cyclin derivatives in *mitotic* extracts that also contained ^{125}I-ubiquitin resulted in rapid and specific degradation of these proteins that was preceded by a transient accumulation of high-molecular-mass ubiquitin conjugates. When a similar experiment was carried out using interphase extracts, the cyclin molecule studied was stable, and the level of cyclin–ubiquitin conjugates was ~10-fold lower than in mitotic extracts. Similarly, when stable mutants of cyclin were incubated in mitotic extracts, the conjugates generated were scarce and of low molecular mass. It is not known whether the Arg residue in the RAALGNISN box is important for recognition by E3 or whether its replacement affects indirectly a different and remote recognition site. As Glotzer and colleagues did not use a specific inhibitor of the ubiquitin system in their studies, a formal possibility existed that the ubiquitin–cyclin conjugates generated are irrelevant side products rather than essential degradation intermediates. Hershko and colleagues added methylated ubiquitin to mitotic clam extracts and showed that this agent efficiently blocks the degradation of both cyclins A and B. Addition of excess unmodified ubiquitin alleviated the inhibition [Hershko et al., 1991]. The modified ubiquitin lacks free amino groups and, therefore, cannot generate poly-ubiquitin chains that are required for the recognition of certain conjugates by the 26S protease [Hershko et al., 1983; Rechsteiner et al., 1993; see above]. These experiments demonstrated directly that formation of such chains is necessary for the degradation of cyclins. With the assumption that cyclins are degraded by the ubiquitin system, it will be interesting to study the regulatory mechanisms. The two main possibilities are that a cyclin-specific ligase is activated during mitosis or that cyclin is modified and converted to a form that is susceptible to the activity of a ubiquitin ligase. At present there is no evidence to support or rule out either mechanism.

Cell surface receptors: The platelet-derived growth factor, the IgE, and the T-cell receptors. Several cell surface receptors, including the lymphocyte homing receptor and the growth hormone receptor, were found to be modified by ubiquitin. However, the function of the modification in these cases remains obscure. Studies of other receptors demonstrate a role for ubiquitin modification in the physiological regulation of the respective proteins. Cenciarelli and colleagues [1992] have shown that the ζ-subunit of the T-cell receptor (TCR) undergoes ubiquitination in response to receptor engagement. It should be noted that the ζ-subunit is involved in TCR-mediated signal transduction and is a substrate for a TCR-activated protein tyrosine kinase. Paolini and Kinet [1993] have shown that the antigen-induced engagement of high-affinity immunoglobulin E receptors (FcεRI) results in immediate multiubiquitination of the β- and γ-chains of the receptor. The modification is independent of receptor phosphorylation, but is tightly linked to activation of the molecule. Disengagement results in rapid de-ubiquitination [Paolini and Kinet, 1993]. Both studies did not demonstrate a relationship between conjugation and degradation of the target molecules. Thus, the activation-induced ubiquitination raises several working hypotheses for the role of the modification in regulation of transmembrane receptor function. Ubiquitination may serve as a means to downregulate receptors by targeting them for proteolysis. Alternatively, the structural alteration conferred by ubiquitination may affect signal-transducing properties and associations of the modified molecules with other effectors. A more direct relationship between ubiquitination and degradation was reported by Mori and colleagues [1992], who found that the platelet-derived growth factor β-receptor undergoes multiple ubiquitination following ligand binding. In this case it appears that the modification also sensitizes the protein for degradation: Ligand-induced *ubiquitination* and *degradation* were partially impaired in a cytosolic tail-truncated receptor that lacked the ubiquitination site [Mori et al., 1992]. In all these cases it appears that the ubiquitination site resides within the cytosolic tail of the molecule. If the modification does indeed target the molecule for degradation, it is important to follow the fate of the extracellular binding domain as well. This segment cannot be ubiquitinated and may be degraded by the vacuolar system. Such a complex proteolytic mechanism must involve signals that are shared by the two proteolytic systems. Indeed, a linkage between the ubiquitin cytosolic system and the lysosomal system has been reported recently [reviewed by Hershko and Ciechanover, 1992].

MHC-restricted class I antigen presentation. MHC class I molecules carry peptide fragments derived from endogenously expressed proteins from the endoplasmic reticulum (ER) to the surface of nucleated cells for display to cytotoxic T lymphocytes. Before peptide presentation in association with MHC molecules can occur, cytoplasmic antigens must be cleaved into peptide fragments and transported into the ER lumen. The proteolytic system that processes cytosolic antigens has not been identified. An interesting finding relates to the possible role of the MCP/26S protease complex/ubiquitin system in the processing of MHC restricted class I antigens. It was found that several subunits of MCP are encoded by the major histocompatibility gene cluster. Furthermore, these subunits are inducible by γ-interferon (IFN-γ), as are other components of the antigen presentation system [reviewed by Goldberg and Rock, 1992; Driscoll and Finley, 1992; Monaco, 1992]. These findings raised the immediate and rather exciting hypothesis that the MCP (alone, within the context of the 26S protease complex, or within the context of the ubiquitin pathway) may be involved in antigen presentation via class I–restricted MHC molecules. More recent studies have shown that MCP subunits encoded by the major histocompatibility complex are not essential for antigen presentation [see, for example, Momburg et al., 1992]. Monaco and colleagues have recently demonstrated that MCP function might be regulated by subunit composition. These researchers have shown that the expression of one of the subunits of MCP, LMP2, is variable among cell lines of different tissue types and can be regulated by IFN-γ [Brown et al., 1993]. In all these studies, researchers investigated the processing of relatively short oligopeptides to the final presented peptides. A recent study by Michalek and colleagues [1993] has shown that a mutant that harbors a thermolabile E1 cannot present peptides derived from ovalbumin following inactivation of the enzyme. The conclusion of the researchers is that the processing of the protein to peptides requires the complete ubiquitin pathway [Michalek et al., 1993]. Further studies are still required to establish the role of the complete system or its proteolytic arm (the MCP or the 26S protease complex) in this important biological process.

General Regulation of the Ubiquitin System

As discussed above, programmed degradation of specific cellular proteins may involve specific post-translational modifications that render substrates susceptible to conjugation and subsequent degradation. In this case, the enzymes of the system remain uncharged. However, under stressed conditions, when general intracellular protein degradation increases, activation of the system may occur. Indeed, Kong and Chock [1992] reported that phosphorylation of E1 and E2 32-kDa results in an approximately twofold activation of the enzymes. The kinase(s) involved in the process have not been identified. The physiological significance of this finding is not clear as the rate-limiting step in the cascade of the ubiquitin pathway in vivo has not been defined. Also, it is not clear whether the relatively small increase in activity is sufficient for adaptation of the cell to the demanding environmental requirements under stress. Medina and colleagues [1991] reported activation of the ubiquitin system in skeletal muscle during fasting and following denervation. These researchers observed an increase in polyubiquitin mRNA and in the steady-state level of ubiquitin-protein conjugates. Genetic studies in yeast have demonstrated that the poly-ubiquitin gene is required for the ability of the cell to withstand stress [reviewed by Finley and Chau, 1991; Jentsch, 1992]. Though the findings suggest that activation/inactivation of the ubiquitin system may play a role in regulation of cellular processes under changing physiological conditions, the signal(s) involved in this type of regulation have not been defined. It may well be that the activation does not involve the enzymes of the system. Rather, post-translational modifications render subpopulations of substrates sensitive to degradation under different physiological conditions, whereas the activity of the system remains constitutive.

CONCLUDING REMARKS

The considerable progress achieved in elucidating the mode of action and the physiological roles of the ubiquitin system is an excellent demonstration for the power of combined biochemical and molecular genetic approaches. Most of the events in this pathway were elucidated by the use of biochemical methods, consisting of the resolution of the components of the reticulocyte cell-free system, purification and characterization of its various enzymes, study of partial reactions, and reconstitution of the complete system from purified components. The use of a molecular genetic approach, mostly in yeast, not only confirmed the conclusions of the biochemical analysis but also added further important dimensions. Cloning the genes of ubiquitin and of other key enzymes along the pathway, the study of their functions by gene disruption and mutational analysis, and the use of genetically engineered substrates were crucial in elucidating the physiological roles of the system as well as in analyzing certain proteolysis targeting signals. It also helped in sorting out important from less important directions of research. Future studies using the two complementing approaches will certainly illuminate important, yet unresolved, problems in this rapidly developing field.

As is often the case, the solving of some problems raises yet additional, and often more complicated, new ones. For example, it is clear now that the N-terminal signal is not involved in the degradation of most cellular proteins, and the identification of other degradation signals in proteins as well as the enzymes involved in their processing is of obvious importance. This last objective can be approached via both biochemical and genetic avenues searching for specific enzymes required for the breakdown of rapidly degraded specific cellular proteins. Among other open questions is the elucidation of the mode of action of the 26S protease complex, including the characterization of the functions of CF-1 and CF-2, the identification of the ubiquitin–protein conjugate binding site, and the elucidation of the roles of ATP in complex assembly and proteolysis. The identification of the functions of different ubiquitin–C-terminal hydrolases in protein degradation, the integration of their action with that of the 26S protease complex, and the possible editing function of some hydrolases to salvage incorrectly ubiquitinated proteins remain to be explored. An important direction for future research is the elucidation of the mechanisms of the regulation of degradation of specific proteins such as oncoproteins. This will not only add to our understanding of signals that target proteins for destruction, but will also shed light on control mechanisms involved in regulation of cellular growth, differentiation, and malignant transformation. Studies of programmed and cell cycle–controlled degradation of cyclins and apoptosis will certainly shed light on the complex mechanisms involved in progression of the cell along the cell cycle and on the irreversibility of temporally controlled processes. As discussed above, such studies should involve the exploration of two equally attractive approaches, post-translational modification of the protein substrates and the modulation of ligases.

ACKNOWLEDGMENTS

Studies in the laboratories of the authors are supported by grants from the US–Israel Binational Science Foundation (BSF), German–Israel Foundation for Scientific Research and Development (GIF), Israeli Academy of Humanities and Sciences, Council for Tobacco Research, Inc. (CTR), Israeli Ministry of Health, Israel Cancer Research Fund, USA (ICRF), Israel Cancer Society, Rappaport Foundation, The Foundation for Promotion of Research in the Technion, The Hedson Fund for Medical Research, Monsanto, Inc., and NIH. Part of this article was published as a review article in the *FASEB Journal*.

REFERENCES

Bachmair A, Finley D, Varshavsky A (1986): In vivo half-life of a protein is a function of its amino-terminal residue. Science 234:179–186.

Bartel B, Wünning I, Varshavsky A (1990): The recognition component of the N-end rule pathway. EMBO J 9:3179–3189.

Brown GM, Driscoll J, Monaco JJ (1993): MHC-linked low molecular mass polypeptide subunits define distinct subsets of proteasomes. J Immunol 151:1193–1204.

Brown JL, Roberts WK (1976): Evidence that approximately eighty percent of the soluble proteins from Ehrlich ascites cells are N-alpha-acetylated. J Biol Chem 251:1009–1014.

Cenciarelli C, Hou D, Hsu K-C, Rellahan BL, Wiest DL, Smith HT, Fried VA, Weissman AM (1992): Activation-induced ubiquitination of the T cell antigen receptor. Science 257:795–797.

Ciechanover A, DiGiuseppe JA, Bercovich B, Orian A, Richter JD, Schwartz AL, Ciechanover A (1991): Degradation of nuclear oncoproteins by the ubiquitin system in vitro. Proc Natl Acad Sci USA 88:139–143.

Ciechanover A, Finley D, Varshavsky A (1984): Ubiquitin dependence of selective protein degradation demonstrated in the mammalian cell cycle mutant ts85. Cell 37:57–66.

Chau V, Tobias JW, Bachmair A, Marriott D, Ecker DJ, Gonda DK, Varshavsky A (1989): A multiubiquitin chain is confined to specific lysine in a targeted short-lived protein. Science 243: 1576–1583.

Chen P, Johnson P, Sommer S, Jentsch S, Hochstrasser M (1993): Multiple ubiquitin-conjugating enzymes participate in the in vivo degradation of the yeast MATα2 repressor. Cell 74:357–369.

Dahlmann B, Kopp F, Kuehn L, Hegerl R, Pfeifer G, Baumeister W (1991): The multicatalytic proteinase (prosome, proteasome): Comparison of the eukaryotic and archaebacterial enzyme. Biomed Biochim Acta 50:465–469.

DeMartino GN, McCullough ML, Reckelhoff JF, Croall DE, Ciechanover A, McGuire MJ (1991): ATP-stimulated degradation of endogenous proteins in cell-free extracts of BHK 21/C13 fibroblasts: A key role for the proteinase, Macropain, in the ubiquitin-dependent degradation of short-lived proteins. Biochim Biophys Acta 1073:29299–29308.

Dohmen RJ, Madura K, Bartel B, Varshavsky A (1991): The N-end rule is mediated by the UBC2(RAD6) ubiquitin-conjugating enzyme. Proc Natl Acad Sci USA 88:7351–7355.

Driscoll J, Finley D (1992): A controlled breakdown: Antigen processing and the turnover of viral proteins. Cell 68:823–825.

Driscoll J, Goldberg AL (1990): The proteasome (multicatalytic protease) is a component of the 1500-kDa proteolytic complex which degrades ubiquitin conjugated proteins. J Biol Chem 265:4789–4792.

Evans T, Rosenthal ET, Youngblom J, Distel D, Hunt T (1983): A protein specified by maternal mRNA in sea urchin eggs that is destroyed at each cleavage division. Cell 33:389–396.

Eytan E, Armon T, Heller H, Beck S, Hershko A (1993): Ubiquitin C-terminal hydrolase activity associated with the 26S protease complex. J Biol Chem 268:4668–4674.

Eytan E, Ganoth D, Armon T, Hershko A (1989): ATP-dependent incorporation of 20S protease into the 26S complex that degrades proteins conjugated to ubiquitin. Proc Natl Acad Sci USA 86:7751–7755.

Ferber S, Ciechanover A (1987): Role of arginine-tRNA in protein degradation by the ubiquitin pathway. Nature 326:808–811.

Finley D, Chau V (1991): Ubiquitination. Annu Rev Cell Biol 7:25–69.

Ganoth D, Leshinsky E, Eytan E, Hershko A (1988): A multicomponent system that degrades proteins conjugated to ubiquitin. J Biol Chem 263:12412–12419.

Glaumann H, Ballard FJ (eds) (1987): Lysosomes: Their Role in Protein Breakdown. London: Academic Press.

Glotzer M, Murray AW, Kirschner MW (1991): Cyclin is degraded by the ubiquitin pathway. Nature 349:132–138.

Goldberg AL, Rock KL (1992): Proteolysis, proteasomes, and antigen presentation. Nature 357:375–379.

Goebl MG, Yochem J, Jentsch S, McGrath JP, Varshavsky A, Byers B (1988): The yeast cell cycle gene CDC34 encodes a ubiquitin-conjugating enzyme. Science 241:1331–1335.

Gonda DK, Bachmair A, Wünning I, Tobias JW, Lane WS, Varshavsky A (1989): Universality and structure of the N-end rule. J Biol Chem 264:16700–16712.

Gonen H, Schwartz AL, Ciechanover A (1991): Purification and characterization of a novel protein that is required for the degradation of N-α-acetylated proteins by the ubiquitin system. J Biol Chem 266:19221–19231.

Gronostajski RM, Goldberg AL, Pardee AB (1984): Energy requirement for degradation of tumor associated p53. Mol Cell Biol 4:442–448.

Hadari T, Warms JV, Rose IA, Hershko A (1992): A ubiquitin–C-terminal isopeptidase that acts on polyubiquitin chains. J Biol Chem 267:719–727.

Heller H, Hershko A (1990): A ubiquitin–protein ligase specific for type III protein substrates. J Biol Chem 265:6532–6535.

Hershko A, Ciechanover A (1992): The ubiquitin system for protein degradation. Annu Rev Biochem 61:761–807.

Hershko A, Ganoth D, Pehrson J, Palazzo RE, Cohen LH (1991): Methylated ubiquitin inhibits cyclin degradation in clam embryo extracts. J Biol Chem 266:376–379.

Hershko A, Heller H, Elias S, Ciechanover A (1983): Components of ubiquitin–protein ligase system: Resolution, affinity purification, and role in protein breakdown. J Biol Chem 258:8206–8214.

Hershko A, Heller H, Eytan E, Kaklij G, Rose IA (1984a): Role of the alpha-amino group of protein in ubiquitin-mediated protein breakdown. Proc Natl Acad Sci USA 81:7201–7205.

Hershko A, Leshinsky E, Ganoth D, Heller H (1984b): ATP-dependent degradation of ubiquitin–protein conjugates. Proc Natl Acad Sci USA 81:1619–1623.

Hershko A, Rose IA (1987): Ubiquitin-aldehyde: A general inhibitor of ubiquitin recycling processes. Proc Natl Acad Sci USA 84:1829–1833.

Hilt W, Enenkel C, Gruhler A, Singer T, Wolf DH (1993): The PRE4 gene codes for a subunit of the yeast proteasome necessary for peptidylglutamyl-peptide-hydrolyzing activity: Mutations link the proteasome to stress- and ubiquitin-dependent proteolysis. J Biol Chem 268:3479–3486.

Hochstrasser M, Varshavsky A (1990): In vivo degradation of a transcriptional regulator: The yeast α2 repressor. Cell 61:697–708.

Hochstrasser M, Ellison MJ, Chau V, Varshavsky A (1991): The short-lived MATα2 transcriptional regulator is ubiquitinated in vivo. Proc Natl Acad Sci USA 88:4606–4610.

Hough R, Pratt G, Rechsteiner M (1987): Purification of two high molecular weight from rabbit reticulocyte lysate. J Biol Chem 262:8303–8313.

Huibregtse JM, Scheffner M, Howley PM (1993): Cloning and expression of the cDNA for E6-AP, a protein that mediates the interaction of the human papillomavirus E6 oncoprotein with p53. Mol Cell Biol 13:775–784.

Ikai A, Nishigai M, Tanaka K, Ichihara A (1991): Electron microscopy of 26S complex containing 20S proteasome. FEBS Lett 292:21–24.

Jentsch S (1992a): Ubiquitin dependent protein degradation: A cellular perspective. Trends Cell Biol 2:98–103.

Jentsch S (1992b): The ubiquitin–conjugation system. Annu Rev Genet 26:179–207.

Jentsch S, Seufert W, Sommer T, Reins H (1990): Ubiquitin-conjugating enzymes: Novel regulators of eukaryotic cells. Trends Biochem Sci 15:195–198.

Johnson ES, Gonda DK, Varshavsky A (1990): Cis-trans recognition and subunit specific degradation of short-lived proteins. Nature 346:287–291.

Kong SK, Chock PB (1992): Protein ubiquitination is regulated by phosphorylation. J Biol Chem 267:14189–14192.

Luca FC, Ruderman JV (1989): Control of programmed cyclin destruction in a cell-free system. J Cell Biol 109:1895–1909.

Luscher B, Eisenman RN (1988): c-myc and c-myb protein degradation: Effect of metabolic inhibitors and heat shock. Mol Cell Biol 8:2504–2512.

Mayer A, Siegel NR, Schwartz AL, Ciechanover A (1989): Degradation of proteins with acetylated amino-termini by the ubiquitin system. Science 244:1480–1483.

Mayer AN, Wilkinson KD (1989): Detection, resolution, and nomenclature of multiple ubiquitin–carboxyl-terminal esterases from bovine calf thymus. Biochemistry 28:166–172.

Medina R, Wing SS, Haas A, Goldberg AL (1991): Activation of the ubiquitin-ATP-dependent proteolytic system in skeletal muscle during fasting and denervation atrophy. Biomed Biochim Acta 50:347–356.

Michalek MT, Grant EP, Gramm C, Goldberg AL, Rock KL (1993): A role for the ubiquitin-dependent proteolytic pathway in MHC class I–restricted antigen presentation. Nature 363:552–554.

Momburg F, Ortiz-Navarrete V, Neefjes J, Goulmy E, van de Val Y, Spits H, Powis SJ, Butcher GW, Howard JC, Walden P, Hämmerling GJ (1992): Proteasome subunits encoded by the major histocompatibility complex are not essential for antigen presentation. Nature 360:174–177.

Monaco JJ (1992): A molecular model of MHC class I–restricted antigen processing. Immunol Today 13:173–179.

Mori S, Heldin C-H, Claesson-Welsch L (1992): Ligand-induced polyubiquitination of the platelet-derived growth factor β-receptor. J Biol Chem 267:6429–6434.

Mueller RD, Yasuda H, Hatch CL, Bonner WM, Bradbury EM (1985): Identification of ubiquitinated histones 2A and 2B in *Physarum polycephalum*. Disappearance of these proteins at metaphase and reappearance at anaphase. J Biol Chem 260:5147–5153.

Murakami Y, Matsufuji S, Kameji T, Hayashi S-I, Igarashi K, Tamura T, Tanaka K, Ichihara A (1992): ornithine decarboxylase is degraded by the 26S proteasome without ubiquitin. Nature 360:597–599.

Murray AW, Kirschner MW (1989): Cyclin synthesis drives the early embryonic cell cycle. Nature 339:275–280.

Paolini R, Kinet J-P (1993): Cell surface control of the multiubiquitination and deubiquitination of high-affinity immunoglobulin E receptors. EMBO J 12:779–786.

Papavassiliou AG, Trier M, Chavrier C, Bohmann D (1992): Targeted degradation of c-Fos, but not v-Fos, by a phosphorylation-dependent signal on c-Jun. Science 258:1941–1944.

Pickart CM, Rose IA (1986): Mechanism of ubiquitin-carboxyl-terminal hydrolase. Borohydride and hydroxylamine inactivate in the presence of ubiquitin. J Biol Chem 261:10210–10217.

Pickart CM, Rose IA (1985): Functional heterogeneity of ubiquitin-carrier proteins. J Biol Chem 260:1573–1581.

Rechsteiner M (1991): Natural substrates of the ubiquitin proteolytic pathway. Cell 66:615–618.

Rechsteiner M, Hoffman L, Dubiel W (1993): The multicatalytic and 26S proteases. J Biol Chem 268:6065–6068.

Reiss Y, Kaim D, Hershko A (1988): Specificity of the binding of NH_2-terminal residue of proteins to ubiquitin-protein ligase. J Biol Chem 263:2693–2698.

Rogers S, Wells R, Rechsteiner M (1986): Amino acid sequences common to rapidly degraded proteins: The PEST hypothesis. Science 234:364–368.

Rose IA (1988): Ubiquitin-carboxyl-terminal hydrolases. In Rechsteiner M (ed): Ubiquitin. New York: Plenum Press, pp 135–155.

Rose IA, Warms JV (1983): An enzyme with ubiquitin-carboxyl-terminal esterase activity from reticulocytes. Biochemistry 22:4234–4237.

Scheffner M, Münger K, Huibregtse JM, Levine AJ, Howley PM (1992): Targeted degradation of the retinoblastoma protein by human papillomavirus E7-E6 fusion proteins. EMBO J 11:2425–2431.

Scheffner M, Werness BA, Huibregtse JM, Levine AJ, Howley PM (1992): Targeted degradation of the retinoblastoma protein by human papillomavirus E7-E6 fusion proteins. EMBO J 11:2425–2431.

Shanklin J, Jabben M, Vierstra RD (1987): Red light-induced formation of ubiquitin-phytochrome conjugates: Identification of a possible intermediate of phytochrome degradation. Proc Natl Acad Sci USA 84:359–363.

Soffer RL (1973): Post-translational modification of proteins catalyzed by aminoacyl-tRNA-protein transferase. Mol Cell Biochem 2:3–14.

Sung P, Berleth E, Pickart C, Prakash S, Prakash L (1991): Yeast RAD6 encoded ubiquitin-conjugating enzyme mediates protein degradation dependent on the N-end recognizing E3 enzyme. EMBO J 10:2187–2193.

Tobias JW, Varshavsky A (1991): Cloning and functional analysis of the ubiquitin-specific protease gene UBP1 of *Saccharomyces cervisiae*. J Biol Chem 266:12021–12028.

Varshavsky A (1992): The N-end rule. Cell 69:725–735.

Waxman L, Fagan JM, Goldberg AL (1987): Demonstration of two distinct high-molecular weight proteases in rabbit reticulocytes, one of which degrades ubiquitin conjugates. J Biol Chem 262:2451–2457.

Wilkinson KD, Lee KM, Deshpande S, Duerksen-Hughes P, Boss JM, Pohl J (1989): The neuron specific protein PGP 9.5 is a ubiquitin carboxyl-terminal hydrolase. Science 246:670–673.

ABOUT THE AUTHORS

AARON J. CIECHANOVER is Professor of Biochemistry and Director of the Rappaport Institute for Research in the Medical Sciences at the Technion Israel Institute of Technology, Haifa, Israel, where he teaches undergraduate, graduate, and medical courses dealing with topics of biochemistry and the regulation of cellular metabolism. After receiving his Bachelor's degree, he received his M.Sc. (1970) and M.D. (1974) from Hadassah and the Hebrew University Medical School, Jerusalem, Israel. Thereafter, Dr. Ciechanover completed his D.Sc. degree in Biochemistry (1981) at the Faculty of Medicine, Technion Israel Institute of Technology, where he discovered the ubiquitin-mediated proteolytic pathway together with Avram Hershko, his graduate advisor. Following a 4-year research fellowship at Massachusetts Institute of Technology, where he began studies on the cell biology of protein sorting, Dr. Ciechanover returned to Israel to continue his studies on the cell and molecular basis of ubiquitin-mediated protein turnover. Dr. Ciechanover has been a Visiting Professor at Washington University School of Medicine with Dr. Schwartz for the past 9 years. In addition, Dr. Ciechanover has been a Fulbright Fellow and Research Career Development Awardee of the Israel Cancer Research Fund.

ALAN L. SCHWARTZ is Alumni-Endowed Professor of Pediatrics, Professor of Molecular Biology and Pharmacology, and Director of the Division of Pediatric Hematology/Oncology at Washington University School of Medicine in St. Louis, MO. He is also Director of the Lucille P. Markey Special Emphasis Pathway in Human Pathobiology. He teaches graduate and medical courses in Molecular Pharmacology and Pathobiology. After receiving his B.A., he pursued his Ph.D. in Pharmacology (1974) and M.D. (1976) at Case Western Reserve University. Following clinical training in Pediatrics, Hematology, and Oncology, Dr. Schwartz began his studies of the cell and molecular biology of receptors and protein sorting/turnover as a Visiting Scientist at the Massachusetts Institute of Technology. Since 1979 his research has focused on the cellular biology of receptor-mediated endocytosis and of intracellular protein turnover. Dr. Schwartz was an American Heart Association Established Investigator and currently is a Distinguished Visiting Professor in the Faculty of Medicine at the University of Utrecht, The Netherlands.

Molecular Genetics of the Ubiquitin System

David Keith Gonda

INTRODUCTION

Ubiquitin-dependent proteolysis in eukaryotes is governed by an elaborate pathway in which covalent attachment of the small, highly conserved protein ubiquitin to other cellular proteins targets them for degradation by a cytoplasmic, ATP-dependent proteolytic activity. It is an essential degradative pathway in eukaryotic cells and accounts for greater than 90% of short-lived protein turnover. Among the substrates of ubiquitin-dependent proteolytic pathways are damaged and abnormal proteins, as well as naturally short-lived proteins such as transcriptional regulators, oncoproteins, and regulators of cell cycle progression [Finley and Chau, 1991; Hershko and Ciechanover, 1992; Jentsch, 1992].

Our understanding of ubiquitin-dependent proteolysis has benefited from the complementary perspectives afforded by biochemical and genetic approaches. Biochemical studies have permitted the identification and characterization of the enzymatic activities involved in ubiquitin–protein conjugation and the degradation of ubiquitinated proteins, while genetic studies have provided insight into the generality of ubiquitin-dependent proteolysis and its in vivo function. More recently, the availability of genes encoding the enzymes of ubiquitin-dependent proteolysis has permitted their detailed comparison and dissection and in the future promises to provide deeper understanding into the function and regulation of this remarkable system.

I present here an overview of the insights provided by genetic analyses of ubiquitin-dependent proteolysis. This work entered its heyday with the identification of the ubiquitin system in the yeast *Saccharomyces cerevisiae*. In many cases, genetic analyses of ubiquitin function in other eukaryotes have been guided by that of yeast, and so much of the discussion below is organized around yeast. Additional discussions of molecular genetics of the ubiquitin system may be found elsewhere [Hochstrasser, 1992; Jentsch, 1992].

UBIQUITIN

Conservation and Organization of Ubiquitin Genes

Virtually every published discussion of ubiquitin and the ubiquitin system has remarked on the extreme sequence conservation of the 76-residue ubiquitin polypeptide among eukaryotes. The cloning of ubiquitin genes from many different organisms, along with the recent identification of archeabacterial ubiquitin [Wolf et al., 1993], has all but secured ubiquitin's place as the most conserved protein on earth. A large number of molecular cloning studies also confirm the conservation of the unusual organization of ubiquitin genes in eukaryotes: Ubiquitin is invariably encoded as a fusion protein that must undergo a post-translational cleavage to produce the mature ubiquitin. Processing of the ubiquitin fusion proteins by ubiquitin-processing proteases (see below) occurs virtually cotranslationally and does not seem to be used to regulate the cellular levels of free, mature ubiquitin [Bachmair et al., 1986; Finley et al., 1989]. This unexpected arrangement was fully characterized first in yeast [Ozkaynak et al., 1984, 1987] but has also been found in every eukaryote examined.

The number of loci from which ubiquitin is expressed varies greatly among different eukayotes; in yeast, ubiquitin is expressed from four loci [Ozkaynak et al., 1987]. UBI1 and UBI2 encode identical polypeptides consisting of a single ubiquitin moiety fused to the N terminus of a ribosomal protein of the large subunit; UBI3 encodes a protein consisting of a single ubiquitin moiety fused to the N terminus of a ribosomal protein of the small subunit [Finley et al., 1989; Redman and Rechsteiner, 1989]. In other eukaryotes, the homologs of the ribosomal proteins found in UBI1–3 are also invariably found expressed as a ubiquitin fusion protein. In yeast, genetically "defusing" the ubiquitin moieties of UBI1–3 from the rest of the coding sequence causes a striking defect in ribosomal biogenesis [Finley et al., 1989]. This defect can be suppressed by increasing the expression of de-ubiquitinated ribosomal protein. Thus, the expression of ribosomal proteins with N-terminal fusions of ubiquitin appears to facilitate the efficiency with which they are incorporated into ribosomes. Expression in both yeast and bacteria of other proteins with engineered N-terminal ubiquitin fusions often increases the efficiency of their expression [Butt et al., 1989; Ecker et al., 1989]. The evolution of ubiquitin–ribosomal fusion proteins may therefore represent a naturally occurring use of ubiquitin to enhance expression and synthesis of the associated ribosomal proteins. It is also possible that the fusion of ubiquitin to these ribosomal proteins provides a connection between the protein synthetic and protein degradative apparatus that facilitates their coordinate regulation.

Ubiquitin is essential. Under normal growth conditions, cellular ubiquitin pools are maintained by UBI1–3 [Finley et al., 1989]. The remaining ubiquitin coding locus in yeast, UBI4, consists of five head-to-tail repeats of ubiquitin and is expressed strongly in response to a variety of environmental and metabolic stresses including heat shock, starvation, and exposure to UV radiation. Mutants with a disrupted UBI4 coding locus are viable, but hypersensitive to a variety of stresses [Finley et al., 1987]. This suggests that UBI4 functions primarily to maintain cellular pools of ubiquitin in times of stress during which conjugation of ubiquitin to intracellular proteins is particularly active. Mutants that lack UBI4 also display a rapid loss of spore viability following sporulation [Finley et al., 1987]. Coding of ubiquitin at this locus as a polyprotein may be a device for allowing efficient expression of ubiquitin in times of need; however, although poly-ubiquitin coding genes are a common motif in other eukaryotes as well, not all are induced by stress stimuli [Graham et al., 1989; Taccioli et al., 1989; Kawalleck et al., 1993].

Ubiquitin Mutants

Considering the extreme conservation of ubiquitin's primary sequence among eukaryotes, surprisingly few residues within the ubiquitin sequence are absolutely essential for ubiquitin function [Breslow et al., 1986; Duerksen et al., 1987; Ecker et al., 1987a,b; Monia et al., 1990]. The C terminus of ubiquitin provides the site of attachment of ubiquitin to other cellular proteins via an isopeptide linkage between ubiquitin's α-carboxyl group and the ϵ-amino group of lysine residues on the target protein. Not surprisingly, ubiquitin function is very sensitive to chemical or genetic alteration of ubiquitin's C terminus: Many mutations at the C-terminal Gly residue of ubiquitin prevent its activation by the E1 enzyme and so prevent ubiquitin conjugation. One mutant, obtained by altering Gly76 to Ala76, is proficient for activation and conjugation to target proteins, but cannot be subsequently released from the target protein by isopeptidases that normally reverse the isopeptide linkage and release free ubiquitin from ubiquitin–protein conjugates [Ecker et al., 1987a; Hodgins et al., 1992]. When this mutant is expressed in yeast, it forms stable multi-ubiquitin chains and induces phenotypes that reflect an apparent deficiency of ubiquitin in the cell [Hodgins et al., 1992]. Since this effect is genetically dominant, this mutant might be used to induce a de facto ubiquitin deficiency in other eukaryotic cells as well [Hodgins et al., 1992].

Critical to ubiquitin function in proteolysis is its ability to conjugate to itself. This activity results in the formation of multi-ubiquitin chains on substrate proteins, in which the C-terminus of one ubiquitin is conjugated to a lysine residue on the next ubiquitin. This feature is critical because only substrates containing a multi-ubiquitin chain are efficiently targeted for degradation [Chau et al., 1989; Gregori et al., 1990]. Lys^{48} is a major acceptor site on ubiquitin for the synthesis of multiubiquitin chains in several ubiquitination reactions [Chau et al., 1989; Haas et al., 1991; Hochstrasser et al., 1991]. While many ubiquitin conjugation reactions appear to only form multi-ubiquitin chains if Lys^{48} is available, others show either increased promiscuity or different specificities in their choice of Lys residues on ubiquitin through which the multi-ubiquitin chain is formed [Haas et al., 1991]. Alternative multi-ubiquitin chains may provide a way of compartmentalizing the action of different ubiquitin conjugation reactions, or they may proscribe different fates within the cell [Finley and Chau, 1991; Hochstrasser, 1992; Jentsch, 1992]. More detailed characterization of ubiquitin mutants in yeast is eagerly awaited.

Ubiquitin-Like Genes

Although the highly conserved, canonical form of ubiquitin has been found in every eukaryote examined, there are increasing numbers of reports that identify ubiquitin-like coding sequences that share from 40% to 80% or more amino acid sequence homology with "true" ubiquitin [Haas et al., 1987; Toniolo et al., 1988; Meyers et al., 1989, 1991; Banerji et al., 1990; Guarino, 1990; Kas et al., 1992; Jones and Candido, 1993; Kumar et al., 1993; Lassalle et al., 1993; Linnen et al., 1993; Michiels et al., 1993; Olvera and Wool, 1993]. Thus, ubiquitin may ultimately come to be viewed as the conserved archetype of a family of ubiquitin-like proteins. Many ubiquitin-like proteins have C-terminal sequences that might permit their conjugation to other cellular proteins, but such an activity has only been clearly demonstrated for one [Loeb and Haas, 1992].

Curiously, ubiquitin homologs are often associated with viral function. Ubiquitin-like proteins have been found encoded in a baculovirus [Guarino, 1990] and a Togavirus [Meyers et al., 1989, 1991], where its presence correlated with the virulence of the strain. The gene sequence of a 15-kDa ubiquitin cross-reacting protein (UCRP) consists of two domains with homology to ubiquitin [Haas et al., 1987]. This protein is constitutively expressed in mammalian cells at low level and induced by interferon. Modification of proteins by UCRP may be among the repertoire of behaviors mounted by the cell in response to viral infection [Haas et al., 1987; Loeb and Haas, 1992]. UCRP is found covalently conjugated to other cellular proteins, suggesting that it exerts its cellular function through attachment to specific target proteins [Loeb and Haas, 1992]. An attractive speculation is that these ubiquitin-like proteins are one aspect of a broader strategy to modify host ubiquitin-dependent pathways in response to viral infection, either for the benefit of the virus in the case of virally encoded proteins or for the benefit of the host in the case of host-encoded proteins [Driscoll and Finley, 1992].

Several ubiquitin-like coding sequences are expressed as N-terminal fusions to other proteins, analogous to the gene fusions between canonical ubiquitin and ribosomal proteins [Toniolo et al., 1988; Banerji et al., 1990; Kas et al., 1992; Jones and Candido, 1993; Linnen et al., 1993; Michiels et al., 1993; Olvera and Wool, 1993]. At least some fusion proteins containing a ubiquitin-like N-terminal domain do not undergo proteolytic processing [Linnen et al., 1993]. The ubiquitin-like domains of these proteins may serve to facilitate their expression.

UBIQUITIN LIGATION SYSTEM

The enzymatic activities of ubiquitin conjugation were first isolated and characterized in the context of an in vitro proteolytic system derived from mammalian reticulocytes [reviewed by Hershko and Ciechanover, 1992]. The first genetic evidence arguing for a direct role of ubiquitin in selective proteolysis in somatic cells came from studies of a mammalian cell line with a temperature-sensitive lesion in ubiquitin conjugation. Shortly thereafter, genes encoding the enzymes that govern ubiquitin conjugation in yeast as well as other organisms were identified. Yeast in particular has provided a powerful context in which to investigate the sometimes subtle role played by specific components of the ubiquitin ligation system in cellular physiology.

Ubiquitin Activating (E1) Enzyme

An irony of ubiquitin history is that the first ubiquitin pathway mutant was identified in one of the least tractable genetic systems. A temperature-sensitive derivative of the mouse mammary carcinoma FM3A, called ts85, arrests at or near the S–G2 boundary and loses its ubiquitinated histones at the nonpermissive temperature. The thermolability of the ts85 strain is due to a mutation within the structural gene for ubiquitin activating (E1) enzyme [Finley et al., 1984]. E1 catalyzes the first step of the ubiquitin ligation pathway by forming a thiol ester with ubiquitin in preparation for its transfer to one of several ubiquitin conjugating (E2) enzymes [Haas et al., 1982]. At nonpermissive temperature, protein ubiquitination in ts85 cells is blocked. This mutant cell line was used to demonstrate that >95% of short-lived proteins were degraded via the ubiquitin system [Ciechanover et al., 1984]. This result confirmed that ubiquitin-dependent proteolysis is a major proteolytic system likely to be found in all eukaryotic cells. This was an important result: since erythroid cells execute an unusual developmental program during their terminal differentiation in which most of their previously long-lived proteins are eliminated, it was not clear prior to this report if the importance of ubiquitin-dependent proteolysis in the degradative processes of reticulocytes was reflected in other cell types.

Genes encoding the ~110-kDa E1 polypeptide have been isolated from several organisms, including yeast, wheat, mouse, and humans [Hatfield et al., 1990; Zacksenhaus and Sheinin, 1990; Handley et al., 1991a,b; Kay et al., 1991; McGrath et al., 1991; Mitchell et al., 1991; Hatfield and Vierstra, 1992; Imai et al., 1992; Tucker et al., 1992; Leyser et al., 1993]. In yeast, only one gene (UBA1) has been definitely shown to encode a bona fide E1 enzyme [McGrath et al., 1991]. UBA1 is essential for vegetative growth and for spore viability. Temperature-sensitive isolates of UBA1 do not yield a uniform arrest phenotype in yeast [McGrath et al., 1991]. In contrast, several other mammalian cell lines with mutations in E1 show specific defects in cell cycle progression, as originally seen with ts85 cells [Kulka et al., 1988; Nishitani et al., 1992; Mori et al., 1993]. E1-encoding genes contain a consensus sequence for a nucleotide-binding site, but comparison of E1 sequences with other enzymes that interact with ubiquitin (such as E2s) failed to identify homologies that might represent consensus sequences for interaction with ubiquitin [McGrath et al., 1991].

Investigations of other eukaryotes reveal that they often contain multiple genes that encode E1s. At least three E1 encoding genes have been characterized from wheat [Hatfield et al., 1990; Hatfield and Vierstra, 1992], and several E1 homologs (including a pseudogene) have been described in mouse [Kay et al., 1991; Mitchell et al., 1991; Imai et al., 1992; Tucker et al., 1992; Disteche et al., 1992]. Whether multiple forms of E1 participate in distinct ubiquitin-dependent functions is unknown. However, an apparent E1 in mouse has been implicated in spermatogenesis [Kay et al., 1991; Mitchell et al., 1991], and an apparent E1 homolog in *Arabidopsis* called AXR1 is required for normal response to the plant hormone auxin [Leyser et al., 1993], though the latter lacks a Cys residue that was identified in wheat E1s as being essential for catalytic activity [Hatfield and Vierstra, 1992].

Ubiquitin Conjugating (E2) Enzymes

The many functions of protein ubiquitination are governed by a multiplicity of ubiquitin conjugating (or E2) enzymes (Table I). E2 enzymes facilitate the transfer of ubiquitin from its activated form as a thiol ester with ubiquitin activating (or E1) enzyme to other cellular proteins via intermediate formation of a ubiquitin thiol ester with the E2 enzyme. Different E2 enzymes can act in distinct cellular functions, apparently reflecting differences in the regulation of their activity and substrate selectivity.

A comparison of E2 genes shows that they constitute a structurally related family of enzymes. All E2s have a homologous (>35%) N-terminal sequence domain of approximately 150–170 residues containing a conserved cysteine residue where the thiol ester with ubiquitin is formed. E2s have been classified into two types: class I E2s consist solely of this N-terminal domain, whereas class II E2s also bear a C-terminal extension [Jentsch et al., 1990]. Recently, the existence of class III E2s that contain divergent domains appended to their N-termini has been noted [Jentsch, 1992].

In *S. cerevisiae,* genes encoding at least 10 different E2 enzymes have been described in the literature [Jentsch et al., 1990]. In most (though not all) cases, the genes encoding E2s in other organisms constitute clear homologs to one of the known species of yeast E2s. Several yeast E2s display a striking degree of functional overlap and pleiotropy of function, and only two yeast E2s are essential for viability under normal growth conditions. These factors can complicate genetic analyses, since the phenotypes resulting from a single mutation may be subtle unless combined with another mutation. Such "synthetic" effects between genetic loci often imply that parallel and redundant functions are shared by the genes involved.

UBC1, UBC4, UBC5, and related genes. *S. cerevisiae* UBC1, 4, and 5 constitute an essential family of ubiquitin conjugating enzymes [Seufert and Jentsch, 1990; Seufert et al., 1990]. UBC1 encodes a class II E2 and is expressed primarily in stationary phase cells [Seufert et al., 1990]. UBC4 and UBC5 both encode class I E2s with better than 95% amino acid sequence homology and show reciprocal expression patterns, with UBC4 being expressed in exponentially growing cells and UBC5 being expressed in stationary cells [Seufert and Jentsch, 1990]. ubc1 mutants show a slow growth phenotype and are defective in growth following spare germination [Seufert et al., 1990]. ubc4 and ubc4ubc5 mutants also display a slow growth phenotype, with the phenotype of the latter being more severe [Seufert and Jentsch, 1990].

UBC1, UBC4, and UBC5 encode overlapping E2 activities that function together in selective proteolysis [Seufert et al., 1990]. UBC4- and UBC5-dependent ubiquitination accounts for the bulk of high-molecular-weight ubiquitin conjugates that are believed to represent intermediates in ubiquitin-dependent proteolysis [Seufert and Jentsch, 1990]. ubc4 mutants are moderately sensitive to amino acid analogs and cadmium [Jungmann et al., 1993a], while ubc4ubc5 double mutants are much more sensitive. Pulse-chase assays directly demonstrate that ubc4ubc5 double mutants display defects in the turnover of short-lived and abnormal proteins [Seufert and Jentsch, 1990]. ubc1 mutants display similar though less severe phenotypes than the ubc4ubc5 double mutant, and overexpression of UBC1 partially suppresses the phenotypes of ubc4ubc5 double mutants [Seufert et al., 1990]. Thus, this gene family appears to govern the degradation of the bulk of abnormal proteins. Genes that encode homologs of UBC4 and UBC5 have been identified in both *Drosophila* and *Caenorhabditis elegans* [Treier et al., 1992; Zhen et al., 1993]. In both cases, demonstration of structural homology based on amino acid sequences was supplemented by demonstration of functional homology based on the ability of the *Drosophila* and *C. elegans* genes to rescue the slow growth and proteolysis defect phenotypes of a ubc4ubc5 deleted yeast [Treier et al., 1992; Zhen et al., 1993].

UBC2(RAD6) and related genes. *S. cerevisiae* UBC2 was known under the guise of the DNA repair gene RAD6 for many years before it was identified as a ubiquitin conjugating (E2) enzyme [Jentsch et al., 1987]. Its functions are accomplished through its role as an E2, since mutagenesis of the single active

TABLE I. Genes of the Ubiquitin Conjugation Pathway

Species	Name	Comments	Sequence reference
Ubiquitin activating (E1) enzymes			
S. cerevisiae	UBA1	Essential gene	McGrath et al. [1991]
Arabidopsis	AXR1	Possible E1	Leyser et al. [1993]
		Required for normal auxin response	
Wheat	UBA1		Hatfield et al. [1990]
	UBA2		Hatfield and Vierstra [1992]
	UBA3		Hatfield and Vierstra [1992]
Mouse	Ube1	X-linked	Disteche et al. [1992]
	A1s9X		Mitchell et al. [1991]
	A1s9Y-1	Y-linked;	Key et al [1991]
	Sby	Candidate spermatogenic gene	Disteche et al. [1992]
	Ube2		Takahashi et al. [1992]
Human	UBE1	X-linked	Handley et al. [1991a,b]
Ubiquitin conjugating (E2) enzymes			
S. cerevisiae	UBC1	Class II E2	Seufert et al. [1990]
		Constitutes essential subfamily with UBC4/UBC5	
		Required for growth after sporulation and germination	
		Required for cadmium resistance	
	UBC2 (RAD6)	Class II E2	Jentsch et al. [1987]
		Required for postreplication repair, induced mutagenesis, sporulation, N-end rule pathway function; mutants perturb retrotransposition	
	UBC3 (CDC34)	Class II E2	Goebl et al. [1988]
		Essential gene; required for transition from G1 to S phase of the cell cycle	
	UBC4/UBC5	Class I E2s with ~95% similarity	Seufert and Jentsch [1990]
		Constitute essential subfamily with UBC1	
		Required for degradation of short-lived and abnormal proteins, sporulation, stress resistance, cadmium resistance	
	UBC6	Class II E2	Sommer and Jentsch [1993]
		Membrane associated with the ER	
		Participates in MATα2 degradation; degradation of membrane proteins	
	UBC7	Class I E2	Jungmann et al. [1993b]
		Required for cadmium resistance	
		Participates in MATα2 degradation	
	UBC8	Class II E2	Qin et al. [1991]
		Function unknown	

	UBC9	Class I E2	sequence unpublished; discussed in Jentsch [1992]
		Essential gene; required for transit through G2/M phases of the cell cycle	
	UBC10 (PAS2)	Class II E2	Haass et al. [1990]
		Required for peroxisome biogenesis	
S. pombe	rhp6	Class I E2	Reynolds et al. [1990]
		Homolog of yeast UBC2(RAD6)	
Plants	E2-23 kDa (UBC4)	Class II E2	Sullivan and Vierstra [1989]
	UBC1	Isolated from wheat	Sullivan and Vierstra [1991b]
		Class I E2	
	UBC7	Isolated from wheat and Arabadopsis	Van Nocker and Vierstra [1991]
		Isolated from wheat	
Drosophila	UbcD1	Class I E2	Treier et al. [1992]
		Functional homolog of yeast UBC4/5	
	Dhr6	Class I E2	Koken et al. [1991a]
		Functional homolog of yeast UBC2(RAD6)	
C. elegans	Ubc-2	Class I E2	Zhen et al. [1993]
		Functional homolog of yeast UBC4/UBC5	
Viral	ASFV-UBC	Class II E2	Hingamp et al. [1992]
		Encoded by African swine flu virus	
Mammals	E2-25 kDa	Class I E2	Chen et al. [1991]
		Isolated from bovine thymus library encodes enzyme involved in multiubiquitin chain formation	
	E2-14 kDa	Class I E2	Wing et al. [1992]
		Isolated from rabbit	
		Encodes apparent homolog of yeast UBC2(RAD6)	
	HHR6-1	Class I E2s;	Koken et al. [1991b]
	HHR6-2	Isolated from human;	
		Functional homologs of yeast UBC2(RAD6)	
Ubiquitin protein ligase (E3)			
S. cerevisiae	UBR1	Nonessential	Bartel et al. [1990]
		Required for N-end rule pathway function	

site Cys residue in UBC2(RAD6) yields the null phenotype [Sung et al., 1990]. Although UBC2(RAD6) defines one of three major epistasis groups that govern DNA repair in yeast, mutations in UBC2(RAD6) show striking pleiotropic defects: rad6 disruption mutants are hypersensitive to UV radiation and various chemical mutagens, are deficient in postreplication repair, display a slow growth phenotype, and are defective in induced mutagenesis, meiotic recombination, and sporulation [Prakash et al., 1990]. They also show enhanced levels and altered target specificity of Ty retrotransposition [Picologlou et al., 1990; Liebman and Newnam, 1993]. Disruption mutants of rad6 display a temperature-sensitive cell cycle defect with arrest around the S–G2 phases of the cell cycle [Ellison et al., 1991] and lack N-end rule pathway function [Dohmen et al., 1991; see below].

Yeast UBC2(RAD6) is a class II E2 and contains a 23-residue highly acidic C-terminal domain [Jentsch et al., 1987; Morrison et al., 1988]. This domain is dispensable for most of its functions except for sporulation and N-end rule pathway function [Morrison et al., 1988; Madura et al., 1993]. Even then, at least partial RAD6 function can be realized if even only a small portion of its tail remains [Morrison et al., 1988; Kolman et al., 1992]. Like UBC4 and UBC5, UBC2(RAD6) defines a highly conserved family of E2 enzymes found in eukaryotes, including fission yeast [Reynolds et al., 1990], *Drosophila* [Koken et al., 1991a], rabbit [Wing et al., 1992], and human [Koken et al., 1991b]. Curiously, all UBC2(RAD6) homologs obtained from eukaryotes other than yeast lack the acidic C-terminal domain of the yeast enzyme. They complement yeast ubc2 mutants for those UBC2(RAD6)-dependent functions that do not require the tail [Reynolds et al., 1990; Koken et al., 1991a,b].

How the above phenotypes can be related to a UBC2(RAD6)-mediated proteolytic event is unclear. UBC2(RAD6) encodes the E2 responsible for degradation of N-terminally targeted substrates via the N-end rule pathway in yeast and the mammals [Dohmen et al., 1991; Sung et al., 1991; Madura et al., 1993]. However, the absence of N-end rule pathway function in ubc2 mutants cannot account for other phenotypes of ubc2(rad6) mutants, since mutations in other genetic loci that eliminate N-end rule pathway function do not display other ubc2-associated phenotypes [Bartel et al., 1990]. There has been much speculation that UBC2 may govern transitions in chromatin structure that may be required for many of the above processes [Jentsch et al., 1987]. UBC2(RAD6) catalyzes the in vitro multi-ubiquitination of histones [Sung et al., 1988], but this may not reflect its true in vivo function since yeast appears to lack ubiquitinated histones [Swerdlow et al., 1990].

UBC3(CDC34) and UBC9. *S. cerevisiae* UBC3, first identified as the cell division cycle gene CDC34, was recognized as encoding a ubiquitin conjugating enzyme by virtue of sequence homology with UBC2(RAD6) [Goebl et al., 1988]. UBC3(CDC34) is essential, with loss of function ubc3/cdc34 mutations causing a defect in the transition from the G1 to S phase of the cell cycle [Goebl et al., 1988]. Ubc3(cdc34) ts mutations at the nonpermissive temperature do not initiate DNA synthesis, and they duplicate but do not separate their spindle pole bodies. The terminal phenotype of these mutants is striking, consisting of cells with multiple, elongated buds.

UBC3(CDC34) encodes a class II E2 that contains an extensive 125-residue C-terminal domain appended to the catalytic N-terminal domain. Within the C-terminal domain is a polyaspartate-containing region that resembles the UBC2(RAD6) C-terminal extension. UBC3(CDC34) also catalyzes the multi-ubiquitination of histones in vitro, via formation of a Lys^{48} multi-ubiquitin chain [Haas et al., 1991]. As for UBC2(RAD6), the in vivo relevance of this histone ubiquitinating activity is questionable. The targets of UBC3(CDC34) whose ubiquitination is required for cell cycle progression are unknown. While there is much speculation in the literature that G1 cyclins may be among the targets of UBC3(CDC34) [Tyers et al., 1992], the arrest phenotype of ubc3(cdc34) mutants is not easily reconciled with the phenotypes of G1 cyclin mutants that are stabilized against degradation, suggesting that they are not the essential targets of ubc3(cdc34) function.

Another E2, designated UBC9, is also essential by virtue of its role in cell cycle function; mutations in UBC9 cause an arrest phenotype consistent with a G2–M phase block in cell cycle progression [discussed by Jentsch, 1992]. UBC9 and UBC3(CDC34) are the only two yeast E2 genes that are individually required for cellular viability.

UBC6. *S. cerevisiae* UBC6 encodes a membrane-associated class II E2. Attachment of UBC6 to membranes is mediated through a hydrophobic membrane anchor sequence at its C terminus, with the N-terminal domain of the E2 residing in the cytosol. Immunofluorescence localizes UBC6 to the endoplasmic reticulum (ER) and nuclear envelope [Sommer and Jentsch, 1993]. Deletion mutants of UBC6 are viable and show no obvious growth defects. Based on synthetic interactions between ubc6 mutants and mutations in the SEC61 component of the secretory apparatus, Sommer and Jentsch [1993] have proposed that the secretory defect of at least some sec61 mutants is exacerbated by UBC6-mediated degradation of secretory apparatus components in the mutant cells. They have further proposed that degradation of short-lived ER membrane proteins may be among the wild-type functions governed by UBC6. UBC6, along with UBC7, also governs the targeting of one of two distinct degradation signals in the MATα2 transcriptional repressor [Chen et al., 1993].

UBC7. The yeast gene UBC7 was cloned by accident in studies unrelated to the ubiquitin system. Null mutants of ubc7 are not sensitive to heat shock, amino acid analogs, or DNA-damaging agents, but are hypersensitive to cadmium [Jungmann et al., 1993a]. Mutants in other components of the ubiquitin system, including UBC1, UBC4, and the proteasome gene PRE1 (Table II) also cause cadmium hypersensitivity, suggesting that cadmium resistance is mediated by abnormal protein degradation [Jungmann et al., 1993a]. Interestingly, ubc7 cells were not similarly sensitive to copper, zinc, or lead. As mentioned above, UBC7 also participates in the degradation of the yeast MATα2 repressor and appears to interact physically with UBC6 in vivo [Chen et al., 1993].

Other E2s. Two additional yeast E2 genes have been identified in other contexts. UBC10 (PAS2), mutations that prevent peroxisome biogenesis, encodes a class II E2 [Wiebel and Kunau, 1992]. Mutations that eliminate its active site Cys residue demonstrate a null phenotype, demonstrating that UBC10(PAS2)'s ubiquitin conjugating activity is required for peroxisome biogenesis. It is not known whether UBC10 function is expressed through proteolysis. UBC8 encodes a class E2 that contains an acidic C-terminal domain similar to that of other class II E2s. Disruption mutants of ubc8 show no obvious phenotype. Although UBC8 has an acidic C-terminal extension, it is apparently unable to catalyze the conjugation of ubiquitin to histones in vitro.

Several other E2s have been identified in a variety of eukaryotes; several are clearly homologs of known yeast E2s. A number of E2 encoding genes have been cloned from plants (both wheat and *Arabidopsis*) that bear varying degrees of homology to the yeast genes [Sullivan and Vierstra, 1989, 1991a; Van Nocker and Vierstra, 1991]. A virally encoded E2 has been described from African swine fever virus [Hingamp et al., 1992; Rodriguez et al., 1992]. The E2s in higher eukaryotes appear to be more structurally diverse than those encoded in yeast: a 230-kDa E2 has been identified in erythroid cells [Klemperer et al., 1989], and the existence of a large, 500-kDa E2 in murine cells has been noted [discussed by Jentsch, 1992]. Cloning and analysis of the genes encoding these E2s will be important milestones in determining what role they play in selective proteolysis.

Structure–Function Studies of Ubiquitin Conjugating Enzymes

The NH_2-terminal (catalytic) domain of E2s. The active site and catalytic activity of E2s reside within their conserved, N-terminal

TABLE II. Proteasome Subunit Genes

Species	Name	Type	Comments	References
A. thermoplasma	Taα	α	Potential tyrosine phosphorylation site	Zwickl et al. [1991]
	Taβ	β	Potential cAMP/cGMP-dependent phosphorylation site	Zwickl et al. [1992]
S. cerevisiae	PRE1	β	Nuclear localization signal-like sequence	Henemeyer et al. [1991]
			Essential gene	
			Pre1-1 mutants show enhanced sensitivity to amino acid analogs and stress	
			Pre1-1/pre1-1 mutants are sporulation deficient	
			Proteasomes isolated from pre1-1 mutants are deficient in chymotrypsin like activity	
	PRE2	β	Essential gene	Friedman et al. [1992]
	PRG1		Suppressor of mutation causing chromosome loss	Heinemeyer et al. [1993]
			Proteasomes isolated from pre2 mutants are deficient in chymotrypsin like activity	
	PRE4	β	Essential gene	Hilt et al. [1993]
			Pre4-1 mutations enhance phenotype of pre1-1 mutants	
			Proteasomes isolated from pre4-1 mutants are deficient in peptidylglutamyl cleaving activity	
	PRS1	α	Encodes subunit YC1	Fujiwara et al. [1990]
			Essential gene	
	PRS2 scl1+	α	Encodes subunit YC7-α	Fujiwara et al. [1990]
			Essential gene	
			Potential tyrosine phosphorylation site	
			Nuclear localization signal-like sequence	
	PRS3	?	Similar to rat C5	Lee et al. [1992]
			Essential gene	
			Lies adjacent to ERD2 on chromosome II	
	Y7	α	Essential gene	Emori et al. [1991]
	Y8		Essential gene	Emori et al. [1991]
	Y13	α	Nonessential gene; disruption mutants display slow-growth phenotypes	Emori et al. [1991]
			Proteasomes isolated from disruption mutant do not lack any proteasome associated proteolytic activity	
	PUP1	β	Essential gene	Haffter and Fox [1991]
	PUP2	α	Essential gene	Georgatsou et al. [1992]
D. melanogaster	PROS-35	α	Potential tyrosine phosphorylation site	Haass et al. [1989]

	PROS-28.1	α	Two potential tyrosine phosphorylation sites and a potential cAMP/cGMP-dependent phosphorylation site	Haass et al. [1990b]
	PROS-29	α	Nuclear localization signal-like sequence	Haass et al. [1990a]
A. thaliana	TAS-g64	α		Genschik et al. [1992]
Xenopus	XC3	α	Isolated as homolog of rat C3 Nuclear localization signal-like sequence	Fujii et al. [1991]
Rat		β		van Riel and Martens [1991]
	C1	β	Homolog of human RING10/murine LMP7; forms apparent subfamily with RING12 and DELTA	Aki et al. [1992]
	C2	α	Nα-acetylated N terminus Apparant homolog of human υ	Fujiwara et al. [1989]
	C3	α	Potential tyrosine phosphorylation site Nα-acetylated N terminus Nuclear localization signal-like sequence	Tanaka et al. [1990a]
	C5	?	Displays low similarity to other proteasome subunits Unmodified N terminus	Tamura et al. [1990]
	C8	α	Nα-acetylated N terminus	Tanaka et al. [1990b]
	C9	α	Potential tyrosine phosphorylation site Nuclear localization signal-like sequence	Kumatori et al. [1990]
	RING12	β	Isolated as homolog of human RING12; forms apparent subfamily with RING10 and DELTA	Tamura et al. [1992]
	DELTA	β	Isolated as homolog of human δ	Tamura et al. [1992]
	IOTA	α	Isolated as homolog of human ι	Tamura et al. [1992]
	ZETA	α	Isolated as homolog of human ζ	Tamura et al. [1992]
	RN3	β?	Tentatively classified as type β based on absence of α-like N-terminal extension Shows equivalent homology to A. thermoplasma α- and β-subunits	Thomsom et al. [1993]
Mouse	LMP2	β	Apparent homolog of human RING12 Encoded in MHC class II region γ-Interferon inducible	Martinez and Monaco [1991]
	LMP7	β	Apparent homolog of human RING10 Encoded in MHC class II region γ-Interferon inducible	Glynne et al. [1993]
	MC13	β	Apparent homolog of human RING10 Encoded in MHC class II region γ-Interferon inducible	Frentzel et al. [1992]

(Continued)

TABLE II. Proteasome Subunit Genes *(Continued)*

Species	Name	Type	Comments	References
Human	C2	α	C2 isolated as homolog of rat C2	DeMartino et al. [1991]
	υ			Tamura et al. [1991]
	C3	α	Isolated as homolog of rat C3 Nuclear localization signal-like sequence	Tamura et al. [1991]
	C5	?	Isolated as homolog of rat C5	Tamura et al. [1991]
	C8	α	Isolated as homolog of rat C8	Tamura et al. [1991]
	C9	α	Isolated as homolog of rat C9	Tamura et al. [1991]
	ι	α		DeMartino et al. [1991]
	ζ	α		DeMartino et al. [1991]
	δ	β		DeMartino et al [1991]
	RING10	β	Encoded in MHC class II region γ-Interferon inducible	Glynne et al. [1991]
	RING12	β	Encoded in MHC class II region γ-Interferon inducible	Kelly et al. [1991]

domain. Sequences surrounding the active site Cys residue are well conserved among the E2s. Alignment of E2 amino acid sequences reveals several residues in addition to those in the active site that are absolutely or near absolutely conserved among all E2s. These residues presumably represent structurally important residues within the E2 enzyme. Two-point mutants isolated in UBC3(CDC34) validate this view. The original temperature-sensitive allele of UBC3(CDC34), cdc34-1, contains a mutation that changes Pro^{71} to Ser. Pro^{71} is one of several Pro residues that are absolutely conserved among E2s [Ellison et al., 1991]. Modification of the analogous residue (Pro^{64}) in UBC2 (RAD6) resulted in the temperature sensitivity of that E2 as well [Ellison et al., 1991]. A second temperature-sensitive allele, cdc34-2, contains a mutation that encodes an Arg residue in the place of a highly conserved Gly residue (Z. Pitluk and D. Gonda, unpublished data). Thus, both of the two known conditional mutations of cdc34 most likely act by destabilizing the structure of the catalytic domain. Identification of these conserved residues as targets for mutagenesis provides a rational, directed way of producing temperature-sensitive alleles of other E2 encoding genes [Ellison et al., 1991].

Because Cys is absolutely necessary for thiol ester formation with ubiquitin, and because in many E2s only the single active site Cys is present, it has been an obvious target for mutagenesis. Mutation of the Cys in UBC2(RAD6) to Ala or Ser yielded native though catalytically inactive enzyme. Failure of this mutant to complement a variety of UBC2(RAD6) mutant phenotypes provided strong evidence that the pleiotropic functions of RAD6 were indeed expressed through its ubiquitin conjugating activity. Similarly, mutant ubc10(pas2) and ubc3(cdc34) that lack their active site Cys do not complement their respective mutants [Wiebel, 1992; D. Gonda, unpublished data]. When the Cys residue at the active site of yeast UBC2(RAD6) or *Arabidopsis* UBC1 is replaced by a Ser, ubiquitin is transferred by E1 to form an ester using the hydroxyl group on the Ser side chain [Sung et al., 1991; Sullivan and Vierstra, 1993]. This adduct is much more stable than the thiol ester formed on the wild-type enzyme and may provide a "trapped" E2–ubiquitin intermediate for structural studies.

Since the mutation of the Cys residue of an E2 presumably does not alter its structure, such a mutant should still interact with its normal partners (such as E3s; see below) in the ubiquitin conjugation pathway. These ubiquitination complexes would be unproductive, since they would involve a catalytically inactive E2. Thus such mutant E2s, when expressed at sufficiently high levels, should have a dominant negative effect on E2-mediated functions. Overexpression of ubc2A88 in yeast does indeed partially inhibit degradation of proteins by the N-end rule pathway, a proteolytic system that requires UBC2(RAD6) E2 function [Madura et al., 1993]. However, overexpression of the corresponding ubc3(cdc34), cdc34A95, has little effect on viability and cell cycle function in a wild-type genetic background (D. Gonda, unpublished data). This negative result can be ascribed to a need for very little UBC3(CDC34) activity for cell cycle function. In contrast, when the cdc34A95 mutation is overexpressed in some mutant ubc3(cdc34) backgrounds, it displays a potent dominant lethal effect (D. Gonda, unpublished data). This "synthetic" dominant lethality (*synthetic* refers to the fact that the phenotype is manifested only in specific genetic backgrounds) presumably arises either because the steady-state level of the endogenous mutant ubc3(cdc34) is reduced due to the ts mutation or because the cdc34A95 mutant more effectively competes away trans-acting factors from the pool of endogenous mutant cdc34 within the cell. These experiments suggest the possibility of using dominant negative ubiquitin conjugating enzyme mutations to explore E2 function in vivo. Since ubiquitin-dependent proteolysis is subject to a kind of combinatorial control in which multiple E2s participate in overlapping functions [Chen et al., 1993], dominant negative mutations may reveal aspects of a specific E2's function that would not be revealed by standard analyses involving the genetic ablation of a single E2 activity.

Although the C-terminal extensions of class II E2s have been proposed to encode the substrate specificity of the E2s in which they reside (see below), determinants of specific E2 function must also reside within the E2 N-terminal "catalytic" domains [Kolman et al., 1992; Silver et al., 1992]. Mutations within an E2 N-terminal domain that do not affect the catalytic activity of an E2 per se, but that eliminate a specific in vivo function of an E2, should indicate sequence determinants that govern interaction with substrates or trans-acting factors (such as E3 proteins; see below) that govern substrate selectivity. Because the N-terminal sequence domains of E2s encode a single, compact structural domain, such mutants are likely to be difficult to identify. A mutant of UBC2(RAD6) that lacks the N-terminal nine residues of the wild-type enzyme does not support the degradation of N-end rule pathway substrates either in vivo or in vitro, displays a sporulation defect, and shows a reduced efficiency of DNA repair. However, it is still proficient in UV-induced mutagenesis [Watkins et al., 1993]. These mutants are deficient in interaction with the UBR1 encoded E3, an auxiliary factor required for UBC2(RAD6) function in the N-end rule pathway [Bartel et al., 1990]. If UBC2(RAD6) function in induced mutagenesis also requires the binding of the E2 to trans-acting factors, they must rely on other determinants within the E2.

The function of E2 C-terminal extensions. The presence of C-terminal extensions in a subset of ubiquitin conjugating enzymes is the most striking feature distinguishing class II E2s from class I E2s and from one another. It was proposed early on that the in vivo role of the C-terminal extensions in class II E2s was to facilitate direct interaction of the E2 with their substrates [Jentsch et al., 1990]. This certainly appears to be their role in several of the known in vitro ubiquitination reactions. Ubiquitination of histones in vitro by UBC2(RAD6) and UBC3(CDC34) requires the presence of their acidic C-terminal extensions. Moreover, addition onto the class I *Arabidopsis* E2 UBC1 (which does not ubiquitinate histones) of an acidic C-terminal domain conferred on the modified E2 the ability to ubiquitinate histones, by virtue of electrostatic interactions between the basic histone substrates and the E2s' acidic tail [Van Nocker and Vierstra, 1991]. The importance of direct interactions between the E2 C-terminal domain and in vivo substrates is not known. It now seems unlikely, as originally suggested, that ubiquitination reactions catalyzed by class I and class II E2s differ significantly in their requirement for auxiliary factors called E3s (see below). Also, as shown by the fact that the C-terminal extension of UBC2(RAD6) is dispensable for many but not all of its in vivo functions [Morrison et al., 1988], the C-terminal domain of a class II E2 may not be necessary for all of its functions.

Despite the difficulties in making sweeping generalizations, however, in specific instances the C-terminal domains of class II E2s do clearly contain determinants involved in specifying in vivo function. The C-terminal domain of UBC3(CDC34) is required for its essential function in the cell cycle: A truncation mutant of UBC3(CDC34) that lacks the tail cannot complement the lethality of a ubc3 disruption mutant or ubc3(cdc34) ts mutants at nonpermissive temperature [Kolman et al., 1992; Silver et al., 1992]. Most of the UBC3(CDC34) tail is dispensable for essential cell cycle function, but a portion of the tail comprising at most 45 residues proximal to the N-terminal domain is required for essential function. Most strikingly, this region of the UBC3(CDC34) tail comprises a portable determinant that confers essential CDC34 function when appended onto the N-terminal domain of UBC2(RAD6) [Kolman et al., 1992; Silver et al., 1992]. The determinant most likely functions by interacting with trans-acting factors that regulate the substrate-specificity of the E2 or with the substrates themselves as previously suggested. It is not known whether the remainder of the CDC34 C-terminal domain is involved in a nonessential UBC3(CDC34)-dependent function or plays a facilitating but nonrate-limiting role in its essential function. Silver et al. [1992] have also argued that the tails of class II E2s, in addition to facilitating ubiquitination reactions, may also serve to restrict the targeting of an E2 to provide greater functional specificity.

The C-terminal domains of class II E2s may also be involved in more subtle regulation of the enzyme. Near the C-terminus of UBC3 (CDC34) are a quartet of Lys residues that can be autoubiquitinated by UBC3(CDC34) in vitro via an intramolecular transfer [Banerjee et al., 1993]. Thus, UBC3(CDC34) may regulate its own degradation through auto-ubiquitination. Given the dispensability of this region, however, such regulation may only play a subtle (though still potentially important) role in regulating UBC3(CDC34) levels.

Ubiquitin Protein Ligases (E3s)

Ubiquitin protein ligases, or E3 proteins, are operationally defined as auxiliary factors required for the E1- and E2-dependent ubiquitination of a target protein [Hershko et al., 1983]. In the best characterized cases, E3s appear to govern the substrate selectivity of protein ubiquitination by virtue of binding of the substrate to the E3 component of an E2–E3 complex [Hershko et al., 1986; Reiss et al., 1989]. Thus, E3s may play a central role in regulating the substrate selectivity of the ubiquitin ligation pathway.

Despite their apparent importance, the E3 proteins remain the least understood components of the ubiquitin system. The precise mechanistic role of E3 proteins in ubiquitination is unclear. Indeed, the paucity of information about known E3 activities makes a definition of E3 based on mechanistic or structural considerations impossible. A minimal view of at least some E3s is that they are simply "molecular docks" that serve to bring together E2 and substrate protein that otherwise would not interact. In this view, any cellular activity that brings together an E2 and a substrate to promote ubiquitination, no matter how fortuitously it does so, could be said to act as an E3. It is also possible, however, that E3s may comprise a more distinct class of enzymes that may actively participate in protein ubiquitination to catalyze or regulate the formation of multi-ubiquitin chains on the target protein.

UBR1. Only one E3 has been defined genetically. The gene UBR1 [Bartel et al., 1990] encodes a 225-kDa protein that appears to be the yeast homolog of the 180-kDa mammalian reticulocyte protein called E3α [Reiss and Hershko, 1990], which governs substrate selection of proteins with basic or hydrophobic N-terminal residues in the N-end rule pathway (see below). Like its mammalian homolog, UBR1 interacts specifically with proteolytic substrates and with its cognate E2 encoded by UBC2(RAD6) [Bartel et al., 1990; Dohmen et al., 1991; Madura et al., 1993]. Interaction between UBR1 and UBC2(RAD6) requires the acidic C-terminal tail [Madura et al., 1993] as well as the first nine residues of UBC2(RAD6) [Watkins et al., 1993]. A portion of the binding site for UBC2(RAD6) appears to reside in the 170 C-terminal residues of UBR1. The UBR1–UBC2 complex regulates the expression of UBR1 via changes in the stability of its mRNA [Madura et al., 1993].

UBR1 is nonessential. Mutants that lack UBR1 function (and which therefore lack a functional N-end rule pathway) display subtle phenotypes, including a very slight growth defect and an increased percentage of aberrant tetrads [Bartel et al., 1990]. Examination of the UBR1 sequence reveals no striking homologies to known motifs or any striking features, except for an unusual richness of Cys residues in one portion of the protein [Bartel et al., 1990]. Low stringency Southern analyses reveal no closely related sequences in mammalian genomes that may represent the mammalian homolog of UBR1. Nor, for that matter, do low stringency Southern analyses reveal any other UBR1–related sequence in the yeast genome. It therefore remains unclear whether UBR1/E3α will define a family of structurally homologous activities that provide E3 function to different ubiquitin-dependent pathways.

PROTEASOMES

Multi-ubiquitination of proteins targets them for destruction by a cytosolic proteolytic activity associated with a large multisubunit complex now termed the *proteasome*. Eukaryotic proteasomes manifest an unusually robust spectrum of proteolytic activities and likely constitute a heterogeneous population whose members differ in subunit composition and vary in response to differentiation and other changes of cell state [reviewed by Goldberg and Rock, 1992; Rivett, 1993].

Molecular genetic analyses of proteasomal subunits over the last several years have seen an intensive but necessary period of "molecular philately," in which the genes from over 40 proteasomal subunits have been cloned and sequenced. These studies reveal a remarkable degree of structural conservation between different proteasomal subunits and between proteasomal subunits of evolutionarily distinct species. They also identify an essential role for proteasomes in eukaryotes.

The 20S Proteasome

A 20S form of the proteasome appears to be the catalytic core of a larger, 26S form that degrades ubiquitin–protein conjugates. In eukaryotes, the 20S proteasome is made up of from 12 to 25 distinct polypeptide subunits (depending on source) ranging in molecular weight from approximately 20 to 35 kDa and has the appearance of four rings (containing six to seven subunits per ring) stacked along their axes to form a cylindrical particle [Goldberg and Rock, 1992; Rivett, 1993]. A simpler form of the 20S proteasome is found in archeabacteria that contain only two types of subunit, α and β, but that also have the appearance of a four-layered cylindrical particle with seven subunits per layer [Puhler et al., 1992; Zwickl et al., 1992].

Sequenced-Based Classification of Proteasome Subunits

A large number of 20S proteasome subunit-encoding genes from organisms as diverse as yeast, *Arabidopsis, Xenopus,* rat, and human have been cloned, including those encoding the α- and β-subunits of the archeabacterial proteasome (Table II). As deduced from their gene sequences, none of the proteasomal subunits show significant homology to previously characterized proteases. All proteasome subunits, however, show significant amino acid sequence homology with one another, indicating that they constitute a novel class of proteolytic enzymes.

Zwickl et al. [1992] have suggested that the two-component archeabacterial proteasome reflects a less complex evolutionary precursor to eukaryotic proteasomes, whose subunits have since evolved into multiple species to provide subtly specialized functions. Consistent with this view, most proteasomal subunits can be classified as α-type subunits or β-type subunits based on their sequence similarity with the archeabacterial proteins. The archeabacterial α- and β-subunits show significant homology with one another (24% identity), indicating that they too evolved from a common precursor [Zwickl et al., 1992]. They differ most strikingly at the N-terminal portion of the molecules, where the α-subunit contains a 26-residue N-terminal extension relative to the N-terminal end of the β-subunit.

Many proteasome subunits can be classified as α-type subunits based on the presence of a similar and homologous N-terminal extension, as well as through sequence similarity throughout the rest of the protein. Other eukaryotic proteasomal subunits are considered to be of β type based on the lack of the N-terminal extension and homology throughout the rest of their sequences. Some members within a specific class may be further classified based on additional internal sequence homologies [Haass et al., 1990a,b; Tamura et al., 1992]. Thus, the rat proteasomal subunits enncoded by RC1, rRING12, and rDELTA appear to constitute a subclass that is distinguishable from other rat β-type subunits [Tamura et al., 1992].

Several eukaryotic proteasomal sequences cannot be unambiguously classified as belonging to either the α or β type of proteasome subunits, however. The RN3 proteasomal subunit from rat lacks the N-terminal extension found in α-type subunits, but shares equivalent (22%) homology with both α- and β-subunits from archeabacteria [Thomson et al., 1993]. The RC5 proteasomal subunit from rat bears only weak homology to other proteasomal sequences [Tamura et al., 1990]. Indeed, its identity as a true component of the 20S proteasome has been questioned [Zwickl et al., 1992].

Immuno-EM studies indicate that the center two discs of the archaebacterial proteasome consist of β-subunits, while the end discs contain the α-subunits [Grziwa et al., 1991; Puhler et al., 1992]. Zwickl et al. [1992] proposed, based on circumstantial evidence, that the β-subunits of the archaebacterial enzyme carry the active sites of the protease, whereas the α-subunits perform regulatory functions. They also posited a similar arrangement of subunits in eukaryotic proteasomes. Congruent with this proposal, potential sites for tyrosine and cAMP/cGMP-dependent phosphorylation and sites with homology to known nuclear localization signal sequences seem to occur primarily within the α-type proteasome subunits (Table II). Whether these sites are functional in the proteasome awaits further study.

In contrast to the wealth of examples of 20S proteasome subunit genes, very little is known about genes that encode the additional subunits presumed to append to the 20S proteasome to form the ATP- and ubiquitin-dependent 26S form. Dubiel et al. [1992] have reported the cloning of a gene that encodes a 51-kDa polypeptide isolated from the 26S particle. Based on sequence homologies, they propose the existence of a family of ATPases that are involved in selective proteolysis in a variety of organisms and contexts.

Genetic Analysis of Proteasome Function

Eukaryotic proteasomes appear to govern both ubiquitin-dependent and -independent proteolysis. Thus, the proteasome might encompass functions beyond the scope of ubiquitin-dependent pathways. However, as ubiquitin-dependent proteolysis comprises greater than 90% of the turnover of short-lived proteins in eukaryotes, the phenotypes of proteasomal mutants are likely to reflect in large part the consequences of obstructing ubiquitin-dependent degradative pathways.

Proteasome function in yeast. Almost all null mutants of proteasome subunit-encoding genes in the yeast *S. cervisiae* are lethal (Table II). This result is not surprising in light of the essentiality of the ubiquitin ligation pathway, but the exact mechanism of lethality in these mutants is unclear. A null mutant may produce an intact proteasome that lacks an associated activity critical for cellular viability. Alternatively, the absence of a proteasomal subunit may prevent assembly of the enzyme. The only known nonessential proteasome gene encodes an α-type subunit called Y13 [Emori et al., 1991]. Mutants deleted for the Y13-encoding gene are viable, but display a slow growth phenotype. Proteasomes isolated from Y13 null mutants possess all three types of proteolytic activities associated with the 20S proteasome [Emori et al., 1991]. Curiously, however, the "latency" of the isolated proteasomes is affected by the absence of Y13. That is, proteasomes isolated from wild-type yeast are relatively inactive unless perturbed in some way, such as by exposure to SDS, whereas the proteasomes isolated from Y13-deleted yeast show strong activity before any perturbing treatments are applied. This may simply be because the absence of the Y13 subunit sufficiently perturbs the proteasome to cause activation without any further treatment. Alternatively, Y13 may serve as a negative regulator of proteasome activity.

Viable strains containing mutant alleles of essential proteasome genes provide insight into subunit function within the proteasome and proteasome function within the cell. Proteasomes isolated from strains carrying mutant alleles of PRE1 (pre1-1) or PRE2 (pre2-1 and pre2-2) are defective in their chymotrypsin-like activity [Heinemeyer et al., 1991, 1993]. Yeast harboring these mutants display enhanced sensitivity to various stresses, including growth on amino acid analogs such as canavanine and nutritional and heat stresses [Heinemeyer et al., 1991, 1993]. In addition, pre 1-1 mutants have been reported to display hypersensitivity to cadmium equivalent to that of ubc7 mutants [Jungmann et al., 1993a]. Moreover, under stress conditions, pre1 and pre2 mutants display defects in bulk proteolysis and a concomitant accumulation of ubiquitin conjugates [Heinemeyer et al., 1991, 1993]. Several proteins known to be substrates of ubiquitin-dependent proteolytic pathways are stabilized in pre1-1 strains [Seufert and Jentsch, 1992]. Taken together, the above observations indicate that proteasome activity is required for the degradation of ubiquitin–protein conjugates. A mutant allele of PRE4 (pre4-1), which results in proteasomes that are deficient in peptidyl-glutamyl-peptide-hydrolyzing activity, shows no stress-related phenotypes and no accumulation of ubiquitin conjugates, though it does enhance the stress-related phenotypes of pre1-1 mutants when combined with them in a single strain [Hilt et al., 1993].

Proteasomes and MHC function. In the last few years, the proteasome has been recognized as the enzyme that digests intracellular antigens to produce antigenic peptides, which are then presented on class I MHC-producing cells [reviewed by Driscoll and Finley, 1992; Goldberg and Rock, 1992]. A major piece of evidence indicating this came from the cloning of genes encoded within the class II MHC region whose homology identified them as proteasome subunits (RING10 and RING12 in humans, corresponding to C1 and RING12 in rats and LMP7 and LMP2 in mouse; see Table II). These particular subunits are apparently not essential for viability, nor are they even required for antigen presentation [Arnold et al., 1992; Momburg et al., 1992]. More likely, they serve to alter subtly the subpopulation of proteosomes in which they reside, possibly to optimize its specificities and activities for the generation of antigenic peptides [Driscoll and Finley, 1992].

UBIQUITIN C-TERMINAL HYDROLASES

Ubiquitin C-terminal hydrolases cleave peptide or isopeptide bonds formed with the C ternminus of ubiquitin [Pickart and Rose, 1985]. Some hydrolases are specialized for cleavage of small adducts from the ubiquitin terminus and may be chiefly responsible for the recycling of ubiquitin following degradation of a ubiquitinated protein. Others can cleave large adducts and are responsible for the processing of ubiquitin–fusion precursor proteins [Ozkaynak et al., 1987], the disassembly of ubiquitin–protein conjugates [Kanda et al., 1986], and the disassembly of multi-ubiquitin chains [Hadari et al., 1992]. The steady-state level of intracellular pools of ubiquitin conjugates and the extent of ubiquitination on a given substrate protein reflect a dynamic balance between the initiation and extension of multi-ubiquitin chains on substrate proteins on the one hand and the disassembly of ubiquitin protein conjugates by isopeptidases on the other [Haas and Bright, 1985, 1987]. C-terminal hydrolase activity thus influences bulk movement of material through intracellular pools of ubiquitinated proteins and may facilitate a kinetic proofreading-like mechanism to minimize the degradation of inappropriately targeted proteins.

Genes encoding four ubiquitin C-terminal hydrolases (YUH1, UBP1, UBP2, and UBP3) have been identified in yeast [Miller et al., 1989; Tobias and Varshavsky, 1991; Baker et al., 1992]. UBP1–3 share little sequence similarity except about their putative active site, and none share significant homology with YUH1 [Baker et al., 1992]. UBP1–3 cleave many large adducts from ubiquitin's C terminus. However, UBP1 and UBP2 can cleave poly-ubiquitin fusion proteins when coexpressed with substrate in *Escherichia coli,* whereas UBP3 cannot [Baker et al., 1992]. YUH1 can also cleave a variety of ubiquitin–protein fusions, but not poly-ubiquitin [Miller et al., 1989]. Yeast mutants that are deleted in all four ubiquitin C-terminal hydrolases are still viable, indicating that additional genes encoding processing activities remain to be found [Baker et al., 1992].

In higher eukaryotes, ubiquitin C-terminal hydrolases can show considerable tissue specificity. PGP 9.5 was originally identified as a neuron-specific protein that constitutes almost 10% of neuronal protein. Comparison of a cDNA clone of human ubiquitin C-terminal hydrolase isozyme L3 with known genes showed its product to be homologous to PGP 9.5, and enzymatic assays subsequently confirmed that PGP 9.5 is a C-terminal hydrolase [Wilkinson et al., 1989]. Its function in neuronal cells and the reason for its abundance are not known. During oogenesis, *Drosophila* nurse cells express a ubiquitin C-terminal hydrolase and transport its transcripts to the embryo [Zhang et al., 1993].

GENETIC ANALYSIS OF SUBSTRATE RECOGNITION

Degradation Signals

Contrary to past speculations that short-lived proteins were subject to intracellular proteolysis because of a global feature of their structure or stability, molecular genetic studies indicate that many short-lived eukaryotic proteins are actively targeted for destruction via degradation signals [Varshavsky, 1991], discrete amino acid sequence determinants that govern the turnover of the protein in which they reside. In substrates of ubiquitin-dependent proteolytic pathways, the degradation signals are bipartite, consisting of a site for recognition by a ubiquitinating enzyme complex (presumably through direct interaction with an E3 protein) and at least one Lys residue onto which a multi-ubiquitin chain is synthesized [Bachmair and Varshavsky, 1989; Johnson et al., 1990; Varshavsky, 1992]. Often, several Lys residues are available for ubiquitination and can serve equally well, even though only one Lys residue may bear a multi-ubiquitin chain on any given target molecule [Bachmair and Varshavsky, 1989; Banerjee et al., 1993]. A short-lived protein may contain multiple degradation signals, with any one signal being sufficient to support at least some degradation of the substrate [Hochstrasser and Varshavsky, 1990].

The two determinants of the bipartite degradation signal need not reside on the same polypeptide chain [Johnson et al., 1990]. A ubiquitination complex (presumably composed of an E2 and an E3) can bind to a multimeric protein by virtue of a determinant on one of the polypeptide chains and ubiquitinate a Lys residue on a different subunit within the multimer. This feature of the ubiquitin-mediated targeting, called *trans-targeting,* may provide a means of coupling the destruction of a protein monomer to its incorporation into a protein complex. Note that under the operational definition of E3, some of the subunits of a multimeric complex that is subject to trans-targeting may themselves be said to comprise a component of an E3 activity.

Genetic Analysis of Degradation Signals

Molecular genetic approaches encourage a linear reductionist view of degradation signals that to date has been successful. However, while elements of known degradation signals reside on compact segments that can be studied in isolation from the protein on which they reside, caution must be exercised in extrapolating from such experiments to the behavior of a degradation signal in the native protein. Many possibilities exist for regulating proteolysis by regulating the context within which a degradation signal resides, since the accessibility of a degradation signal to ubiquitination enzymes will influence its efficacy. For example, phosphorylation near a degradation signal or local structure may affect whether the signal will be recognized by the targeting apparatus. Degradation signal activity may thus be affected by other elements of the protein that are distant in linear sequence but close in three-dimensional space.

A common tool in characterizing degradation signals is the use of test proteins in which the short-lived protein under investigation is fused to a readily assayable reporter construct. A popular reporter has been *E. coli* β-galactosidase, because the steady-state level of β-galactosidase activity, which reflects the rate of turnover of the test protein, can be readily assayed both in culture and through colorimetric plate assays [Bachmair et al., 1986]. Genetic analyses can verify that the proteolytic systems examined using the test proteins are indeed the same systems that govern the targeting of the original substrate, by showing that mutations within the degradation signal or genes governing the targeting pathway affect degradation of the native substrate and the test protein in a congruent fashion. Experience with several systems has validated the use of fusion test proteins as tools for examining selective proteolysis in vivo.

The N-end rule pathway. The first degradation signals to be genetically defined were those of the N-end rule pathway [Varshavsky, 1992]. This pathway governs the degradation of substrates based on the identity of their N-terminal residue. It was discovered serendipitously when a set of ubiquitin–β-galactosidase fusion proteins was expressed in yeast. The ubiquitin moiety is rapidly and precisely removed by the same enzymes responsible for the processing of naturally occurring ubiquitin fusion precursors. This cleavage occurs regardless of which amino acid immediately follows the C-terminal Gly residue of the ubiquitin (except for Pro, which slows the cleavage). Thus, these constructs provide a "trick" for generating proteins with any desired N-terminal residue, circumventing the normal cellular pathways that generate protein N termini and that restrict cytosolic protein termini to only a subset of residues [Bachmair et al., 1986]. Depending on the identity of the residue on its N terminus, the half-life of the resulting protein varies from less than 2 minutes to greater than 20 hours [Bachmair et al., 1986]. Bacterially expressed ubiquitin–β-galactosidase fusion proteins are processed in an analogous manner in a reticulocyte in vitro system and reveal the presence of the N-end rule pathway here as well [Gonda et al., 1989]. In fact, many of the substrates used to characterize ubiquitin-dependent proteolysis in vitro turn out to be targeted via the mammalian N-end rule [Reiss et al., 1988].

Classification of the βgal test proteins and comparison of the N-end "rule books" of yeast and mammalian reticulocytes suggests that the reticulocyte N-end rule pathway constitutes an expanded version of the yeast pathway [Gonda et al., 1989]. βgal test proteins that possess a hydrophobic amino termini (called type II substrates), or those that possess (or can be modified to possess) a basic N terminus (called type I substrates) are short lived in yeast. Both classes of substrates interact with the UBR1-encoded E3, and degradation of both classes is eliminated in UBR1 mutants, suggesting that the substrate recognition site for both types of substrates resides on the single UBR1 polypeptide [Bartel et al., 1990]. UBR1 therefore appears to be the yeast homolog of reticulocyte E3α, which also binds substrates with hydrophobic or basic N termini in two distinct and functionally separable binding sites [Reiss et al., 1988; Reiss and Hershko, 1990]. An additional set of substrates that is targeted in the mammalian system, defined by Ser, Ala, and Thr-βgal test proteins, is not short-lived in yeast, apparently because yeast lacks the mammalian E3 activity that recognizes them [Gonda et al., 1989]. As already noted, the UBC2 (RAD6)-encoded E2 is necessary for N-end rule pathway function in yeast and physically interacts with UBR1 [Dohmen et al., 1991; Madura et al., 1993].

Substrates with basic or hydrophobic N termini are recognized directly by UBR1; thus, basic or hydrophobic amino acid residues have been termed *primary destabilizing residues.* Other N-end rule pathway substrates undergo post-translational modification prior to targeting for degradation. Proteins with acidic N-terminal residues (Asp and Glu; also Cys in mammalian cells) have an Arg residue post-translationally conjugated onto their N termini by the enzyme arginyl-tRNA protein transferase [Ferber and Ciechanover, 1987; Gonda et al., 1989]. Asp and Glu (and Cys) are therefore said to be secondary destabilizing residues [Gonda et al., 1989]. In yeast this enzyme is encoded by a single gene called ATE1, mutations that prevent the modification and thus the degradation of proteins bearing secondary destabilizing residues [Balzi et al., 1990]. Proteins with Asn or Gln at their N termini are first deamidated to form the secondary residues Asp or Glu (respectively) and then modified by the transferase [Gonda et al., 1989]. A yeast gene called DEA1 encodes the N-terminal deamidase required for conversion of tertiary destabilizing residues to secondary residues [discussed by Varshavsky, 1992].

The prototypic N-end rule degradation signal resides within an approximately 40-residue extension contained at the N-terminus of the βgal test proteins. The signal is portable and able to confer UBR1-dependent targeting and degradation on heterologous proteins to which it is appended [Bachmair and Varshavsky, 1989; Park et al., 1992]. This signal consists of the destabilizing N-terminal residue that is recognized by E3α/UBR1 and either one of two lysines that lie 15 and 17 residues from the N terminus. These two Lys residues are the only sites within βgal and DHFR-derived test proteins that can be efficiently conjugated by ubiquitin, and their elimination by mutagenesis stabilizes the substrates [Bachmair and Varshavsky, 1989; Chau et al., 1989; Johnson et al., 1990]. It is not clear how the sequence context directly around the ubiquitinatable Lys residues affect the efficiency of targeting, but one requirement appears to be that the Lys residues lie in a "segmentally mobile" region [Bachmair and Varshavsky, 1989]. This can be rationalized by the need for the two determinants of the degradation signal to accommodate themselves to the fixed, relative position of the N-terminal and Lys-binding sites within an E2–E3 ubiquitination complex. Substrates that are subject to other targeting pathways (e.g., damaged and abnormal proteins) may have degradation signals with similar conformational requirements.

The N-end rule pathway must perform a nonessential function in yeast, since mutants lacking N-end rule pathway function are viable. The mature N termini of almost all cytosolic proteins are of the stabilizing class, whereas the majority of secreted and compartmentalized proteins have N-terminal residues that are destabilizing [Bachmair et al., 1986]. Thus, this pathway may catalyze the degradation of miscompartmentalized proteins within the cytosol. In light of this speculation, it is worth noting that UBI4, the poly-ubiquitin encoding locus, was isolated as a high copy suppresser of the lethality caused by clathrin deficiency in some yeast strains [Nelson and Lemmon, 1993].

PEST sequences. Rogers et al. [1986] suggested that sequences enriched in proline, glutamic acid, serine, and threonine (P, E, S, and T) act as degradation signal, based on their association with short-lived proteins. Subsequent studies have not upheld the generality of the PEST hypothesis, as the degradation signals of several short-lived proteins do not map to the PEST regions [Hochstrasser and Varshavsky, 1990; Glotzer et al., 1991]. PEST regions may be associated with short-lived proteins because short-lived proteins are likely to be under several different modes of regulation, including reversible phosphorylation, whose sites will be enriched in PEST-like sequences.

Cell cycle regulators. The degradation of cyclins is a key element in the regulation of the eukaryotic cell cycle. The first examples of cyclins were originally identified in marine invertebrates as proteins whose levels oscillated with the passage through the cell cycle, being synthesized continuously but degraded abruptly upon exit from mitosis. Cyclins comprise a large family of related proteins that bind to cdc2 and cdc2-like kinases. In turn these kinases, regulated by reversible phosphorylations and their association with cyclin, govern progression through the cell cycle. The timed degradation of mitotic cyclins that are bound to cdc2 is one of the suite of mechanisms that inactivates M-phase promoting factor kinase activity, allowing exit from mitosis [Nasmyth, 1993; Solomon, 1993].

The degradation of mitotic cyclins has been examined most intensively in *Xenopus* extracts, which reproduce many of the biochemical events of the *Xenopus* embryonic cell cycle [Murray and Kirschner, 1989; Murray et al., 1989]. Indirect but persuasive evidence indicates that cyclin is proteolyzed by a ubiquitin-dependent pathway [Glotzer et al., 1991; Hershko et al., 1991]. A degradation signal that can target a heterologous test protein for degradation during exit from mitosis maps to the N-terminal 90 residues of cyclin. Key components of the signal appear to be a nine-residue sequence motif referred to as the *destruction box* and a region of Lys residues that lies downstream of the destruction box and may serve as a ubiquitin attachment site [Glotzer et al., 1991]. The destruction box may represent the site of recognition by ubiquitination enzymes. Thus, the general design of the cyclin degradation signal is similar to that of the N-end rule pathway. Similar sequence motifs representing homologous degradation signals are found in several *S. cerevisiae* cyclins, and mutations that remove or alter the destruction boxes of the yeast mitotic cyclins encoded by CLB1 and CLB2 appear to prevent their degradation [Ghiara et al., 1991; Surana et al., 1993]. It remains an outstanding problem how the temporal component of cyclin degradation is imposed on the targeting of cyclins for destruction.

Other cell cycle regulators are targets for temporally regulated degradation. The c-mos proto-oncogene product, Mos, functions in both early and late steps during meiotic maturation in *Xenopus* oocytes. Mos is unstable during germinal vesicle breakdown but is later stabilized. The instability of Mos in the earlier steps is governed by the identity of its penultimate residue. Later, autophosphorylation at Ser^3 of Mos stabilizes it. Mutation of a specific Lys residue blocked ubiquitination of Mos and its degradation, indicating that Mos is targeted by the ubiquitin system. Thus, substrate phosphorylation may be used to govern its targeting by the ubiquitin system and may be involved in regulating the degradation of other cell cycle proteins (such as cyclins) as well.

MATα2 repressor. The steady-state level of a short-lived protein can be rapidly adjusted through changes in rates of its synthesis. This is an especially important feature of regulators that must act within a time frame less than that encompassing cell division. The MATα2 transcriptional repressor, which functions in mating type control in yeast, is short-lived, with a half-life of approximately 5 minutes [Hochstrasser and Varshavsky, 1990]. The MATα2 protein contains two degradation signals named Deg1 and Deg2, each of which can confer a short lifetime on a β-galactosidase test protein on which it is fused. The MATα2 protein is degraded via ubiquitin-dependent pathways, though the sites of ubiquitin attachment have not been defined [Hochstrasser et al., 1991]. Deg1 and Deg2 are targeted by pathways defined by distinct subsets of ubiquitin conjugating enzymes. Test proteins containing the Deg1 signal are degraded via a UBC6/UBC7-dependent pathway. UBC4 and UBC5 also define an epistasis group of E2s that participate in MATα2 degradation [Chen et al., 1993]. Based on the involvement of multiple E2s in MATα2 degradation, and their physical interaction with one another, Chen et al. [1993] propose that targeting of proteolytic substrates may be governed by a kind of "combinatorial control," in which the target selectivity of ubiquitination complexes is a function of the E2s and E3s that make up the complexes.

FUTURE PROSPECTS

Genetic analyses of ubiquitin-dependent proteolysis have shown it to be both highly conserved among eukaryotes and capable of fascinating variation. The scope of its variation will be revealed as more genes of the ubiquitin system are cloned and characterized. Through genetics, additional elements can often be easily recognized based on sequence homologies, which offers in many cases a means simpler than direct biochemistry for candidate components of the ubiquitin system to be identified. The availability of cloned genes of ubiquitin system enzymes will permit detailed dissection of their structure and function through direct mutagenesis and provide insight into their mechanism. Ubiquitin protein ligases, the least understood component of the ubiquitin system, should soon yield their secrets through a combination of genetic and biochemical strategies and complete the outline of how substrates are targeted for destruction. And, finally, exhaustive dissection of degradation signals in a variety of short-lived proteins will hopefully provide a coherent picture of how substrate proteins are distinguished from nonsubstrate proteins in the cell. Analyses of the gene encoding components of the ubiquitin system should continue to play an important part in exploring its mechanisms, function, and regulation.

ACKNOWLEDGMENTS

I thank the members of the Department of Molecular Biophysics and Biochemistry, my laboratory, and my family for their support. Work in my laboratory has been funded by grants from the NIH (GM45314), the ACS (CB-80), and the Donaghe Foundation. D.G. is a Scholar of the Leukemia Society of America.

REFERENCES

Aki M, Tamura T, Tokunaga F, Iwanaga S, Kawamura Y, Shimbara N, Kagawa S, Tanaka K, Ichihara A (1992): cDNA cloning of rat proteasome subunit RC1, a homologue of RING10 located in the human MHC class II region. FEBS Lett 301:65–68.

Arnold D, Driscoll J, Androlewicz M, Hughes E, Cresswell P, Spies T (1992): Proteasome subunits encoded in the MHC are not generally required for the processing of peptides bound by MHC class I molecules. Nature 360:171–174.

Bachmair A, Finley D, Varshavsky A (1986): In vivo half-life of a protein is a function of its amino-terminal residue. Science 234:179–186.

Bachmair A, Varshavsky A (1989): The degradation signal in a short-lived protein. Cell 56:1019–1032.

Baker RT, Tobias JW, Varshavsky A (1992): Ubiquitin-specific proteases of *Saccharomyces cerevisiae*. Cloning of UBP2 and UBP3, and functional analysis of the UBP gene family. J Biol Chem 267:23364–23375.

Balzi E, Choder M, Chen W, Varshavsky A, Goffeau A (1990): Cloning and functional analysis of the arginyl-tRNA-protein transferase gene ATE1 of *Saccharomyces cerevisiae*. J Biol Chem 265:7464–7471.

Banerjee A, Gregori L, Xu Y, Chau V (1993): The bacterially expressed yeast CDC34 gene product can undergo autoubiquitination to form a multiubiquitin chain-linked protein. J Biol Chem 268:5668–5675.

Banerji J, Sands J, Strominger JL, Spies T (1990): A gene pair from the human major histocompatibility complex encodes large proline-rich proteins with multiple repeated motifs and a single ubiquitin-like domain. Proc Natl Acad Sci USA 87:2374–2378.

Bartel B, Wunning I, Varshavsky A (1990): The recognition component of the N-end rule pathway. EMBO J 9:3179–3189.

Breslow E, Chauhan Y, Daniel R, Tate S (1986): Role of methionine-1 in ubiquitin conformation and activity. Biochem Biophys Res Commun 138:437–444.

Butt TR, Jonnalagadda S, Monia BP, Sternberg EJ, Marsh JA, Stadel JM, Ecker DJ, Crooke ST (1989): Ubiquitin fusion augments the yield of cloned gene products in *Escherichia coli:* Proc Natl Acad Sci USA 86:2540–2544.

Chau V, Tobias JW, Bachmair A, Marriott D, Ecker DJ, Gonda DK, Varshavsky A (1989): A multiubiquitin chain is confined to specific lysine in a targeted short-lived protein. Science 243:1576–1583.

Chen P, Johnson P, Sommer T, Jentsch S, Hochstrasser M (1993): Multiple ubiquitin-conjugating enzymes participate in the in vivo degradation of the yeast MAT alpha 2 repressor. Cell 74:357–369.

Chen ZJ, Niles EG, Pickart CM (1991): Isolation of a cDNA encoding a mammalian multiubiquitinating enzyme (E225K) and overexpression of the functional enzyme in *Escherichia coli*. J Biol Chem 266:15698–15704.

Ciechanover A, Finley D, Varshavsky A (1984): The ubiquitin-mediated proteolytic pathway and mechanisms of energy-dependent intracellular protein degradation. J Cell Biochem 24:27–53.

DeMartino GN, Orth K, McCullough ML, Lee LW, Munn TZ, Moomaw CR, Dawson PA, Slaughter CA (1991): The primary structures of four subunits of the human, high-molecular-weight proteinase, macropain (proteasome), are distinct but homologous. Biochim Biophys Acta 1079:29–38.

Disteche CM, Zacksenhaus E, Adler DA, Bressler SL, Keitz BT, Chapman VM (1992): Mapping and expression of the ubiquitin-activating enzyme E1 (Ube1) gene in the mouse. Mamm Genome 3:156–161.

Dohmen RJ, Madura K, Bartel B, Varshavsky A (1991): The N-end rule is mediated by the UBC2(RAD6) ubiquitin-conjugating enzyme. Proc Natl Acad Sci USA 88:7351–7355.

Driscoll J, Finley D (1992): A controlled breakdown: Antigen processing and the turnover of viral proteins. Cell 68:823–825.

Dubiel W, Ferrell K, Pratt G, Rechsteiner M (1992): Subunit 4 of the 26S protease is a member of a novel eukaryotic ATPase family. J Biol Chem 267:22699–2702.

Duerksen HP, Xu XX, Wilkinson KD (1987): Structure and function of ubiquitin: Evidence for differential interactions of arginine-74 with the activating enzyme and the proteases of ATP-dependent proteolysis. Biochemistry 26:6980–6987.

Ecker DJ, Butt TR, Marsh J, Sternberg EJ, Margolis N, Monia BP, Jonnalagadda S, Khan MI, Weber PL, Mueller L, et al. (1987a): Gene synthesis, expression, structures, and functional activities of site-specific mutants of ubiquitin. J Biol Chem 262:14213–14221.

Ecker DJ, Khan MI, Marsh J, Butt TR, Crooke ST (1987b):

Chemical synthesis and expression of a cassette adapted ubiquitin gene. J Biol Chem 262:3524–3527.

Ecker DJ, Stadel JM, Butt TR, Marsh JA, Monia BP, Powers DA, Gorman JA, Clark PE, Warren F, Shatzman A, et al. (1989): Increasing gene expression in yeast by fusion to ubiquitin. J Biol Chem 264:7715–7719.

Ellison KS, Gwozd T, Prendergast JA, Paterson MC, Ellison MJ (1991): A site-directed approach for constructing temperature-sensitive ubiquitin-conjugating enzymes reveals a cell cycle function and growth function for RAD6. J Biol Chem 266:24116–24120.

Emori Y, Tsukahara T, Kawasaki H, Ishiura S, Sugita H, Suzuki K (1991): Molecular cloning and functional analysis of three subunits of yeast proteasome. Mol Cell Biol 11:344–353.

Ferber S, Ciechanover A (1987): Role of arginine-tRNA in protein degradation by the ubiquitin pathway. Nature 326:808–811.

Finley D, Bartel B, Varshavsky A (1989): The tails of ubiquitin precursors are ribosomal proteins whose fusion to ubiquitin facilitates ribosome biogenesis. Nature 338:394–401.

Finley D, Chau V (1991): Ubiquitination. Annu Rev Cell Biol 7:25–69.

Finley D, Ciechanover A, Varshavsky A (1984): Thermolability of ubiquitin-activating enzyme from the mammalian cell cycle mutant ts85. Cell 37:43–55.

Finley D, Ozkaynak E, Varshavsky A (1987): The yeast polyubiquitin gene is essential for resistance to high temperatures, starvation, and other stresses. Cell 48:1035–1046.

Frentzel S, Graf U, Hammerling GJ, Kloetzel PM (1992): Isolation and characterization of the MHC linked beta-type proteasome subunit MC13 cDNA. FEBS Lett 302:121–125.

Friedman H, Goebel M, Snyder M (1992): A homolog of the proteasome-related RING10 gene is essential for yeast cell growth. Gene 122:203–206.

Fujii G, Tashiro K, Emori Y, Saigo K, Tanaka K, Shiokawa K (1991): Deduced primary structure of a *Xenopus* proteasome subunit XC3 and expression of its mRNA during early development. Biochem Biophys Res Commun 178:1233–1239.

Fujiwara T, Tanaka K, Kumatori A, Shin S, Yoshimura T, Ichihara A, Tokunaga F, Aruga R, Iwanaga S, Kakizuka A, et al. (1989): Molecular cloning of cDNA for proteasomes (multicatalytic proteinase complexes) from rat liver: Primary structure of the largest component (C2). Biochemistry 28:7332–7340.

Fujiwara T, Tanaka K, Orino E, Yoshimura T, Kumatori A, Tamura T, Chung CH, Nakai T, Yamaguchi K, Shin S, et al. (1990): Proteasomes are essential for yeast proliferation. cDNA cloning and gene disruption of two major subunits. J Biol Chem 265:16604–16613.

Genschik P, Philipps G, Gigot C, Fleck J (1992): Cloning and sequence analysis of a cDNA clone from *Arabidopsis thaliana* homologous to a proteasome alpha subunit from *Drosophila*. FEBS Lett 309:311–315.

Georgatsou E, Georgakopoulos T, Thireos G (1992): Molecular cloning of an essential yeast gene encoding a proteasomal subunit. FEBS Lett 299:39–43.

Ghiara JB, Richardson HE, Sugimoto K, Henze M, Lew DJ, Wittenberg C, Reed SI (1991): A cyclin B homolog in *S. cerevisiae:* Chronic activation of the Cdc28 protein kinase by cyclin prevents exit from mitosis. Cell 65:163–174.

Glotzer M, Murray AW, Kirschner MW (1991): Cyclin is degraded by the ubiquitin pathway. Nature 349:132–138.

Glynne R, Kerr LA, Mockridge I, Beck S, Kelly A, Trowsdale J (1993): The major histocompatibility complex-encoded proteasome component LMP7: Alternative first exons and post-translational processing. Eur J Immunol 23:860–866.

Glynne R, Powis SH, Beck S, Kelly A, Kerr LA, Trowsdale J (1991): A proteasome-related gene between the two ABC transporter loci in the class II region of the human MHC. Nature 353:357–360.

Goebl MG, Yochem J, Jentsch S, McGrath JP, Varshavsky A, Byers B (1988): The yeast cell cycle gene CDC34 encodes a ubiquitin-conjugating enzyme. Science 241:1331–1335.

Goldberg AL, Rock KL (1992): Proteolysis, proteasomes and antigen presentation. Nature 357:375–379.

Gonda DK, Bachmair A, Wunning I, Tobias JW, Lane WS, Varshavsky A (1989): Universality and structure of the N-end rule. J Biol Chem 264:16700–16712.

Graham RW, Jones D, Candido EP (1989): UbiA, the major polyubiquitin locus in *Caenorhabditis elegans*, has unusual structural features and is constitutively expressed. Mol Cell Biol 9:268–277.

Gregori L, Poosch MS, Cousins G, Chau V (1990): A uniform isopeptide-linked multiubiquitin chain is sufficient to target substrate for degradation in ubiquitin-mediated proteolysis. J Biol Chem 265:8354–8357.

Grziwa A, Baumeister W, Dahlmann B, Kopp F (1991): Localization of subunits in proteasomes from Thermoplasma acidophilum by immunoelectron microscopy. FEBS Lett 290:186–190.

Guarino LA (1990): Identification of a viral gene encoding a ubiquitin-like protein. Proc Natl Acad Sci USA 87:409–413.

Haas AL, Ahrens P, Bright PM, Ankel H (1987): Interferon induces a 15-kilodalton protein exhibiting marked homology to ubiquitin. J Biol Chem 262:11315–11323.

Haas AL, Bright PM (1985): The immunochemical detection and quantitation of intracellular ubiquitin-protein conjugates. J Biol Chem 260:12464–12473.

Haas AL, Bright PM (1987): The dynamics of ubiquitin pools within cultured human lung fibroblasts. J Biol Chem 262:345–351.

Haas AL, Reback PB, Chau V (1991): Ubiquitin conjugation by the yeast RAD6 and CDC34 gene products. Comparison to their putative rabbit homologs, E2(20K) AND E2(32K). J Biol Chem 266:5104–5112.

Haas AL, Warms JV, Hershko A, Rose IA (1982): Ubiquitin-activating enzyme. Mechanism and role in protein-ubiquitin conjugation. J Biol Chem 257: 2543–2548.

Haass C, Pesold HB, Kloetzel PM (1990a): The *Drosophila* PROS-29 gene is a new member of the PROS-gene family. Nucleic Acids Res 18:4018.

Haass C, Pesold HB, Multhaup G, Beyreuther K, Kloetzel PM (1989): The PROS-35 gene encodes the 35 kd protein subunit of *Drosophila melanogaster* proteasome. EMBO J 8:2373–2379.

Haass C, Pesold HB, Multhaup G, Beyreuther K, Kloetzel PM (1990b): The Drosophila PROS-28.1 gene is a member of the proteasome gene family. Gene 90:235–241.

Hadari T, Warms JV, Rose IA, Hershko A (1992): A ubiquitin C-terminal isopeptidase that acts on polyubiquitin chains. Role in protein degradation. J Biol Chem 267:719–727.

Haffter P, Fox TD (1991): Nucleotide sequence of PUP1 encoding a putative proteasome subunit in *Saccharomyces cerevisiae*. Nucleic Acids Res 19:5075.

Handley PM, Mueckler M, Siegel NR, Ciechanover A, Schwartz AL (1991a): Correction: Molecular cloning, sequence, and tissue distribution of the human ubiquitin-activating enzyme E1. Proc Natl Acad Sci USA 88:7456.

Handley PM, Mueckler M, Siegel NR, Ciechanover A, Schwartz AL (1991b): Molecular cloning, sequence, and tissue distribution of the human ubiquitin-activating enzyme E1. Proc Natl Acad Sci USA 88:258–262.

Hatfield PM, Callis J, Vierstra RD (1990): Cloning of ubiquitin activating enzyme from wheat and expression of a functional protein in *Escherichia coli*. J Biol Chem 265:15813–15817.

Hatfield PM, Vierstra RD (1992): Multiple forms of ubiquitin-activating enzyme E1 from wheat. Identification of an essential cysteine by in vitro mutagenesis. J Biol Chem 267:14799–14803.

Heinemeyer W, Gruhler A, Mohrle V, Mahe Y, Wolf DH (1993): PRE2, highly homologous to the human major histocompatibility complex-linked RING10 gene, codes for a yeast proteasome subunit necessary for chymotryptic activity and degradation of ubiqutinated proteins. J Biol Chem 268:5115–5120.

Heinemeyer W, Kleinschmidt JA, Saidowsky J, Escher C, Wolf DH (1991): Proteinase yscE, the yeast proteasome/multicatalytic-multifunctional proteinase: Mutants unravel its function in stress induced proteolysis and uncover its necessity for cell survival. EMBO J 10:555–562.

Hershko A, Ciechanover A (1992): The ubiquitin system for protein degradation. Annu Rev Biochem 61: 761–807.

Hershko A, Ganoth D, Pehrson J, Palazzo RE, Cohen LH (1991): Methylated ubiquitin inhibits cyclin degradation in clam embryo extracts. J Biol Chem 266:16376–16379.

Hershko A, Heller H, Elias S, Ciechanover A (1983): Components of ubiquitin–protein ligase system. Resolution, affinity purification, and role in protein breakdown. J Biol Chem 258:8206–8206.

Hershko A, Heller H, Eytan E, Reiss Y (1986): The protein substrate binding site of the ubiquitin–protein ligase system. J Biol Chem 261:11992–11999.

Hilt W, Enenkel C, Gruhler A, Singer T, Wolf DH (1993): The PRE4 gene codes for a subunit of the yeast proteasome necessary for peptidylglutamyl-peptide-hydrolyzing activity. Mutations link the proteasome to stress- and ubiquitin-dependent proteolysis. J Biol Chem 268:3479–3486.

Hingamp PM, Arnold JE, Mayer RJ, Dixon LK (1992): A ubiquitin conjugating enzyme encoded by African swine fever virus. EMBO J 11:361–366.

Hochstrasser M (1992): Ubiquitin and intracellular protein degradation. Curr Opin Cell Biol 4:1024–1031.

Hochstrasser M, Ellison MJ, Chau V, Varshavsky A (1991): The short-lived MAT alpha 2 transcriptional regulator is ubiquitinated in vivo. Proc Natl Acad Sci USA 88:4606–4610.

Hochstrasser M, Varshavsky A (1990): In vivo degradation of a transcriptional regulator: The yeast alpha 2 repressor. Cell 61:697–708.

Hodgins RR, Ellison KS, Ellison MJ (1992): Expression of a ubiquitin derivative that conjugates to protein irreversibly produces phenotypes consistent with a ubiquitin deficiency. J Biol Chem 267: 8807–8812.

Imai N, Kaneda S, Nagai Y, Seno T, Ayusawa D, Hanaoka F, Yamao F (1992): Cloning and sequence of a functionally active cDNA encoding the mouse ubiquitin-activating enzyme E1. Gene 118:279–282.

Jentsch S (1992): The ubiquitin-conjugation system. Annu Rev Genet 26:179–207.

Jentsch S, McGrath JP, Varshavsky A (1987): The yeast DNA repair gene RAD6 encodes a ubiquitin-conjugating enzyme. Nature 329:131–134.

Jentsch S, Seufert W, Sommer T, Reins HA (1990): Ubiquitin-conjugating enzymes: Novel regulators of eukaryotic cells. Trends Biochem Sci 15:195–198.

Johnson ES, Gonda DK, Varshavsky A (1990): cis-trans recognition and subunit-specific degradation of short-lived proteins. Nature 346:287–291.

Jones D, Candido EP (1993): Novel ubiquitin-like ribosomal protein fusion genes from the nematodes *Caenorhabditis elegans* and *Caenorhabditis briggsae*. J Biol Chem 268:19545–19551.

Jungmann J, Reins HA, Schobert C, Jentsch S (1993a): Resistance to cadmium mediated by ubiquitin-dependent proteolysis. Nature 361:369–371.

Jungmann J, Reins HA, Schobert C, Jentsch S (1993b):

Resistance to cadmium mediated by ubiquitin-dependent proteolysis. Nature 361:369–371.

Kanda F, Sykes DE, Yasuda H, Sandberg AA, Matsui S (1986): Substrate recognition of isopeptidase: Specific cleavage of the epsilon-(alpha-glycyl)lysine linkage in ubiquitin-protein conjugates. Biochim Biophys Acta 870:64–75.

Kas K, Michiels L, Merregaert J (1992): Genomic structure and expression of the human fau gene: Encoding the ribosomal protein S30 fused to a ubiquitin-like protein. Biochem Biophys Res Commun 187:927–933.

Kawalleck P, Somssich IE, Feldbrugge M, Hahlbrock K, Weisshaar B (1993): Polyubiquitin gene expression and structural properties of the ubi4-2 gene in *Petroselinum crispum*. Plant Mol Biol 21:673–684.

Kay GF, Ashworth A, Penny GD, Dunlop M, Swift S, Brockforff N, Rastan S (1991): A candidate spermatogenesis gene on the mouse Y chromosome is homologous to ubiquitin-activating enzyme E1. Nature 354:486–489.

Kelly A, Powis SH, Glynne R, Radley E, Beck S, Trowsdale J (1991): Second proteasome-related gene in the human MHC class II region. Nature 353:667–668.

Klemperer NS, Berleth ES, Pickart CM (1989): A novel, arsenite-sensitive E2 of the ubiquitin pathway: purification and properties. Biochemistry 28:6035–6041.

Koken M, Reynolds P, Bootsma D, Hoeijmakers J, Prakash S, et al. (1991a): Dhr6, a *Drosophila* homolog of the yeast DNA-repair gene RAD6. Proc Natl Acad Sci USA 88:3832–3836.

Koken MHM, Reynolds P, Jaspers-Dekker I, Prakash L, Prakash S, Bootsma D, HJHJ (1991b): Structural and functional conservation of two human homologs of the yeast DNA repair gene RAD6. Proc Natl Acad Sci USA 88:8865–8869.

Kolman CJ, Toth J, Gonda DK (1992): Identification of a portable determinant of cell cycle function within the carboxyl-terminal domain of the yeast CDC34 (UBC3) ubiquitin conjugating (E2) enzyme. EMBO J 11:3081–3090.

Kulka RG, Raboy B, Schuster R, Parag HA, Diamond G, Ciechanover A, Marcus M (1988): A Chinese hamster cell cycle mutant arrested at G2 phase has a temperature-sensitive ubiquitin-activating enzyme, E1. J Biol Chem 263:15726–15731.

Kumar S, Yoshida Y, Noda M (1993): Cloning of a cDNA which encodes a novel ubiquitin-like protein. Biochem Biophys Res Commun 195:393–399.

Kumatori A, Tanaka K, Tamura T, Fujiwara T, Ichihara A, Tokunaga F, Onikura A, Iwanaga S (1990): cDNA cloning and sequencing of component C9 of proteasomes from rat hepatoma cells. FEBS Lett 264:279–282.

Lassalle F, Lassegues M, Roch P (1993): Serological evidence and amino acid sequence of ubiquitin-like protein isolated from coelomic fluid and cells of the earthworm *Eisenia fetida andrei*. Comp Biochem Physiol [B] 104:623–628.

Lee DH, Tanaka K, Tamura T, Chung CH, Ichihara A (1992): PRS3 encoding an essential subunit of yeast proteasomes homologous to mammalian proteasome subunit C5. Biochem Biophys Res Commun 182:452–460.

Leyser HM, Lincoln CA, Timpte C, Lammer D, Turner J, Estelle M (1993): *Arabidopsis auxin*–resistance gene AXR1 encodes a protein related to ubiquitin-activating enzyme E1. Nature 364:161–164.

Liebman SW, Newnam G (1993): A ubiquitin-conjugating enzyme, RAD6, affects the distribution of Ty1 retrotransposon integration positions. Genetics 133:499–508.

Linnen JM, Bailey CP, Weeks DL (1993): Two related localized mRNAs from Xenopus laevis encode ubiquitin-like fusion proteins. Gene 128:181–188.

Loeb KR, Haas AL (1992): The interferon-inducible 15-kDa ubiquitin homolog conjugates to intracellular proteins. J Biol Chem 267:7806–7813.

Madura K, Dohmen RJ, Varshavsky A (1993): N-recognin/Ubc2 interactions in the N-end rule pathway. J Biol Chem 268:12046–12054.

Martinez CK, Monaco JJ (1991): Homology of proteasome subunits to a major histocompatibility complex-linked LMP gene. Nature 353:664–667.

McGrath JP, Jentsch S, Varshavsky A (1991): UBA 1: An essential yeast gene encoding ubiquitin-activating enzyme. EMBO J 10:227–236.

Meyers G, Rumenapf T, Thiel HJ (1989): Ubiquitin in a togavirus [Letter]. Nature 341:491.

Meyers G, Tautz N, Dubovi EJ, Thiel HJ (1991): Viral cytopathogenicity correlated with integration of ubiquitin-coding sequences. Virology 180:602–616.

Michiels L, Van der Rauwelaert E, Van HF, Kas K, Merregaert J (1993): fau cDNA encodes a ubiquitin-like–S30 fusion protein and is expressed as an antisense sequence in the Finkel-Biskis-Reilly murine sarcoma virus. Oncogene 8:2537–2546.

Miller HI, Henzel WJ, Ridgeway JB, Kuang W-J, Chisholm V (1989): Cloning and expression of a yeast ubiquitin-protein cleaving activity in *Escherichia coli*. Bio/Technology 7:698–704.

Mitchell MJ, Woods DR, Tucker PK, Opp JS, Bishop CE (1991): Homology of a candidate spermatogenic gene from the mouse Y chromosome to the ubiquitin-activating enzyme E1. Nature 354:483–486.

Momburg F, Ortiz NV, Neefjes J, Goulmy E, van de Wal Y, Spits H, Powis SJ, Butcher GW, Howard JC, Walden P, et al. (1992): Proteasome subunits encoded by the major histocompatibility complex are not essential for antigen presentation. Nature 360:174–177.

Monia BP, Ecker DJ, Crooke ST (1990): New perspectives on the structure and function of ubiquitin. Bio/Technology 8:209–215.

Mori M, Eki T, Takahashi KM, Hanaoka F, Ui M, Enomoto

T (1993): Characterization of DNA synthesis at a restrictive temperature in the temperature-sensitive mutants, tsFT5 cells, that belong to the complementation group of ts85 cells containing a thermolabile ubiquitin-activating enzyme E1. Involvement of the ubiquitin-conjugating system in DNA replication. J Biol Chem 268:16803–16809.

Morrison A, Miller EJ, Prakash L (1988): Domain structure and functional analysis of the carboxyl-terminal polyacidic sequence of the RAD6 protein of Saccharomyces cerevisiae. Mol Cell Biol 8:1179–1185.

Murray AW, Kirschner MW (1989): Cyclin synthesis drives the early embryonic cell cycle. Nature 339:275–280.

Murray AW, Solomon MJ, Kirschner MW (1989): The role of cyclin synthesis and degradation in the control of maturation promoting factor activity. Nature 339:280–286.

Nasmyth K (1993): Control of the yeast cell cycle by the Cdc28 protein kinase. Curr Opin Cell Biol 5:166–179.

Nelson KK, Lemmon SK (1993): Suppressors of clathrin deficiency: overexpression of ubiquitin rescues lethal strains of clathrin-deficient *Saccharomyces cerevisiae*. Mol Cell Biol 13:521–532.

Nishitani H, Goto H, Kaneda S, Yamao F, Seno T, Handley P, Schwartz AL, Nishimoto T (1992): tsBN75 and tsBN423, temperature-sensitive x-linked mutants of the BHK21 cell line, can be complemented by the ubiquitin-activating enzyme E1 cDNA. Biochem Biophys Res Commun 184:1015–10121.

Olvera J, Wool IG (1993): The carboxyl extension of a ubiquitin-like protein is rat ribosomal protein S30. J Biol Chem 268:17967–17974.

Ozkaynak E, Finley D, Solomon MJ, Varshavsky A (1987): The yeast ubiquitin genes: A family of natural gene fusions. EMBO J 6:1429–1439.

Ozkaynak E, Finley D, Varshavsky A (1984): The yeast ubiquitin gene: Head-to-tail repeats encoding a polyubiquitin precursor protein. Nature 312:663–666.

Park EC, Finley D, Szostak JW (1992): A strategy for the generation of conditional mutations by protein destabilization. Proc Natl Acad Sci USA 89:1249–1252.

Pickart CM, Rose IA (1985): Ubiquitin carboxyl-terminal hydrolase acts on ubiquitin carboxyl-terminal amides. J Biol Chem 260:7903–7910.

Picologlou S, Brown N, Liebman SW (1990): Mutations in RAD6, a yeast gene encoding a ubiquitin-conjugating enzyme, stimulate retrotransposition. Mol Cell Biol 10:1017–1022.

Prakash S, Sung P, Prakash L (1990): The Eukaryotic Nucleus. Caldwell, NJ: Telford Press, pp 275–292.

Puhler G, Weinkauf S, Bachmann L, Muller S, Engel A, Hegerl R, Baumeister W (1992): Subunit stoichiometry and three-dimensional arrangement in proteasomes from *Thermoplasma acidophilum*. EMBO J 11:1607–1616.

Qin S, Nakajima B, Nomura M, Arfin SM (1991): Cloning and characterization of a *Saccharomyces cerevisiae* gene encoding a new member of the ubiquitin-conjugating protein family. J Biol Chem 266:15549–15554.

Redman KL, Rechsteiner M (1989): Identification of the long ubiquitin extension as ribosomal protein S27a. Nature 338:438–440.

Reiss Y, Heller H, Hershko A (1989): Binding sites of ubiquitin–protein ligase. Binding of ubiquitin–protein conjugates and of ubiquitin-carrier protein. J Biol Chem 264:10378–10383.

Reiss Y, Hershko A (1990): Affinity purification of ubiquitin–protein ligase on immobilized protein substrates. Evidence for the existence of separate NH2-terminal binding sites on a single enzyme. J Biol Chem 265:3685–3690.

Reiss Y, Kaim D, Hershko A (1988): Specificity of binding of NH2-terminal residue of proteins to ubiquitin–protein ligase. Use of amino acid derivatives to characterize specific binding sites. J Biol Chem 263:2693–2698.

Reynolds P, Koken MH, Hoeijmakers JH, Prakash S, Prakash L (1990): The rhp6+ gene of *Schizosaccharomyces pombe:* A structural and functional homolog of the RAD6 gene from the distantly related yeast *Saccharomyces cerevisiae*. EMBO J 9:1423–1430.

Rivett JA (1993): Proteasomes: Multicatalytic proteinase complexes. Biochem J 291:1–10.

Rodriguez JM, Salas ML, Vinuela E (1992): Genes homologous to ubiquitin-conjugating proteins and eukaryotic transcription factor SII in African swine fever virus. Virology 186:40–52.

Rogers S, Rechsteiner M (1986): Amino acid sequences common to rapidly degraded proteins: the PEST hypothesis. Science 234:364–368.

Seufert W, Jentsch S (1990): Ubiquitin-conjugating enzymes UBC4 and UBC5 mediate selective degradation of short-lived and abnormal proteins. EMBO J 9:543–550.

Seufert W, Jentsch S (1992): In vivo function of the proteasome in the ubiquitin pathway. EMBO J 11:3077–3080.

Seufert W, McGrath JP, Jentsch S (1990): UBC1 encodes a novel member of an essential subfamily of yeast ubiquitin-conjugating enzymes involved in protein degradation. EMBO J 9:4535–4541.

Silver ET, Gwozd TJ, Ptak C, Goebl M, Ellison MJ (1992): A chimeric ubiquitin conjugating enzyme that combines the cell cycle properties of CDC34 (UBC3) and the DNA repair properties of RAD6 (UBC2): Implications for the structure, function and evolution of the E2s. EMBO J 11:3091–3098.

Solomon MJ (1993): Activation of the various cyclin/cdc2 protein kinases. Curr Opin Cell Biol 5:180–186.

Sommer T, Jentsch S (1993): A protein translocation defect linked to ubiquitin conjugation at the endoplasmic reticulum. Nature 365:176–179.

Sullivan ML, Vierstra RD (1989): A ubiquitin carrier protein from wheat germ is structurally and functionally similar to the yeast DNA repair enzyme encoded by RAD6. Proc Natl Acad Sci USA 86:9861–9865.

Sullivan ML, Vierstra RD (1991a): Cloning of a 16-kDa ubiquitin carrier protein from wheat and *Arabidopsis thaliana*. Identification of functional domains by in vitro mutagenesis. J Biol Chem 266:23878–23885.

Sullivan ML, Vierstra RD (1991b): Cloning of a 16-kDa ubiquitin carrier protein from wheat and *Arabidopsis thaliana*. Identification of functional domains by in vitro mutagenesis. J Biol Chem 266:23878–23885.

Sullivan ML, Vierstra RD (1993): Formation of a stable adduct between ubiquitin and the *Arabidopsis* ubiquitin-conjugating enzyme, AtUBC1+. J Biol Chem 268:8777–8780.

Sung P, Berleth E, Pickart C, Prakash S, Prakash L (1991): Yeast RAD6 encoded ubiquitin conjugating enzyme mediates protein degradation dependent on the N-end recognizing E3 enzyme. EMBO J 10:2187–2193.

Sung P, Prakash S, Prakash L (1988): The RAD6 protein of *Saccharomyces cerevisiae* polyubiquitinates histones, and its acidic domain mediates this activity. Genes Dev 2:1476–1485.

Sung P, Prakash S, Prakash L (1990): Mutation of cysteine-88 in the Saccharomyces cerevisiae RAD6 protein abolishes its ubiquitin-conjugating activity and its various biological functions. Proc Natl Acad Sci USA 87:2695–2699.

Sung P, Prakash S, Prakash L (1991): Stable ester conjugate between the *Saccharomyces cerevisiae* RAD6 protein and ubiquitin has no biological activity. J Mol Biol 221:745–749.

Surana U, Amon A, Dowzer C, McGrew J, Byers B, Nasmyth K (1993): Destruction of the CDC28/CLB mitotic kinase is not required for the metaphase to anaphase transition in budding yeast. EMBO J 12:1969–1978.

Swerdlow PS, Schuster T, Finley D (1990): A conserved sequence in histone H2A which is a ubiquitination site in higher eukaryotes is not required for growth in Saccharomyces cerevisiae. Mol Cell Biol 10:4905–4911.

Taccioli GE, Grotewold E, Aisemberg GO, Judewicz ND (1989): Ubiquitin expression in *Neurospora crassa*: Cloning and sequencing of a polyubiquitin gene. Nucleic Acids Res 17:6153–6165.

Takahashi E, Ayusawa D, Kaneda S, Itoh Y, Seno T, Hori T (1992): The human ubiquitin-activating enzyme E1 gene (UBE1) mapped to band Xp11.3—p11.23 by fluorescence in situ hybridization. Cytogenet Cell Genet 59:268–269.

Tamura T, Lee DH, Osaka F, Fujiwara T, Shin S, Chung CH, Tanaka K, Ichihara A (1991): Molecular cloning and sequence analysis of cDNAs for five major subunits of human proteasomes (multi-catalytic proteinase complexes). Biochim Biophys Acta 1089:95–102.

Tamura T, Shimbara N, Aki M, Ishida N, Bey F, Scherrer K, Tanaka K, Ichihara A (1992): Molecular cloning of cDNAs for rat proteasomes: Deduced primary structures of four other subunits. J Biochem (Tokyo) 112:530–534.

Tamura T, Tanaka K, Kumatori A, Yamada F, Tsurumi C, Fujiwara T, Ichihara A, Tokunaga F, Aruga R, Iwanaga S (1990): cDNA cloning and sequencing of component C5 of proteasomes from rat hepatoma cells. FEBS Lett 264:91–94.

Tanaka K, Fujiwara T, Kumatori A, Shin S, Yoshimura T, Ichihara A, Tokunaga F, Aruga R, Iwanaga S, Kakizuka A, et al. (1990a): Molecular cloning of cDNA for proteasomes from rat liver: Primary structure of component C3 with a possible tyrosine phosphorylation site. Biochemistry 29:3777–3785.

Tanaka K, Kanayama H, Tamura T, Lee DH, Kumatori A, Fujiwara T, Ichihara A, Tokunaga F, Aruga R, Iwanaga S (1990b): cDNA cloning and sequencing of component C8 of proteasomes from rat hepatoma cells. Biochem Biophys Res Commun 171:676–683.

Thomson S, Balson DF, Rivett AJ (1993): cDNA cloning of a new type of subunit of mammalian proteasomes. FEBS Lett 322:135–138.

Tobias JW, Varshavsky A (1991): Cloning and functional analysis of the ubiquitin-specific protease gene UBP1 of *Saccharomyces cerevisiae*. J Biol Chem 266: 12021–12028.

Toniolo D, Persico M, Alcalay M (1988): A "housekeeping" gene on the X chromosome encodes a protein similar to ubiquitin. Proc Natl Acad Sci USA 85:851–855.

Treier M, Seufert W, Jentsch S (1992): *Drosophila* UbcD1 encodes a highly conserved ubiquitin-conjugating enzyme involved in selective protein degradation. EMBO J 11:367–372.

Tucker PK, Phillips KS, Lundrigan B (1992): A mouse Y chromosome pseudogene is related to human ubiquitin activating enzyme E1. Mammal Genome 3:28–35.

Tyers M, Tokiwa G, Nash R, Futcher B (1992): The Cln3-Cdc28 kinase complex of *S. cerevisiae* is regulated by proteolysis and phosphorylation. EMBO J 11:1773–1784.

Van Nocker S, Vierstra RD (1991): Cloning and characterization of a 20-kDa ubiquitin carrier protein from wheat that catalyzes multiubiquitin chain formation in vitro. Proc Natl Acad Sci USA 88:10297–10301.

van Riel M, Martens GJ (1991): Cloning and sequence analysis of pituitary cDNA encoding the beta-subunit of *Xenopus* proteasome. FEBS Lett 291:37–40.

Varshavsky A (1991): Naming a targeting signal. Cell 64:13–15.

Varshavsky A (1992): The N-end rule. Cell 69:725–735.

Watkins JF, Sung P, Prakash S, Prakash L (1993): The extremely conserved amino terminus of RAD6 ubiquitin-conjugating enzyme is essential for amino-end rule–dependent protein degradation. Genes Dev 7:250–261.

Wiebel FF, Kunau WH (1992): The Pas2 protein essential for peroxisome biogenesis is related to ubiquitin-conjugating enzymes. Nature 359:73–76.

Wilkinson KD, Lee KM, Deshpande S, Duerksen HP, Boss JM, Pohl J (1989): The neuron-specific protein PGP 9.5 is a ubiquitin carboxyl-terminal hydrolase. Science 246:670–673.

Wing SS, Dumas F, Banville D (1992): A rabbit reticulocyte ubiquitin carrier protein that supports ubiquitin-dependent proteolysis (E214k) is homologous to the yeast DNA repair gene RAD6. J Biol Chem 267:6495–6501.

Wolf S, Lottspeich F, Baumeister W (1993): Ubiquitin found in the archaebacterium *Thermoplasma acidophilum*. FEBS Lett 326:42–44.

Zacksenhaus E, Sheinin R (1990): Molecular cloning, primary structure and expression of the human X-linked A1S9 gene cDNA which complements the ts A1S9 mouse L cell defect in DNA replication. EMBO J 9:2923–2929.

Zhang N, Wilkinson K, Bownes M (1993): Cloning and analysis of expression of a ubiquitin carboxyl terminal hydrolase expressed during oogenesis in *Drosophila melanogaster*. Dev Biol 157:214–223.

Zhen M, Heinlein R, Jones D, Jentsch S, Candido EP (1993): The ubc-2 gene of *Caenorhabditis elegans* encodes a ubiquitin-conjugating enzyme involved in selective protein degradation. Mol Cell Biol 13:1371–1377.

Zwickl P, Grziwa A, Puhler G, Dahlmann B, Lottspeich F, Baumeister W (1992): Primary structure of the *Thermoplasma* proteasome and its implications for the structure, function, and evolution of the multicatalytic proteinase. Biochemistry 31:964–972.

Zwickl P, Lottspeich F, Dahlmann B, Baumeister W (1991): Cloning and sequencing of the gene encoding the large (alpha-) subunit of the proteasome from *Thermoplasma acidophilum*. FEBS Lett 278:217–221.

ABOUT THE AUTHOR

DAVID K. GONDA is Assistant Professor of Molecular Biophysics and Biochemistry at Yale University in New Haven, CT, where he teaches undergraduate, graduate, and medical courses dealing with topics in biochemistry, molecular biology, and biophysics. After receiving a B.S. from Harvey Mudd College in Claremont, California, he pursued doctoral research at Yale University in the laboratory of Charles Radding, where he earned a Ph.D. in 1986 for work on in vitro studies of *E. coli* recA protein. This was followed by postdoctoral research as a Jane Coffin Childs Fellow at the Massachusetts Institute of Technology, where he worked with Alexander Varshavsky on studies of the ubiquitin system. Since joining the faculty at Yale University in 1990, his research has focused on biochemical and genetic studies of the ubiquitin system. In 1993 he was named as a Scholar of the Leukemia Society of America.

LYSOSOMAL/VACUOLAR SYSTEMS

Selective Degradation of Cytosolic Proteins by Lysosomes

J. Fred Dice and Stanley R. Terlecky

INTRODUCTION

Lysosomes are able to internalize cellular proteins in a variety of ways. One pathway is selective for cytosolic proteins containing peptide sequences biochemically related to Lys-Phe-Glu-Arg-Gln (KFERQ). This pathway is activated in confluent monolayers of cultured cells in response to deprivation of serum growth factors and applies to approximately 30% of cytosolic proteins. Intact animals also activate this proteolytic pathway in tissues such as liver, kidney, and heart in response to fasting.

We have reconstituted this lysosomal degradation pathway in vitro using highly purified lysosomes. Uptake and degradation of substrate proteins are stimulated by ATP and a member of the heat shock 70-kDa protein family, the 73-kDa constitutive heat shock protein (hsc73). This pathway is selective, since ribonuclease A and ribonuclease S-peptide are good substrates, but ribonuclease S-protein and dihydrofolate reductase are not. Furthermore, the uptake mechanism is saturable, and at 4°C substrate proteins bind specifically to a protein component of lysosomal membranes, presumably a receptor or polypeptide transport channel. A portion of cellular hsc73 is associated with lysosomes both on the cytoplasmic face of the lysosomal membrane and within the lysosomal lumen. Hsc73 within the lumen is required for polypeptide import into the organelle. These results indicate that the lysosomal polypeptide import process is strikingly similar to those for import of proteins for residence in other organelles.

LYSOSOMAL PATHWAYS OF PROTEOLYSIS

Both cytosolic and lysosomal pathways of proteolysis operate in most cells, and the contribution of the various proteolytic pathways depends on the cell type and its developmental and physiological status [Dice, 1987; Olson et al., 1992]. Intracellular proteins can be taken up and degraded by lysosomes by multiple pathways (Fig. 1).

Endocytosis and Exocytosis

The lysosomal degradation of endocytosed extracellular proteins has been extensively studied [Schmidt, 1992]. Many plasma membrane proteins and certain other proteins within the vacuolar apparatus are also degraded by lysosomes through endocytic pathways [Hare, 1990] (Fig. 1). Proteins traveling through exocytic pathways to reside on the cell surface or to be secreted from cells can also be diverted to lysosomes for degradation [Marzella and Glaumann, 1987; Olson et al., 1992].

Microautophagy

In well-nourished cells lysosomes appear to be able to internalize cytosolic proteins by a poorly understood process called *microautophagy* (Fig. 1) in which the lysosomal membrane invaginates at multiple locations to form

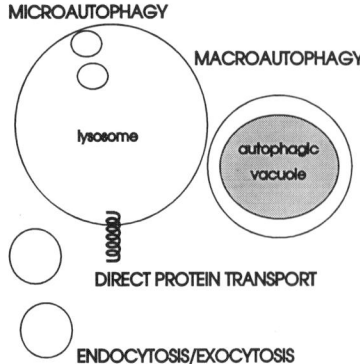

Fig. 1. *Schematic representations of pathways of lysosomal proteolysis. See text for descriptions of the pathways.*

intralysosomal vesicles [Ahlberg et al., 1982; Dice, 1987]. These vesicles presumably disintegrate whereupon their contents can be accessed by lysosomal hydrolases. Microautophagy appears to be nonselective in that several proteins and inert particles are internalized at similar rates [Ahlberg et al., 1982].

Macroautophagy

When cultured cells reach confluence and in certain tissues of fasted animals an additional lysosomal pathway of proteolysis is stimulated [Cockle and Dean, 1982; Kominami et al., 1983; Knecht et al., 1984; Mortimore, 1987]. Macroautophagy (Fig. 1) involves formation of double-membrane autophagic vacuoles from ribosome-free regions of the rough endoplasmic reticulum [Dunn, 1990a], the smooth endoplasmic reticulum [Ueno et al., 1991], or a preexisting organelle called the *phagophore* [Seglen et al., 1990]. These autophagic vacuoles then acquire lysosomal membrane proteins, acidify, and finally acquire lysosomal hydrolases. Macroautophagy also appears to be nonselective in that many different organelles and proteins are sequestered at approximately the same rates [Kominami et al., 1983; Kopitz et al., 1990].

Direct Protein Transport

Specific cytosolic proteins also enter lysosomes under conditions of stress, including serum growth factor deprivation of cultured confluent cells and in certain animal tissues in response to starvation. This pathway is restricted to cytosolic proteins that contain peptide sequences biochemically related to Lys-Phe-Glu-Arg-Gln (KFERQ). The mechanism by which proteins with KFERQ-like peptide regions are targeted to lysosomes for degradation is similar in many respects to the import of newly synthesized proteins into the mitochondrion, endoplasmic reticulum, nucleus, or peroxisome. Therefore, proteins with KFERQ-like peptide regions may enter lysosomes by directly crossing a membrane bilayer rather than by vesicular pathways (Fig. 1).

Activation of Lysosomal Proteolytic Pathways

Endocytosis and exocytosis appear to be active under many different conditions with the notable exception of the mitotic phase of the cell cycle [Warren, 1993]. However, the other pathways of lysosomal proteolysis are more acutely regulated and are active during different growth conditions of cultured cells (Fig. 2). Microautophagy may be the primary pathway of lysosomal proteolysis in cells during rapid cell division. Macroautophagy is maximally stimulated when cultured cells become confluent and is not further stimulated by serum withdrawal [Knecht et al., 1986]. Instead, serum withdrawal activates the direct protein transport pathway of lysosomal proteolysis [Dice, 1987].

Different lysosomal proteolytic pathways are also activated in liver in response to feeding and fasting (Fig. 3). Microautophagy appears to be the major lysosomal proteolytic pathway immediately after a meal, but macroautophagy is activated within a few hours following feeding [Mortimore, 1987]. In response to longer periods of starvation, the direct protein transfer pathway is activated [Simon et al., 1991; A.M. Cuervo, S.R. Terlecky, J.F. Dice, and E. Knecht, unpublished results].

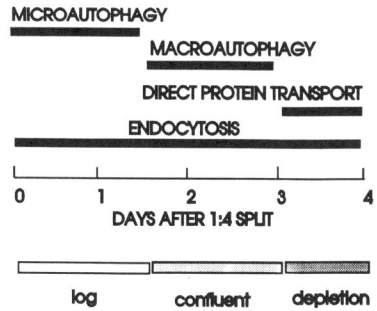

Fig. 2. *Lysosomal proteolytic pathways in cultured cells. Solid bars represent activity of the indicated pathways of lysosomal proteolysis.*

RIBONUCLEASE A AS A PROBE FOR SELECTIVE LYSOSOMAL PROTEOLYSIS

We used red cell–mediated microinjection to introduce specific radiolabeled proteins into the cytosol of confluent cultures of human fibroblasts [Neff et al., 1981; McElligott and Dice, 1984]. Degradation of the injected protein could be followed by monitoring either the loss of radioactivity from the cell monolayer or the appearance of acid-soluble radioactivity in the culture medium [McElligott and Dice, 1984]. These two measurements do not give identical half-lives because some acid-insoluble peptide degradation fragments as well as intact protein are also released from cells [Isenman and Dice, 1989]. However, conclusions regarding serum-regulated degradation are evident using either measurement.

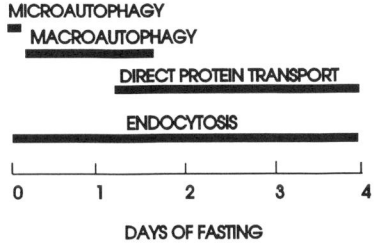

Fig. 3. *Lysosomal proteolytic pathways in liver. Solid bars indicate activity of the indicated pathways of lysosomal proteolysis in rat liver.*

TABLE I. Degradation of Long-Lived Proteins Microinjected Into Human Cells*

Protein	$t_{1/2}$ (hours)	
	± Serum	– Serum
Group 1: Regulated degradation		
RNase A	100	50
Aspartate amino transferase	80	40
Pyruvate kinase	296	121
Hemoglobin	213	106
Group 2: Nonregulated degradation		
RNase S-Protein	100	100
Ovalbumin	55	49
Lysozyme	59	59
Insulin α-chain	104	104

*IMR-90 human diploid fibroblasts were recipient cells in all cases except for pyruvate kinase, which was injected into HeLa cells. References to half-life values have been previously cited in Dice [1990].

Selectivity of the Proteolytic Pathway

Radiolabeled bovine pancreatic ribonuclease A (RNase A) is degraded with a half-life of approximately 100 hours in serum-supplemented cells and 50 hours in serum-deprived cells [Backer et al., 1983]. The half-lives of certain other microinjected proteins are also reduced in response to serum withdrawal, but half-lives of other microinjected proteins are unaffected by serum (Table I).

This selectivity in proteolysis is particularly striking in experiments in which ^{125}I-RNase and ^{131}I-RNase S-protein (residues 21–124 of RNase A) are coinjected into the same cells [Backer et al., 1983]. ^{125}I-RNase A is degraded more rapidly in response to serum withdrawal, while ^{131}I-RNase S-protein is degraded at the same rate in the presence or absence of serum (Fig. 4).

Targeting Signal for Selective Proteolysis

The N-terminal 20 amino acids of RNase A (RNase S-peptide) microinjected by itself is degraded in a serum-dependent manner [Backer et al., 1983] (Fig. 5), confirming that it contains the essential information for serum-regulated degradation. Furthermore, covalent attachment of RNase S-peptide to heterologous

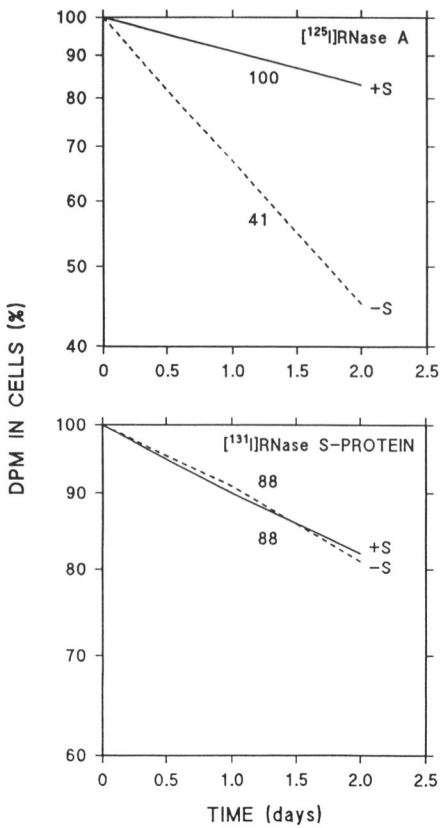

Fig. 4. Degradation of ^{125}I-RNase A and ^{131}I-RNase S-protein microinjected into the same cells. S = 10% calf serum. The numbers refer to half-lives in hours.

PEPTIDE	$T_{1/2}$ +S	−S
KETAAAKFERQHMDSSTSSA	62	30
KETAAAKFERQHMD	51	25
AAKFERQHM	45	25
ETAAAKF	72	73
KETAAAKFER	46	45
KFERQ	85	43

Fig. 5. Degradation characteristics of RNase S-peptide and smaller peptides. The first peptide listed is RNase S-peptide. Half-lives ($t_{1/2}$) were measured in the presence (+S) and absence (−S) of serum.

TABLE 2. Effect of RNase S-Peptide Sequences on Serum-Regulated Degradation of Test Proteins*

| | $t_{1/2}$ (hours) | |
Protein	± Serum	− Serum
Lysozyme	59	59
RNase S-peptide-lysozyme	47	21
Insulin α-chain	104	104
RNase S-peptide-insulin α-chain	100	44
β-Galactosidase**	134	154
RNase S-peptide-β-galactosidase**	178	78

*Half-life values can be found in Backer and Dice [1986].
**L. Jeffreys-Terlecky, M. Kirven-Brooks, and J.F. Dice (unpublished data).

proteins using water-soluble carbodiimides also causes their degradation rates to increase in response to serum withdrawal [Backer and Dice, 1986] (Table II).

To define further the crucial amino acid residues within RNase S-peptide, we microinjected various synthesized fragments of RNase S-peptide and determined whether or not their degradation is serum regulated (Fig. 5). Degradation of certain fragments (amino acids 1–14 and 3–13) were serum regulated while degradation of other fragments (amino acids 2–8 and 1–10) were not [Dice et al., 1986]. We then took advantage of the observation that red blood cell proteases cleave [^{3}H]RNase S-peptide during loading. A radiolabeled pentapeptide derived from RNase S-peptide is degraded faster in response to serum withdrawal, and we were able to identify this pentapeptide as residues 7–11, KFERQ. ^{3}H-KFERQ microinjected into fibroblasts was degraded in a serum-regulated fashion [Dice et al., 1986] (Fig. 5). Additionally, comicroinjection of ^{125}I-RNase A and excess nonradioactive KFERQ blocked the enhanced degradation of ^{125}I-RNase A in response to serum withdrawal [Chiang and Dice, 1988].

Further attempts to analyze the degradation characteristics of peptides related to KFERQ were not successful because many of such peptides were very rapidly degraded after microinjection perhaps because they are good substrates for cytosolic proteases or peptidases. For example, we attempted to assess the importance of the terminal glutamine by synthesizing and microinjecting ^{3}H-KFERA. However, this pentapeptide was degraded with a half-life of 4 hours, so whether or not it could also be degraded by the slower serum-dependent pathway could not be established.

To determine experimentally the importance of individual amino acids within the KFERQ region, we turned to a recombinant DNA approach. We synthesized an oligonucleotide that encoded a slightly modified RNase S-peptide [Goff et al., 1987] and fused it in frame to the *Escherichia coli* β-galactosidase gene. β-Galactosidase introduced into human fibroblasts is degraded at the same rate in the presence and absence of serum. However, an RNase S-peptide–β-galactosidase fusion protein is degraded more rapidly in the absence of serum (L. Jeffreys-Terlecky, M. Kirven-Brooks, and J.F. Dice, unpublished results; Table II). We are currently examining the effects of mutations in each of the codons of the KFERQ region to determine experimentally the types of peptides that will lead to enhanced degradation in response to serum withdrawal.

Lysosomal Degradation of Microinjected RNase A

Several lines of evidence indicate that the pathway of degradation of microinjected RNase A is lysosomal. For example, degradation of RNase A in the absence of serum is partially inhibited by the lysosomotropic agent ammonium chloride [McElligott et al., 1985]. In addition, after microinjection into the cytosol, a small amount of RNase A fractionates with lysosomes; other microinjected proteins known to be degraded by cytosolic pathways of proteolysis show no lysosomal association.

As mentioned earlier, degradation of microinjected RNase A results in secretion not only of free amino acids but also peptide degradation products from cells. The same peptides derived from RNase A are released from cells after microinjection and after lysosomal hydrolysis following endocytosis [Isenman and Dice, 1989]. Since the lysosomal degradation of proteins following endocytosis is well-established [Schmidt, 1992], lysosomes are also likely to be responsible for the degradation of microinjected RNase A.

More compelling evidence in support of lysosomal degradation of microinjected RNase A comes from studies in which RNase A is labeled with ^{3}H-raffinose, an inert trisaccharide. The degradation products of ^{3}H-raffinose–tagged proteins are trapped in the compartment where they are generated, so their location serves as a cumulative marker of the site of degradation of a protein. All of the degradation products from microinjected ^{3}H-raffinose–RNase A accumulate within lysosomes for cells grown in both the presence [McElligott et al., 1985] and the absence of serum (M.A. McElligott and J.F. Dice, unpublished results). However, the pathway of entry of RNase A into lysosomes differs in cells maintained in the presence or absence of serum [Chiang and Dice, 1988]. In the presence of serum, RNase A and RNase S-protein are both taken up by lysosomes probably by micro- or macroautophagy. In the absence of serum, RNase A is taken up by lysosomes by the direct protein transport pathway.

GENERALITY OF THIS PATHWAY OF PROTEIN DEGRADATION

To examine whether peptide regions similar to KFERQ exist in intracellular proteins, we raised polyclonal antibodies to KFERQ and affinity-purified IgGs specifically directed toward the pentapeptide [Chiang and Dice, 1988]. The anti-KFERQ IgGs immunoprecipitate approximately 30% of ^{3}H-leucine–labeled cystosolic proteins from human fibroblasts.

We followed loss of radioactivity from total, immunoreactive, and nonimmunoreactive cytosolic proteins by incubating radiolabeled cells in medium containing excess unlabeled leucine to suppress reutilization of the isotope (Fig. 6). The immunoprecipitable proteins are preferentially degraded in response to serum withdrawal. In contrast, nonimmunoprecipitable proteins are degraded at the same rate in the presence and absence of serum. These results

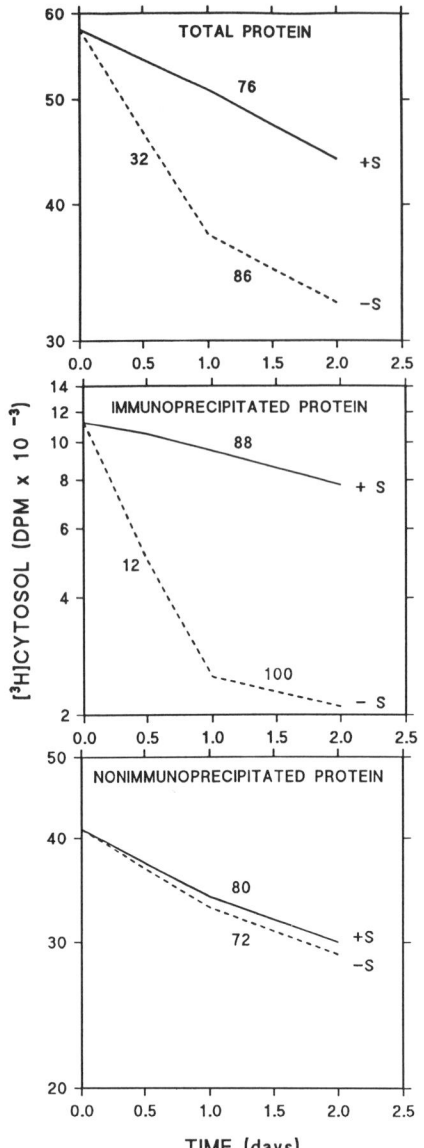

Fig. 6. *Selective degradation of proteins immunoreactive with anti-KFERQ IgGs during serum withdrawal. S = 10% serum. The numbers refer to half-lives in hours.*

show that the approximately twofold enhanced degradation in total cytosolic protein in response to serum withdrawal is actually due to more strikingly enhanced degradation of 30% of the cytosolic proteins containing peptide regions biochemically related to KFERQ.

As mentioned earlier, a similar pathway of proteolysis appears to be active in the liver of fasted animals. Certain additional tissues (kidney and heart), but not others (brain, testes, skeletal muscle) also exhibit this pathway of proteolysis in response to fasting [Chiang and Dice, 1988; Wing et al., 1991].

THE ROLE OF HEAT SHOCK 70 kDa PROTEINS (HSP70s) IN SELECTIVE LYSOSOMAL PROTEOLYSIS

We considered the possibility that an intracellular protein may recognize KFERQ-like peptide regions in proteins that are degraded more rapidly in response to serum withdrawal. We isolated a protein of 73 kDa from an RNase S-peptide affinity column and called it prp73 for *peptide recognition protein of 73 kDa* [Chiang et al., 1989].

A monoclonal antibody that recognizes all of the members of the hsp70 family [Kurtz et al., 1986] reacted with prp73 purified from human fibroblast cytosol. In addition, sequence data obtained from purified prp73 tentatively identified it as the constitutively expressed heat shock cognate protein of 73 kDa (hsc73) [Chiang et al., 1989].

Hsc73 and prp73 were functionally equivalent in several assays. Both proteins bound to RNase A, RNase S-peptide, KFERQ, aspartate aminotransferase, and pyruvate kinase—proteins and peptides that contain KFERQ-like peptide motifs. Neither hsc73 nor prp73 binds to ovalbumin, lysozyme, or ubiquitin—proteins that lack KFERQ-like peptide motifs [Terlecky et al., 1992]. In addition, hsc73 and prp73 functioned identically in stimulating degradation of RNase S-peptide by isolated lysosomes under conditions in which three other members of the hsp70 family had no activity [Terlecky et al., 1992].

SELECTIVE DEGRADATION OF PROTEINS BY ISOLATED LYSOSOMES

To try to elucidate the mechanism of degradation of KFERQ motif-containing proteins, we developed an in vitro assay using lysosomes isolated from human fibroblasts over two consecutive discontinuous density gradients [Chiang et al., 1989; Terlecky et al., 1992; Terlecky and Dice, 1993]. Maximal degradation of ^{3}H-RNase S-peptide and ^{3}H-RNase A by isolated lysosomes requires both ATP and hsc73 [Chiang et al., 1989; Terlecky et al., 1992]. This degradation is selective because ^{3}H-RNase S-protein and ^{3}H-dihydrofolate reductase, proteins that do not contain KFERQ motifs, are degraded little, if at all, under the same conditions [Terlecky and Dice, 1993] (Fig. 7).

Degradation of ^{3}H-RNase S-peptide can be inhibited by reducing the temperature, and degradation appears to occur within lysosomes because it is inhibited by ammonium chloride [Chiang et al., 1989] and leupeptin [Terlecky and Dice, 1993]. Hydrolases released from damaged lysosomes cannot account for the proteolysis because the strong incubation buffer maintains the pH at 7.2 and there is no detectable activity of released proteases at this pH [Terlecky and Dice, 1993].

Lysosomal uptake of ^{3}H-RNase S-peptide is saturable ($K_m = 5$ μM). At 4°C and in the presence of hsc73 ^{3}H-RNase S-peptide specifically binds to lysosomal membranes, and this binding is reduced by prior mild trypsinization [Terlecky and Dice, 1993]. No such binding is observed for RNase S-protein. Presumably, this RNase S-peptide binding component is a receptor or a polypeptide transport channel. A 39-kDa protein within the lysosomal membrane specifically binds to RNase S-peptide, and we are currently determining its identity (R. Skurat and J.F. Dice, unpublished results).

Lysosomes isolated from serum-deprived cells are twice as active in protein uptake in vitro as are lysosomes derived from serum-supplemented cells. Correlated with this increased activity is an increased amount of cellular hsc73 in the lysosomal fraction [Terlecky and Dice, 1993]. Some of this hsc73 appears to be associated with the lysosome surface because it can be removed with trypsin. This hsc73 is probably engaged in the transport of proteins into lysosomes. However, most of the lysosomal hsc73 appears to be in the lumen of the lysosome, since it is not digested by trypsin unless the lysosomal membrane is disrupted. Indirect immunofluorescence confirms the colocalization of a portion of cellular hsc73 with lysosomes (S.R. Terlecky, F. Agarraberes, and J.F. Dice, unpublished results).

Lysosomes isolated from rat liver have recently been reported to take up and degrade glyceraldehyde-3-phosphate dehydrogenase (GAPDH) selectively [Aniento et al., 1993]. This process appears to be very similar to the uptake and degradation of RNase A by fibroblast lysosomes in that it is selective and stimulated by hsc73 and ATP. Indeed, uptake of GAPDH by rat liver lysosomes can be competed with RNase A or RNase S-peptide, and uptake of RNase A can be competed by GAPDH. RNase S-protein and ovalbumin show no competition in these assays. Interestingly, in this rat liver lysosome system an import intermediate of RNase A has been detected in which most of the molecule has entered the lysosome but a small portion remains outside (A.M. Ceurvo, S.R. Terlecky, J.F. Dice, and E. Knecht, unpublished results).

The lumenal hsp70 plays a critical role in this selective lysosomal protein degradation

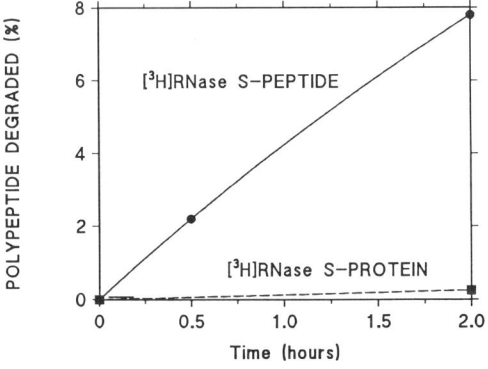

Fig. 7. *Import and degradation of polypeptides by isolated lysosomes. Degradation was measured in the presence of hsc73 and ATP.*

pathway. Cells that had been radiolabeled with ^{3}H-leucine were allowed to endocytose a monoclonal antibody (13D3) [Terlecky et al., 1992] that recognizes native and denatured hsc73 but not other hsp70 family members [Terlecky et al., 1992; S.R. Terlecky, F. Agarreberes, and J.F. Dice, unpublished results]. Protein degradation in cells incubated in the presence or absence of serum indicated that the enhanced degradation due to serum withdrawal is completely blocked while degradation in the presence of serum was unaffected (S.R. Terlecky, F. Agarraberes, and J.F. Dice, unpublished results).

The intralysosomal hsc73 is likely to be required to pull the substrate proteins across the lipid bilayer. Such an action has been shown for other hsp70 family members that reside in the endoplasmic reticulum or in the mitochondrial matrix. Depletion of these intraorganellar hsp70s impaired the import of precursor proteins [Kang et al., 1990; Vogel et al., 1990; Nicchita and Blobel, 1993]. The similarities of these polypeptide transport pathways into different organelles are highlighted in Figure 8.

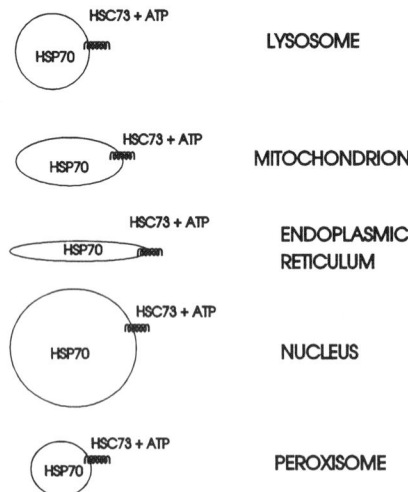

Fig. 8. *Protein import into organelles. Hsc73 and ATP in the cytosol stimulate these protein transport pathways [Hendrick and Hartl, 1993; Walton et al., 1993], and various hsp70 family members within organelles appear to be required for protein import.*

FUTURE DIRECTIONS

The mechanisms by which hsc73 promotes lysosomal degradation of proteins containing KFERQ-like peptide regions are not yet known. Hsc73 stimulates import of precursor proteins into mitochondria at least in part by preventing misfolding into a transport-incompetent conformation [Hendrick and Hartl, 1993]. Such a role seems unlikely for the stimulation of lysosomal uptake of mature, folded proteins such as RNase A. Perhaps hsc73 is also able to unfold proteins prior to lysosomal import. Alternatively, hsc73 may participate in binding of the substrate proteins to the receptor or protein translocation channel in the lysosomal membrane. Hsc73 may even enter lysosomes along with substrate proteins.

Another important issue to resolve is how hsc73 actions are regulated. Perhaps post-translational modifications or cofactors modulate hsc73's peptide-binding properties or subcellular localization so that in certain conditions hsc73 promotes protein import while in others it promotes lysosomal proteolysis. However, it is also possible that other components of the various polypeptide import pathways are regulated. For example, activation of the lysosomal degradation pathway may be due to changes in the substrate proteins that expose the KFERQ-like regions. Another possibility is that the putative receptor or peptide transporter on the lysosome surface may be regulated. Answers to these intriguing questions promise to reveal important new insights about targeting of proteins to intracellular organelles in addition to furthering our understanding of intracellular protein breakdown.

ACKNOWLEDGMENTS

Research in the authors laboratory was supported by NIH grant AG06116.

REFERENCES

Ahlberg J, Marzella L, Glaumann H (1982): Uptake and degradation of proteins by isolated rat liver lysosomes.

Suggestion of a microautophagic pathway of proteolysis. Lab Invest 47:523–532.

Aniento F, Roche E, Cuervo AM, Knecht E (1993): Uptake and degradation of glyceraldehyde-3-phosphate dehydrogenase by rat liver lysosomes. J Biol Chem 268:10463–10470.

Backer JM, Bourret L, Dice JF (1983): Regulation of catabolism of microinjected ribonuclease A requires the amino terminal twenty amino acids. Proc Natl Acad Sci USA 80:2166–2170.

Backer JM, Dice JF (1986): Covalent linkage of ribonuclease S-peptide to microinjected proteins causes their intracellular degradation to be enhanced during serum withdrawal. Proc Natl Acad Sci USA 83:5830–5834.

Chiang H-L, Dice JF (1988): Peptide sequences that target proteins for enhanced degradation during serum withdrawal. J Biol Chem 263:6797–6805.

Chiang H-L, Terlecky SR, Plant CP, Dice JF (1989): A role for a 70-kilodalton heat shock protein in lysosomal proteolysis of intracellular proteins. Science 246:282–285.

Chirico WJ, Waters MG, Blobel G (1988): 70K heat shock related proteins stimulate protein translocation into microsomes. Nature 332:805–810.

Cockle SM, Dean RT (1982): The regulation of proteolysis in normal fibroblasts as they approach confluence. Evidence for participation of the lysosomal system. Biochem J 208:243–249.

Deshaies RJ, Koch BD, Werner-Washburne M, Craig EA, Schekman R (1988): 70 kD stress protein homologues facilitate translocation of secretory and mitochondrial precursor polypeptides. Nature 332:800–805.

Dice JF (1987): Molecular determinants of protein half-lives in eukaryotic cells. FASEB J 1:349–357.

Dice JF (1990): Peptide sequences that target cytosolic proteins for lysosomal proteolysis. Trends Biochem Sci 15:305–309.

Dice JF, Chiang H-L, Spenser EP, Backer JM (1986): Regulation of catabolism of microinjected ribonuclease A: Identification of residues 7-11 as the essential pentapeptide J Biol Chem 262:6853–6859.

Dunn WA (1990a): Studies on the mechanisms of autophagy: Formation of the autophagic vacuole. J Cell Biol 110:193–1933.

Dunn WA (1990b): Studies on the mechanisms of autophagy: Maturation of the autophagic vacuole. J Cell Biol 110:1935–1945.

Goff SA, Short-Russell SR, Dice JF (1987): Efficient saturation mutagenesis of a pentapeptide coding sequence using mixed oligonucleotides. DNA 6:381–388.

Hare JF (1990): Mechanisms of membrane protein turnover. Biochim Biophys Acta 1031:71–90.

Hendrick JP, Hartl F-U (1993): Molecular chaperone functions of heat shock proteins. Annu Rev Biochem 62:349–384.

Isenman LD, Dice JF (1989): Secretion of intact proteins and peptide fragments by lysosomal pathways of protein degradation. J Biol Chem 264:21591–21596.

Kang P-J, Ostermann J, Shilling J, Neupert W, Craig EA, and Pfanner N (1990): Requirement for hsp70 in the mitochondrial matrix for translocation and folding of precursor proteins. Nature 348:137–143.

Knecht E, Hernandez-Yago J, Grisolia S (1984): Regulation of lysosomal autophagy in transformed and nontransformed mouse fibroblasts under several growth conditions. Exp Cell Res 154:224–232.

Kominami E, Hashida E, Khairallah E, Katunuma N (1983): Sequestration of cytoplasmic enzymes in an autophagic vacuole–lysosomal system induced by injection of leupeptin. J Biol Chem 258:6093–6100.

Kopitz J, Kisen GO, Gordon PB, Bohley P, Seglen PO (1990): Non-selective autophagy of cytosolic enzymes in isolated rat hepatocytes. J Cell Biol 111:941–954.

Kurtz S, Rossi J, Petko L, Lindquist S (1986): An ancient developmental induction: Heat shock proteins induced in sporulation and oogenesis. Science 231:1154–1157.

Marzella L, Glaumann H (1987): Autophagy, microautophagy, and crinophagy as mechanisms for protein degradation. In Glaumann H, Ballard FJ (eds): Lysosomes: Their Role in Protein Breakdown. New York: Academic Press, pp 319–367.

McElligott MA, Dice JF (1984): Microinjection of cultured cells using red cell-mediated fusion and osmotic lysis of pinosomes: A review of methods and applications. Biosci Rep 4:451–466.

McElligott MA, Miao P, Dice JF (1985): Lysosomal degradation of ribonuclease A and ribonuclease S–protein microinjected into human fibroblasts. J Biol Chem 260:11986–11993.

Mortimore GE (1987): Mechanism and regulation of induced and basal protein degradation in liver. In Glaumann H, Ballard FJ (eds): Lysosomes: Their Role in Protein Breakdown. New York: Academic Press, pp 415–444.

Neff NT, Bourret L, Miao P, Dice JF (1981): Degradation of proteins microinjected into IMR-90 human diploid fibroblasts. J Cell Biol 91:184–194.

Nicchitta CV, Blobel G (1993): Lumenal proteins of the mammalian endoplasmic reticulum are required to complete protein translocation. Cell 73:989–998.

Olson TS, Terlecky SR, Dice JF (1992): Pathways of intracellular protein degradation in eukaryotic cells. In Ahern TJ, Manning MC (eds): Stability of Protein Pharmaceuticals: In Vivo Pathways of Degradation and Strategies for Protein Stabilization. New York: Plenum Publishing Corporation, pp 89–118.

Pfeifer U (1987): Functional morphology of the lysosomal apparatus. In Glaumann H, Ballard FJ (eds): Lysosomes: Their Role in Protein Breakdown. New York: Academic Press, pp 3–59.

Schmidt SL (1992): The mechanism of receptor-mediated endocytosis: More questions than answers. BioEssays 14:589–596.

Seglen PO, Gordon PB, Holen I (1990): Non-selective autophagy. Semin Cell Biol 1:441–448.

Terlecky SR, Chiang H-L, Olson TS, Dice JF (1992): Protein and peptide binding and stimulation of in vitro lysosomal proteolysis by the 73-kDa heat shock cognate protein. J Biol Chem 267:9202–9209.

Terlecky SR, Dice JF (1993): Polypeptide import and degradation by isolated lysosomes. J Biol Chem (in press).

Ueno T, Muno D, Kominami E (1991): Membrane markers of endoplasmic reticulum preserved in autophagic vacuolar membranes from leupeptin-administered rat liver. J Biol Chem 266:18995–18999.

Vogel JP, Misra LM, Rose MD (1990): Loss of BiP/GRP78 function blocks translocation of secretory proteins in yeast. J Cell Biol 110:1855–1895.

Walton PA, Morello JP, Welch WJ (1993): Inhibition of the peroxisomal import of a microinjected protein by coinjection of antibodies to members of the 70 kD heat shock protein family. J Cell Biochem Suppl 17C:22.

Warren G (1993): Membrane partitioning during cell division. Annu Rev Biochem 62:323–348.

ABOUT THE AUTHORS

J. FRED DICE is Professor of Physiology at Tufts University School of Medicine in Boston, MA, where he teaches graduate and medical students on topics related to protein degradation, protein targeting, aging, and endocrinology. After receiving a B.A. from the University of California at Santa Cruz, he pursued doctoral research at Stanford University in the laboratory of Dr. Robert T. Schimke, where he earned a Ph.D. in 1973 for work on protein degradation in rat liver. This was followed by postdoctoral research at Harvard Medical School where he worked with Dr. Alfred Goldberg on protein degradation in skeletal muscle and other tissues. Since 1980 Dr. Dice's research has involved protein degradation studies in human fibroblasts, especially lysosomal pathways of proteolysis, and how proteolysis is altered in aged cells. Another research interest concerns the mechanism of accumulation of ATP synthase subunit 9 in Batten disease. Dr. Dice is on the Editorial Board of the *Journal of Biological Chemistry* and is a member of the Molecular Cytology Study Section at NIH.

STANLEY R. TERLECKY is a postdoctoral fellow at the University of California at San Diego in the laboratory of Suresh Subramani, where he is currently studying peroxisome biogenesis. After receiving a B.A. from New York University, he pursued doctoral research at Tufts University School of Medicine in the laboratory of Dr. Dice. He earned his Ph.D. in 1993 for reconstituting the selective lysosomal proteolytic pathway in vitro. His picture hangs in the Physiology Conference Room.

Autophagy: Its Mechanism and Regulation

Glenn E. Mortimore and Motoni Kadowaki

INTRODUCTION

Autophagy, also termed *autophagocytosis*, is an intracellular vacuolar process in eukaryotic cells that sequesters and degrades the macromolecular constituents of cytoplasm. It is highly conserved and found in virtually all lower plants and animals as well as higher forms. Autophagy, for example, plays a major role in endogenous amino acid supply in fungi/yeast (see chapter by Jones and Murdock, this volume) and germinating seeds [Nishimura and Beevers, 1979] and in intracellular protein turnover in protoplasts of cultured plant cells [Canut et al., 1985]. It is also utilized in cell restructuring [Marty, 1978; Paavola, 1978a,b] and as a source of substrate for gluconeogenesis in such diverse cells as the amoeba *Tetrahymena pyriformis* [May et al., 1982] and the mammalian hepatocyte [reviewed by Mortimore and Pösö, 1987].

Cytoplasmic sequestration is the step that distinguishes autophagy from other intracellular degradative processes. As a volume uptake mechanism it is relatively nonselective and able to capture most classes of macromolecules and organelles. Therein lies its uniqueness. No other degradative mechanism of which we are aware can balance the coordinated synthesis of structural elements in a way that maintains their cytoplasmic proportions. In cells such as the hepatocyte the balance can be tipped strongly in favor of degradation, thus providing an important source of amino acids for energy needs and other uses early in starvation. In this instance, the fact that autophagy is an ongoing process and closely regulated by complex feedback mechanisms attests to its fundamental role in cellular homeostasis.

The aims of this chapter are twofold: 1) to review the general features of intracellular protein and RNA turnover and the major classes of autophagy and 2) to present a current overview of the mechanism and regulation of autophagy, focusing largely on the mammalian hepatocyte.

GENERAL PROTEIN AND RNA TURNOVER IN CELLS
Protein Degradation

In the evaluation of autophagy the most useful experimental approach for determining general rates of protein degradation in tissues and cells over short intervals is one that employs the release of labeled amino acids from previously labeled cells as an end point. Measurements based on decreases of resident protein content would be too small for the degree of accuracy required, and estimates derived from post-transcriptionally modified amino acids, such as 3-methylhistidine in muscle [Young and Munro, 1978], are useful only for specific proteins or protein groups. All methods, however, share two strict requirements: First, the reutilization of released label by protein synthesis must be decreased to negligible values either by flooding the cells (and their precursor sites) with a massive dose of unlabeled amino acid [Mortimore et al., 1972] or inhibiting synthesis with an agent such as cycloheximide [Khairallah and Mortimore, 1976]; second, the treatment should not perturb the underlying proteolytic mechanism. The particular amino acid chosen as a marker depends on the type of cell to be studied. Ideally, the amino acid should be widely distributed in cellular proteins and metabolically stable. Valine fulfills these requirements in liver [Mortimore and Mondon, 1970; Seglen and Solheim, 1978] as does phenylalanine and tyrosine in muscle [Rannels et al., 1975; Fulks et al., 1975]. Leucine has also been widely employed as a marker. However, because it is an effective inhibitor of autophagy [Seglen et al., 1980; Pösö et al., 1982b], its use should be restricted to experiments not involving regulation.

Classes of General Protein Degradation

Although degradation rates of individual cellular proteins vary widely, only two classes of turnover have been seen in instances where breakdown is determined from amino acid release [reviewed by Mortimore and Pösö, 1987]. The first is short lived; the second is termed *long-lived* or *resident* protein degradation. Separation between the two classes is wide, and no intermediate components have been found. In the perfused liver, for example, the half-life of the short-lived fraction is about 10 minutes, representing a turnover that is ≥100-fold faster than long-lived breakdown [Hutson and Mortimore, 1982]. Analogous results have been obtained in cultured cells [Poole and Wibo, 1973; Epstein et al., 1975; Auteri et al., 1983].

The nature of the short-lived fraction is not clear, but there are indications that it involves the breakdown of nascent peptides [Solheim and Seglen, 1980; Wheatley, 1984], mediated by the ubiquitin–ATP-dependent proteolytic system [Ciechanover et al., 1984]. Although its turnover is believed to represent approximately one-third of total protein synthesis in liver [Hutson and Mortimore, 1982; Wheatley, 1984], calculations indicate that the pool size is very small, amounting to less than 0.3% of total cell protein. Because of its rapid turnover, the short-lived fraction in labeled cells can be effectively depleted of label during a 60-minute chase [Hutson and Mortimore, 1982]. Of relevance to autophagic function (see discussion later in this chapter) is the fact that only long-lived breakdown, which is generated from more than 99% of the remaining label in resident proteins [Schworer et al., 1981], is physiologically regulated [Poole and Wibo, 1973; Epstein et al., 1975; Knowles and Ballard, 1976; Vandenburgh and Kaufman, 1980; Hutson and Mortimore, 1982; Auteri et al., 1983].

RNA Degradation

A method has been devised for measuring RNA degradation in the perfused rat liver [Lardeux et al., 1987; Lardeux and Mortimore, 1987] and isolated hepatocyte [Balavoine et al., 1990] that is similar in principle to the use of labeled valine in proteolytic determinations. The procedure takes advantage of the absence of cytidine deaminase in rat liver, which is normally required in its oxidation. Hence, in the orotic acid-labeled rat liver or isolated hepatocyte, labeled cytidine will accumulate in direct proportion to the release of CMP from RNA when label reutilization is prevented by unlabeled cytidine [Lardeux et al., 1987]. With this technique it has been possible to determine rates of RNA breakdown on a moment-to-moment basis, an indispensible asset in evaluating rapid regulatory responses [Lardeux and Mortimore, 1987]. In contrast to the two classes of turnover that characterize general protein breakdown, RNA exhibits only the long-lived component [Lardeux et al., 1987].

MECHANISM OF AUTOPHAGY

Autophagy is a morphologically diverse process. Apart from its primary role in sequestering and digesting cytoplasmic macromolecules, which is remarkably similar in both plant and animal cells, the process is difficult to categorize mechanistically. Major differences arise from the nature of intracellular membranes that form the vacuoles and the manner by which acid hydrolases are acquired. In addition, rates of vacuole formation and their mode of regulation can vary widely depending on the particu-

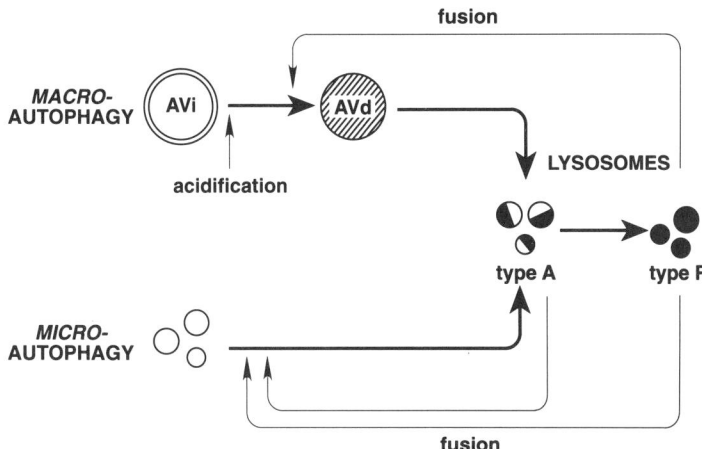

Fig. 1. *Scheme depicting major lysosomal–vacuolar compartments and routes of acid hydrolase flow in macro- and microautophagocytosis in the rat hepatocyte. AVi, autophagosome; AVd, autolysosome [de Duve and Wattiaux, 1966]; ER, endoplasmic reticulum; type A and R secondary lysosomes (see Figs. 2 and 6). AVi obtain their forming membranes from rough ER [Dunn, 1990a], while the limiting membranes of microautophagosomes (except some invagination forms, not shown) are probably derived from smooth ER [Mortimore et al., 1988]. Newly synthesized enzymes enter via Golgi-derived primary lysosomes, but most of the vacuolar acid hydrolase is acquired by fusion with preexisting lysosomes [Surmacz et al., 1987; Lawrence and Brown, 1992]. Acidification of AVi occurs before fusion [Dunn, 1990b]. The point (or points) where endosomes join the lysosomal–vacuolar pathway is unsettled and, for this reason, was omitted (see text).*

lar needs of the cell or organism. Rather than attempt merely to catalog this information, we have chosen to use the mammalian animal cell as a model for integrating autophagic structure and function. Since considerable morphologic and biochemical information has been obtained in liver, muscle, and isolated cells that is relevant to autophagically mediated macromolecular turnover, we hope this choice will be of value in suggesting how autophagy might fit into an overall scheme of cellular regulation.

In cells such as the hepatocyte, where protein turnover is especially large and autophagy comparatively easy to define in morphologic terms, two major classes of autophagy, macro- and microautophagy, have been recognized (Fig. 1). The first constitutes the classic overt variety, while the second is characterized by smaller, less conspicuous vacuoles [Mortimore et al., 1988a]. Both arise spontaneously in the normal, untreated cell. The former, however, is controlled by amino acids and induced by deprivation [Schworer et al., 1981], while the latter predominates under basal conditions [Mortimore et al., 1988a]. Together, these two vacuolar systems account for nearly all cytoplasmic sequestration in the hepatocyte [Hutson and Mortimore, 1982; Mortimore et al., 1983]. These distinctions provide the framework for the following discussion of autophagic structure and function.

Macroautophagy

Vacuole formation. Macroautophagy is an intrinsic cellular process that is normally restrained by amino acids and, in some cases, by serum or other growth-promoting factors in cell media [Ballard and Gunn, 1982]. Removal of one or more has been shown to induce autophagic responses in liver [Neely et al., 1974; Mortimore and Schworer, 1977; Neely et al., 1977; Ward et al., 1977], heart [Jefferson et al., 1974], and cultured cells [Mitchener et al., 1976; Amenta and Brocher, 1981]. Amino acids

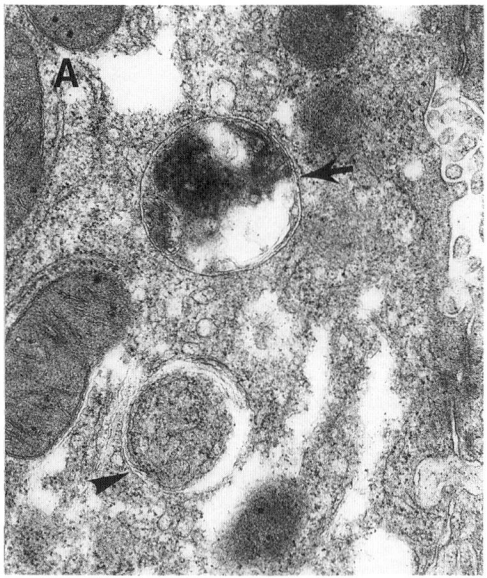

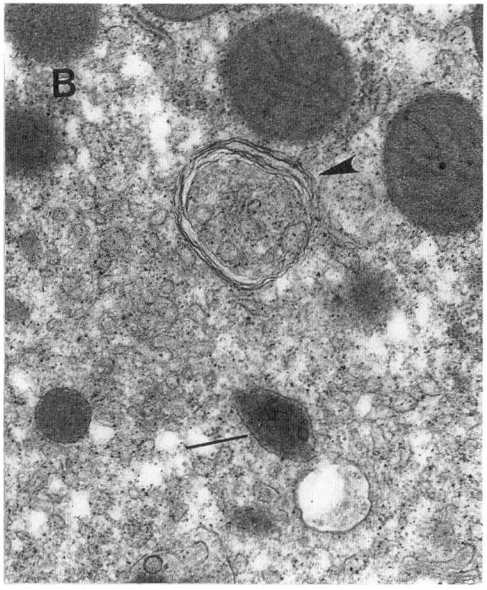

Fig. 2. Electron micrographs of lysosomal–vacuolar components in hepatocytes of a rat liver perfused 40 minutes without added amino acids. **A:** AVi (arrowhead) with sequestered membranes and ribosomes; note the space between the inner and outer membranes and extra membrane at the top. The AVd (arrow) contains disrupted membranes and glycogen (white); the limiting membrane is slightly thicker than that of AVi. The particle at the top of the panel is a peroxisome. ×28,140. **B:** AVi (arrowhead) with sequestered smooth ER, surrounded by multiple membranes. Below the AVi lies a type R secondary lysosome (line) with its distinctive halo beneath the limiting membrane. The body to its left is unidentified. ×28,140.

are the prime regulators of macroautophagy, and maximal rates of vacuole formation can be achieved by stringent amino acid depletion alone in both perfused liver [Mortimore and Schworer, 1977; Schworer and Mortimore, 1979; Schworer et al., 1981] and isolated hepatocyte [Kovács et al., 1981]. In addition, heat stress augments the nutrient-deprivation response in temperature-sensitive CHO mutant cells [Schwartz et al., 1992]. Vacuole formation is clearly ATP dependent, although the steps involved have not been identified [Plomp et al., 1989; Kadowaki et al., 1994]. It is also probable that one (or more) GTP-binding proteins is required in vacuole formation, since GTPγS and other nonhydrolyzable GTP analogs have been shown to inhibit the vacuolar uptake of ^{125}I-tyramine-cellobiitol in hepatocytes permeabilized by α-toxin [Kadowaki et al., 1994].

Cycloheximide and other inhibitors of protein synthesis strongly suppress macroautophagy, possibly through increases in intracellular amino acids [reviewed in Mortimore and Pösö, 1987]. In addition, 3-methyladenine, which has little inhibitory activity on synthesis, is also effective [Seglen and Gordon, 1982]; its mechanism is not known.

Vacuole formation takes place by movement of a double-walled membrane around a portion of cytoplasm, isolating it from the surrounding matrix. As depicted in Figure 2, separation of the sequestered cytoplasm appears to occur as though it were surgically excised, without disturbance to the local cytoarchitecture. The results of recent immunocytochemical studies using antibodies to proteins specific for the rough endoplasmic reticulum have demonstrated that the smooth-surfaced enveloping membranes of the initial vacuole AVi (see Fig. 1 for definition of terms) in the perfused rat liver are derived from rough endoplasmic reticulum denuded of ribosomes [Dunn, 1990a]. It is not known, though, whether ribosomal detachment is a requirement for AVi formation.

The foregoing ATP requirement is also unclear, but it is probable that it is required for membrane movement. In this regard it is of interest that the microfilament inhibitors cytochalasins B and D have been reported to decrease vacuole formation in cultured rat kidney cells [Aplin et al., 1992]. The effect, though, could be cell specific, since results from an earlier study in Ehrlich ascites tumor cells failed to reveal inhibition by cytochalasin B of AVi formation induced by vinblastine despite evidence of microfilament disorganization [Hirsimäki and Hirsimäki, 1984].

Because macroautophagic vacuoles are widely distributed in the cell, their contents tend to reflect the overall composition of cytoplasm. In the hepatocyte, for example, the endoplasmic reticulum and presumably their associated proteins constitute most of the sequestered volume while mitochondria, peroxisomes, glycogen, and free ribosomes occupy smaller fractions [Pfeifer, 1973, 1978; Schworer et al., 1981]. Since the space between membranes is also sequestered, one may presume that free cytosol is taken up as well. Indeed, this prediction is supported by the nonselective uptake of such cytosolic enzymes as ornithine decarboxylase, tyrosine aminotransferase, and lactate dehydrogenase into autophagic vacuoles after proteolytic suppression by the cysteine proteinase inhibitor leupeptin [Kominami et al., 1983; Henell and Glaumann, 1984; Kopitz et al., 1990].

One exception to this apparent nonselective sequestration should be mentioned. When maximal or near-maximal rates of autophagic sequestration are induced in the perfused liver by removal of alanine at normal amino acid concentrations, the character of sequestration changes. Since alanine is a coregulator that is specifically required for inhibition by the regulatory amino acids at normal plasma levels (see Regulation of Autophagy, below), its removal results in an immediate deprivation response. When compared with complete amino acid deprivation, though, the vacuoles contain a 4.5-fold greater proportion of smooth to rough endoplasmic reticulum and significantly fewer mitochondria and peroxisomes [Mortimore et al., 1987]. The reason for these differences is not known. It is possible that amino acids are capable of modulating the cellular distribution of autophagic activity in addition to regulating the rate of sequestration.

Vacuole maturation and dynamics of turnover. As shown in Figures 2 and 3, the autophagic response to amino acid deprivation in the perfused liver is rapid [Schworer et al., 1981; Dunn, 1990a]. Initially, only AVi are seen, but after a lag of 7–8 minutes degradative forms (AVd/autolysosomes) appear [Schworer et al., 1981]. Typically, AVd are characterized by 1) a single limiting membrane that is somewhat thicker than the outer membranes of AVi, 2) cytochemical evidence of acid hydrolase activity within the vacuoles, and 3) varying degrees of disruption of the vacuolar contents [Mortimore and Schworer, 1977; Schworer et al., 1981; Dunn, 1990a]. While AVd acquire most of their acid hydrolases following the fusion of AVi with type R lysosomes and primary lysosomes (Fig. 1) [Lawrence and Brown, 1992], the transformation of AVi to AVd does not involve just a single event, but requires multiple steps, with acidification appearing before lysosomal fusion [Dunn, 1990b]. Although Dunn [1990a] obtained no immunocytochemical evidence that plasma membrane markers are transferred to AVi membranes, Seglen and coworkers concluded from indirect evidence in electroinjected hepatocytes that fluid phase endocytosis converges on the autophagic pathway before lysosomal fusion [Gordon and Seglen, 1988; Gordon et al., 1992]; Tooze et al. [1990] arrived at a similar conclusion from morphologic investigations in pancreatic exocrine cells. On the other hand, in a detailed study with colloidal gold Lawrence and Brown [1992] observed that AVi in cultured hepatocytes rapidly fuse with preexisting lysosomes, but seldom with a prelysosomal compartment. Clearly, additional studies are needed to clarify these inconsistencies.

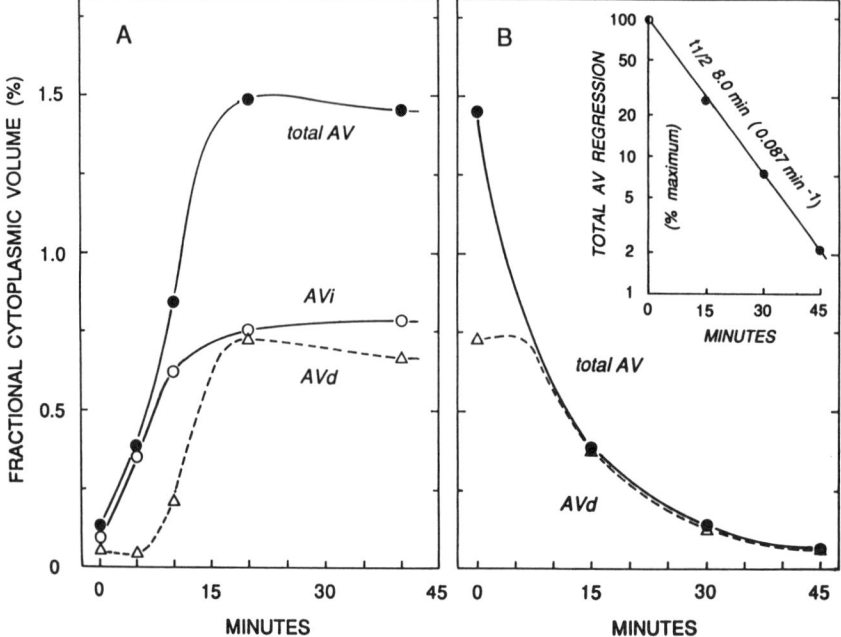

Fig. 3. *Time course of macroautophagic vacuole (AV) formation and regression.* **A:** *Livers from normal fed rats were perfused in parallel without added amino acids and, at intervals, fixed for electron microscopy. Volume densities of AVi, AVd, and total AV were determined stereologically [modified from Schworer et al., 1981].* **B:** *Livers were perfused in parallel for 20 minutes without amino acids; at time zero 10× plasma amino acids were added. Volume densities were determined as in A.* **Inset:** *Semilogarithmic plot of the regression of total AV volume. [Modified from Schworer et al., 1981, with permission of the publisher.]*

The lysosomal/vacuolar circuit is complete with digestion of the sequestered cytoplasm and reappearance of secondary lysosomes as end products (Figs. 1, 2). With ongoing autophagy, a large part of the acid hydrolase activity acquired by AVi is derived by the recycling of preexisting enzymes [Lawrence and Brown, 1992]. At the same time, newly processed enzymes may enter the lysosomal/vacuolar pathway via primary lysosomes that are ultimately derived from Golgi vesicles.

While the formation of lysosomal end products was thought to be a straightforward process of intravacuolar digestion and disposal of excess membrane, new findings have appeared indicating that the ubiquitin system plays an important role in autophagic/lysosomal function [Gropper et al., 1991; Lenk et al., 1992; Schwartz et al., 1992]. Of particular relevance to autophagy are studies with temperature-sensitive CHO mutant cells that restrict ubiquitin conjugate formation at the initial step (E1) when the temperature is elevated above permissive levels [Gropper et al., 1991]. The results showed that the conversion of AVd to lysosomes is strongly inhibited at restrictive temperatures, although AVi formation and their maturation to AVd are not affected [Lenk et al., 1992; Schwartz et al., 1992]. It is of interest that the effects are limited to macroautophagy; basal protein turnover (microautophagy) is apparently not involved [Gropper et al., 1991].

Based on the time course of autophagic vacuole formation in the perfused liver (Fig. 3), cytoplasmic sequestration is an ongoing process in which a steady state between AVi for-

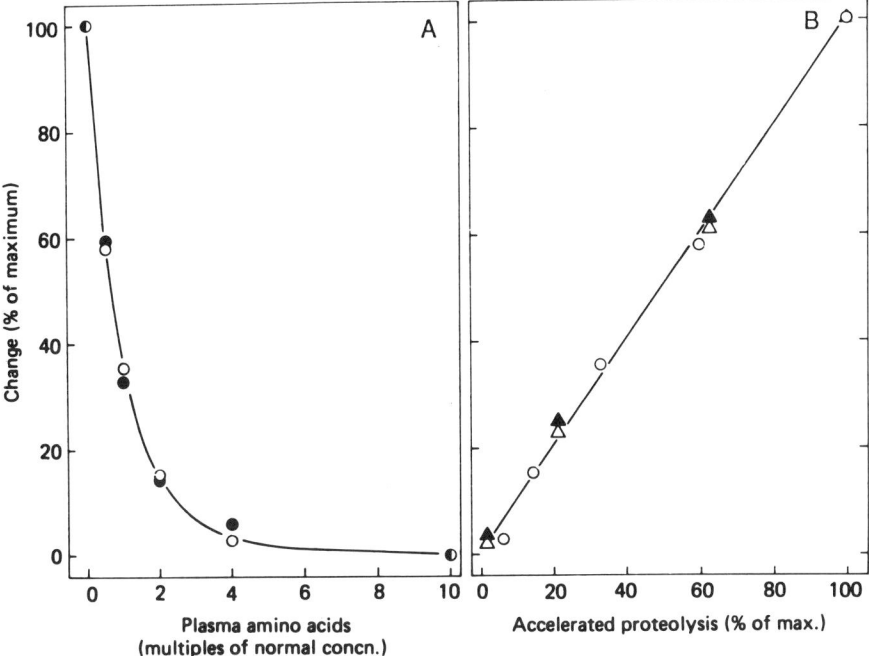

Fig. 4. *Correlations between proteolysis, aggregate volumes of AVi and AVd, and the shift of β-hexosaminidase from lysosomes to AVi during deprivation-induced macroautophagy in perfused livers of fed rats.* **A:** *Relationship between accelerated long-lived proteolysis (●) and the above shift of β-hexosaminidase (○) at various levels of plasma amino acids. The shift, which was determined in lysosomal–vacuolar fractions separated in colloidal silica gradients, is a measure of lysosomal–AVi fusion.* **B:** *Correlation of AVi (△), AVd (▲), and the β-hexosaminidase shift (○) versus rates of deprivation-induced proteolysis.* [From Surmacz et al., 1987, with permission of the publisher.]

mation and their maturation to AVd and lysosomes is attained within 20 minutes [Schworer et al., 1981]. In the same experiments, the addition of an amino acid load after maximal levels of macroautophagy had been reached caused an immediate cessation of AVi formation. The existing population of AVd regressed exponentially with a half-life of 8 minutes (Fig. 3). In other hepatocyte studies, estimates of vacuole half life determined from volume regression and osmotic sensitivity have agreed extremely well, with most results averaging 7.9–8.0 minutes [Neely et al., 1974; Pfeifer, 1978; Papadopoulos and Pfeifer, 1986]. Similar values have been obtained in mouse pancreatic acinar and seminal vesicle cells [Kovács et al., 1987].

Correlation between macroautophagy and the deprivation-induced breakdown of protein and RNA.

In livers perfused with graded levels of plasma amino acids, the aggregate volumes of AVi and Avd decrease in direct proportion to deprivation-induced rates of long-lived proteolysis over the full range of macroautophagy [Schworer et al., 1981]. The results are depicted in Figure 4, together with measurements of lysosomal fusion based on the proportion of β-hexosaminidase in colloidal silica density gradients that had shifted from lysosomes to macroautophagic vacuoles. It is thus evident that deprivation-induced changes in lysosomal fusion as well as in aggregate volumes of AVi and AVd are directly proportional to induced rates of proteolysis.

This relationship suggests that the proportion of type R lysosomes that fuse with AVi on the average is strictly maintained despite wide differences in individual vacuole size and shape [Surmacz et al., 1987]. Because the volume/number ratio of AVi is not constant but rises sharply with increasing deprivation [Schworer et al., 1981], it is probable that the direct correlation in Figure 4 would not exist without a mechanism that regulates the number of fusions in vacuoles according to the volume of the individual vacuoles [Surmacz et al., 1987]. The nature of the signal is not known, but intravacuolar pH is a possibility [Surmacz et al., 1987].

Since autophagy induced by plasma amino acid deprivation is relatively nonselective in nature, one would expect comparable increases in the fractional rates of protein and RNA degradation. Results in Table I show that this is so when rates are corrected for basal turnover. The increases, which are directly attributable to macroautophagy, proved to be the same within experimental error over the full deprivation-induced range. Since RNA degradation, unlike that of protein, lacks a short-lived fraction (see General Protein and RNA Turnover in Cells, below), the agreement underscores the contention that proteolysis reflects the breakdown of long-lived or resident proteins. Although induced rates were the same for both components, basal rates differed strikingly. With RNA, the value was approximately 20% of that for protein. The reason for the difference is not entirely clear, but it is probable that membrane-bound ribosomes, which make up at least two-thirds of rRNA, are not sequestered in the basal state. On the other hand, free ribosomes have been observed in microautophagic vacuoles under basal conditions [Mortimore et al., 1983, 1988a].

Intralysosomal pools of degradable protein and RNA. Apart from vacuolar volumes and their turnover, the quantity of substrate undergoing hydrolysis and its relationship to rates of degradation in the intact cell represent additional elements in the development of a unified mechanism of autophagic function. These were first explored in experiments in which the content of degradable protein within lysosomal particles (including autophagic vacuoles) was correlated with rates of long-lived proteolysis [Mortimore and Ward, 1981]. Pool size was ascertained from the total release of valine in isotonic liver homogenates during incubations lasting 90–120 minutes. All but 4%–5% was derived from protein previously internalized within lysosomes, and none could be attributed to interactions between cytosolic proteins and lysosomes during the incubations; lysosomal latency was well maintained over the course of the experiments. The quantity of sequestered protein was determined from the total quantity of free valine released and corrected for an intralysosomal pH of 4.5 [Mortimore and Pösö, 1984] and a valine content of rat liver protein of 0.591 μmol/mg [Schworer et al., 1981].

TABLE I. Effects of Amino Acids on the Fractional Turnover of Protein and RNA in Livers of Fed Rats*

	Fractional turnover (Percent h^{-1})			
	Protein		RNA	
Amino acids (×)	Total	Minus 10×	Total	Minus 10×
0	4.57 ± 0.13	3.16	3.48 ± 0.23	3.19
0.5	3.25 ± 0.09	1.84	2.24 ± 0.25	1.95
4	1.64 ± 0.10	0.23	0.45 ± 0.06	0.16
10	1.41 ± 0.08	0	0.29 ± 0.04	0

*Livers were perfused for 40 minutes in the single-pass mode with multiples/fractions (×) of a normal plasma amino acid mixture. Rates of protein and RNA degradation were determined as described elsewhere [Pösö et al., 1982b; Lardeux et al., 1987]. The columns headed "Minus 10×" represent accelerated turnover, corrected for basal or 10× rates. Results are means ± of 6–18 experiments.

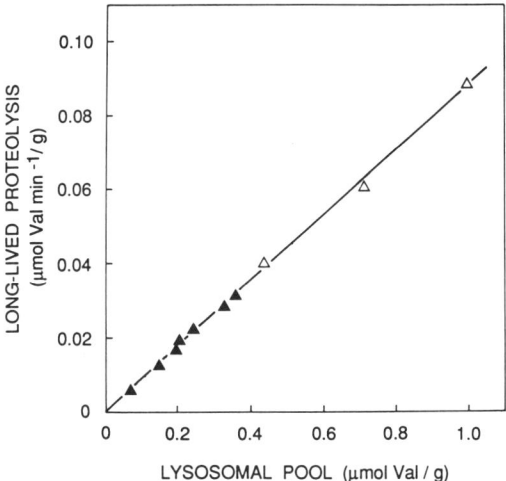

Fig. 5. *Relationship between long-lived proteolysis and pools of intralysosomal degradable protein in rat liver over the full range of protein turnover. The quantity of valine residues in the protein pools from livers in various accelerated (open triangles) and basal (filled triangles) states of protein degradation was correlated with corresponding rates of proteolysis monitored from valine release. The above regression equation was $y = -0.0002 + 0.0882x$ ($r = 0.999$), based on data from three studies [Mortimore and Ward, 1981; Hutson and Mortimore, 1982; Mortimore et al., 1988a], all corrected for an intralysosomal pH of 4.5 [Mortimore and Pösö, 1984].*

Degradable intralysosomal protein in liver from three studies [Mortimore and Ward, 1981; Hutson and Mortimore, 1982; Mortimore et al., 1988a] was found to correlate directly with long-lived protein breakdown ($r = 0.999$) over the full range of turnover (Fig. 5). The regression exhibits three important features. First, the slope of 0.088 min^{-1} ($t_{½} = 7.9$ minutes) is nearly identical to the apparent rate constant of macroautophagic turnover in Figure 3, thus independently confirming the initial estimate. Second, since valine is derived from at least two major classes of autophagic vacuoles, macro and micro, it follows that the slope must be the same in each class. This is illustrated by the fact that the lower seven points represent conditions in which macroautophagic vacuoles are lacking [Hutson and Mortimore, 1982; Mortimore et al., 1988a]. It is probable that the rate constant of vacuole turnover is a general feature of autophagy, independent of the size and type of vacuole, a point also emphasized by Papadopoulos and Pfeifer [1986]. Third, the absence of a significant zero intercept suggests that, within the limits of experimental error, long-lived proteolysis in the adult hepatocyte is a function of autophagic sequestration. A similar direct correlation has been demonstrated between hepatic RNA breakdown and lysosomal pools of degradable of RNA [Heydrick et al., 1991].

The third point is not inconsistent with the coexistence of important nonautophagic degradative mechanisms. It should be emphasized that the relevance of these mechanisms is not determined by their impact on protein turnover generally but rather by their effect on specific (frequently regulatory) proteins whose pool sizes may be small relative to the total quantity of resident cellular proteins.

Microautophagy

Because of the large variation in size of autophagic vacuoles, de Duve and Wattiaux [1966] coined the term *microautophagy* to advance the notion that the cytoplasmic "bite" could extend below the limit of macroautophagy into the molecular range. In the hepatocyte, the term has been used to denote the process of sequestration in the basal state, the existence of which is demonstrated by the lower seven points in Figure 5. It is probable that most cells possess similar mechanisms, although most of the evidence has come from the liver parenchymal cell. By the criterion of de Duve and Wattiaux, the lysosomal uptake of specific proteins should fall into this category. Such a mechanism does in fact exist, but, because of its uniqueness, has been dealt with separately (see chapter by Dice and Terlecky, this volume).

That cytoplasmic proteins are sequestered by lysosomal elements under basal conditions was initially demonstrated from the time course of valine release in liver homogenates [Mortimore et al., 1973]; similar findings were obtained by Ahlberg et al. [1985]. The results of later stud-

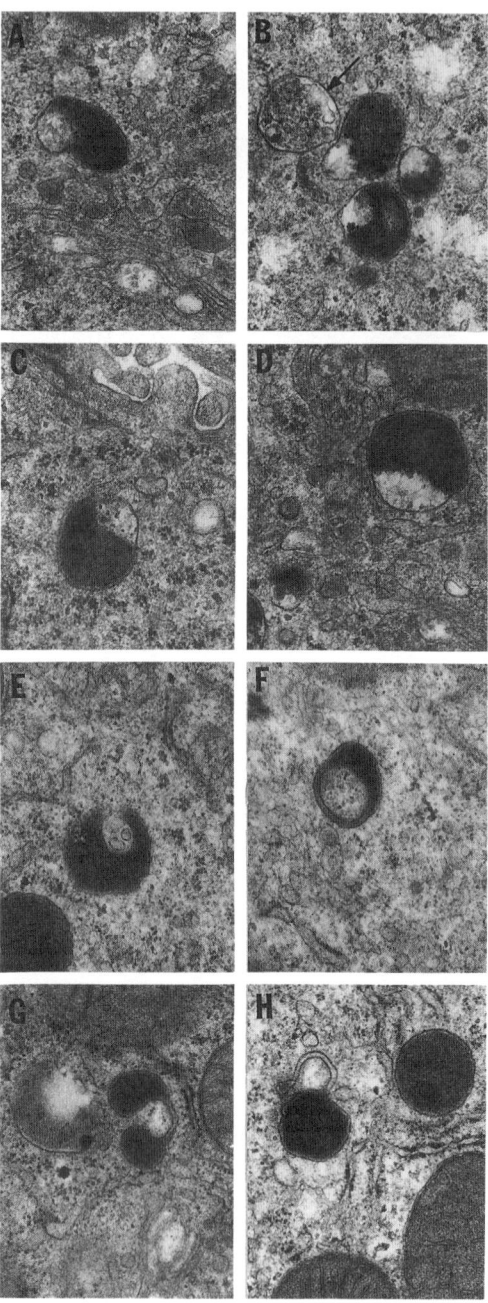

ies with the starved–refed mouse pointed to the possible role of a secondary lysosome, termed type A, as the site of sequestration [Hutson and Mortimore, 1982; Mortimore et al., 1983]. A direct correlation between microautophagy and long-lived proteolysis was established in perfused rat livers in which macroautophagy was suppressed by the addition of amino acids [Mortimore et al., 1988a]. In this study, basal proteolytic rates decreased >50% during 48 hours of starvation in association with a fall in the aggregate volume of type A lysosomes (Fig. 6A–D). By contrast, a small (~10%) volume fraction comprising various types of lysosomal invaginations [Marzella et al., 1980; Pfeifer, 1981; de Waal et al., 1986] did not change (see Fig. 6).

Close correlations were found between proteolysis and 1) total volumes of lysosomal elements and 2) all microautophagic elements after the latter were corrected for underestimation (Fig. 7). It may be appreciated from Figure 6 that the sharply demarcated, electron-lucent and dense zones in type A lysosomes will unavoidably generate some misinterpretation. For example, a section through a dense zone will be viewed as a type R, not type A, lysosome. Thus, for reasons of particle geometry alone, type A lysosomes will be consistently underestimated [Mortimore et al., 1988a]. The underestimation

Fig. 6. *Electron micrographs of secondary lysosomes containing sequestered cytoplasmic material. Livers from fed rats were perfused for 40 minutes with 10× plasma amino acids to suppress macroautophagy and then were fixed for electron microscopy.* ***A–D:*** *Examples of type A lysosomes. These elements together with type R lysosomes constitute more than 95% of secondary lysosomes. Type R components (see H, right) are similar to type A except for the lack of electron-lucent zones; note the distinctive halo beneath the limiting membrane of both types. All electron-lucent areas are sharply demarcated from dense regions, but no limiting membranes are evident. A and D, membrane remnants resembling smooth ER are visible in the lucent zones in association with glycogen (white); free ribosomes are seen in C. The single-walled vesicle in B (arrow), with profiles similar to smooth ER, could represent a nascent type A vesicle. ×31,000. [From Mortimore et al., 1988a, with permission of the publisher.]* ***E–H:*** *Rare examples of lysosomal invaginations and particles with flap-like extensions (H) that may (F) or may not (E, G, H) sequester cytoplasm. As a group they compose only 10% of all microsequestrational elements and, in contrast to type A elements, are not regulated (suppressed) by short-term starvation. ×31,000. [From Mortimore et al., 1988a, with permission of the publisher.]*

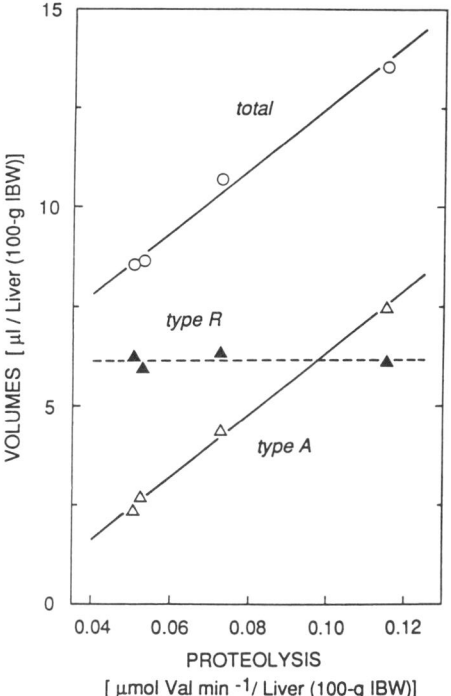

Fig. 7. *Relationship between basal protein turnover and volume densities of type A and type R lysosomes during short-term starvation in rat liver. Livers from fed and 48-hour starved rats were perfused with 10× plasma amino acids for the measurement of basal long-lived proteolysis and stereological analysis of secondary lysosomes [Mortimore et al., 1988a]. Under- and overestimation of type A and type R elements, respectively, were corrected by multiplying the observed type A volumes in Mortimore et al. [1988a] by 1.285, which equalized its slope to that of the total (see figure); R was then determined by subtracting type A from the total. Type A lysosomes were also assumed to be underestimated ~2% because of random sections through electron-lucent zones; this correction increased the total ~1% but did not affect type R. The type A volumes shown include closed invagination forms (Fig. 6).*

was corrected in Figure 7 and the correction validated in Table II.

How type A lysosomes are formed is not clear. While some may represent end products of macroautophagy, most are generated de novo since they persist in the presence of amino acid concentrations high enough to inhibit macroautophagy by 95%–97% [Schworer and Mortimore, 1979; Schworer et al., 1981]. The fact that they appear in close association with the smooth endoplasmic reticulum and frequently contain glycogen [Amherdt et al., 1974] suggests that they could arise by fusion between dense bodies and small vesicles derived from the latter [Mortimore et al., 1988a]. If so, distinctive precursor vacuoles might be difficult to identify (Fig. 6B). Novikoff and Shin [1978] have suggested that membranes of the smooth endoplasmic reticulum are in constant motion and could isolate bits of cytoplasm. Such a notion is attractive because it fits with the foregoing and could explain the decline in microautophagic activity with starvation; endoplasmic reticulum is rapidly lost early in starvation [Pfeifer, 1973], and microsequestration could decrease in parallel with the loss.

A Unified Mechanism for Autophagic Proteolysis

Because both macro- and microautophagy are volume uptake processes and have the same apparent turnover constants, it should be possible to predict rates of long-lived proteolysis from the product of the autophagic flux and the concentration of protein in the degradative compartment. The former can be computed for macroautophagy from the aggregate volume of AVd and the turnover constant of 0.087–0.088 min^{-1}. However, the corresponding value for microautophagy is less certain, as it is necessary to assume that the proteolytic compartment is limited to type A lysosomes [Mortimore et al., 1988a]. The assumption seems reasonable, though, since type A, but not type R, volumes correlate directly with long-lived proteolysis (Fig. 7). In either case, however, we have little direct information of the actual protein concentration at sites of degradation (see below).

Studies of autophagy in isolated hepatocytes permeabilized with α-toxin from *Staphylococcus aureus* [Kadowaki et al., 1994] have shown that the rate of sequestration (autophagic flux) can be determined from the steady-state uptake of externally added ^{125}I-iodotyramine-cellobiitol into macro- and microautophagic vacuoles separated from other organelles in colloidal silica gradients [Surmacz et al., 1983a,b]. Of

TABLE II. Prediction of Long-Lived Proteolysis From Cytoplasmic Protein Concentration and the Flux Through Autophagic Vacuoles: A Unified Mechanism of Autophagy (AV)*

AV/cell status	AV volume (μl) A	AV flux (μl/h) B	Protein concentration (nmol Val/μl)		Long-lived proteolysis (μmol Val/h)	
			Cytoplasm C	AV D/B	Pred. [B.C]	Obs. D
1. Macro fed	16.802	88.71 (3.27)	171.6	170.4	15.22	15.12 (3.25)
2. Micro fed	7.444	39.30 (1.40)	177.5	175.6	6.98	6.90 (1.37)
3. Micro 48-hour starved	3.128	16.52 (1.00)	209.2	213.1	3.46	3.52 (1.01)
4. Macro + micro α-toxin	—	12.18 (1.88)	163.7	164.2	2.00	2.00 (1.89)

*__Line 1:__ __A__ was determined from the maximal deprivation-induced increase in fractional volumes of AVd in perfused rat livers [Schworer and Mortimore, 1979], multiplied by the total cytoplasmic space of 2.71 ml per 100-g rat as calculated in Mortimore et al. [1988a] and Blouin et al. [1977]. __B__ is the product of __A__ and the rate constant of vacuole turnover 0.088 min^{-1} (Fig. 5). __C__ was computed from the valine content of liver protein, 465 μmol per 100-g rat, divided by the cytoplasmic volume. __D__ is the maximal deprivation-induced rate of hepatic protein degradation per 100-g rat [Schworer and Mortimore, 1979]. The number in parentheses under __B__ and __D__ in lines 1 through 4 are percentage fractional rates of the corresponding absolute rates. Note: Although otherwise appropriate, data from Schworer et al. [1981] were not used in this summary as AVd volumes are available only from the figures.

__Lines 2 and 3:__ Values corresponding to those above were taken from rat liver perfusion studies of Mortimore et al. [1988a] with type A volumes corrected for underestimation as detailed in Figure 7 and related text; values from the three 48-hour starved rats were averaged. Valine in liver protein was computed from the proportion 0.591 μmol Val per mg protein [Schworer et al., 1981]; the same AV rate constant 0.088 min^{-1} was employed, and all hepatic volumes and rates were calculated per 100-g initial body weight.

__Line 4:__ The data were obtained from experiments with isolated rat hepatocytes permeabilized with *S. aureus* α-toxin [Kadowaki et al., 1994]. __B__ was determined directly from the AV uptake of ^{125}I-iodotyramine-cellobiitol; the cytoplasmic volume and valine content of protein were 646.5 ml and 105.8 μmol; except for concentration, the values are expressed per 10^8 cells.

interest was the finding that the ratio of proteolysis/autophagic flux, which is an indirect measure of the protein concentration in degradative autophagic vacuoles, is higher than average intracellular values and equivalent to the apparent cytoplasmic protein concentration, computed from total cell protein divided by the cytoplasmic space (Table II). As summarized in Table II, remarkably close agreement between the ratio and the computed cytoplasmic protein concentrations was found in all previous studies for which complete data are available. And, as would be expected, equally close agreement was obtained between predicted and observed rates of proteolysis.

Several conclusions may be drawn from these findings. The agreements between predicted and observed cytoplasmic protein and proteolytic rates validate the determination of macroautophagic flux by the use of stereology and the rate constant of vacuole turnover (0.088 min^{-1}). They also confirm the method used for correcting the underestimation of type A lysosomes (Fig. 7). Because the correlations between both sets of predicted and observed values are optimal (maximal) only when constancy of the type R lysosomal volume is assumed over the 48-hour period of starvation, any upward or downward deviation from a slope of zero during the period would adversely affect them. Why protein concentrations in degradative vacuoles are higher than intracellular concentrations is not entirely clear, although it could be linked to vacuolar acidification and the attending net loss of K^+, Na^+, and water [Schworer et al., 1981]. Since the aggregate volume of AVd is about 18% smaller than that of AVi [Schworer et al., 1981], the change is sufficient to explain the increase in protein concentration. Finally, it is important to mention that these cellular adjustments in the volume and protein content of AVd result in almost exact correspondence between the fractional turnover of cytoplasm by autophagic sequestration and the turnover of total cell protein (Table II). With regard to the foregoing features, the two classes of autophagy, macro and micro, appear to be identical.

REGULATION OF AUTOPHAGY
Amino Acid Control of Macroautophagy

Amino acids are considered to be the prime regulators of macroautophagy as they are capable of evoking responses in the breakdown of long-lived proteins over the full range of control without hormonal intervention (Fig. 4). The fact that optimal regulation in liver [Woodside and Mortimore, 1972; Seglen et al., 1980; Pösö et al., 1982b] and in cultured kidney cells [Rabkin et al., 1991] requires the concerted action of several amino acids rather than just leucine as in skeletal and cardiac muscle [Buse and Reid, 1975; Fulks et al., 1975; Chua et al., 1979] may be an indication of the multiple roles that protein breakdown in liver and kidney play in supplying amino acids for a variety of metabolic needs.

Regulatory and nonregulatory amino acids. Studies with the isolated hepatocyte and perfused liver have shown that eight amino acids (Leu, Gln, Tyr, Phe, Pro, Met, His, and Trp) contribute to the direct suppression of proteolysis [Woodside and Mortimore, 1972; Hopgood et al., 1980; Seglen et al., 1980; Sommercorn and Swick, 1981; Pösö et al., 1982b]. Inhibition has been observed with asparagine alone [Seglen et al., 1980]. However, its action may differ from others of the regulatory group, since its effectiveness is lost when it is added to physiological mixtures [Pösö et al., 1982b]. Whether all of the eight amino acids are true inhibitors is not easily determined. In many instances individual effects are too small to measure or are obscured by differences between test systems. In elucidating regulatory mechanisms one useful strategy has been to determine the smallest number of amino acids that can evoke a maximal response. In perfused livers of synchronously fed rats, a combination of leucine and glutamine was found to be completely effective [Mortimore et al., 1991], and in livers of normally fed animals, leucine, glutamine, and either tyrosine or phenylalanine has produced comparable responses [Kadowaki et al., 1992]. The choice between the two aromatic amino acids raises an interesting problem that will be discussed below. Although the original number of regulatory amino acids may have been overestimated, it is unlikely that an important regulator was overlooked, since the 12 complementary amino acids as a group are totally ineffective at plasma levels as high as 10× [Pösö et al., 1982a].

Multiphasic control by regulatory amino acids and coregulation by alanine. In skeletal and cardiac muscle [Tischler et al., 1982] and cultured kidney cells [Rabkin et al., 1991], proteolytic dose responses to leucine display a single phase of inhibition. By contrast, dose responses to regulatory amino acids in the perfused rat liver [Mortimore et al., 1987, 1991] and isolated rat hepatocyte [Venerando et al., 1994] are mediated via two alternate, concentration-dependent mechanisms or modes. One (L) elicits direct inhibition at low and high plasma levels, but requires the coregulator alanine to express inhibition at normal concentrations. The second (H) is inactive at normal levels and below, but is fully effective at higher physiological concentrations. Studies with synchronously fed rats have shown that the H and L modes evolve in a preset pattern during each feeding cycle [Mortimore et al., 1991]. H appears first after food intake. Shortly thereafter, H begins to alternate randomly with L. With omission of the next feeding, H disappears and L becomes constant during early deprivation and starvation.

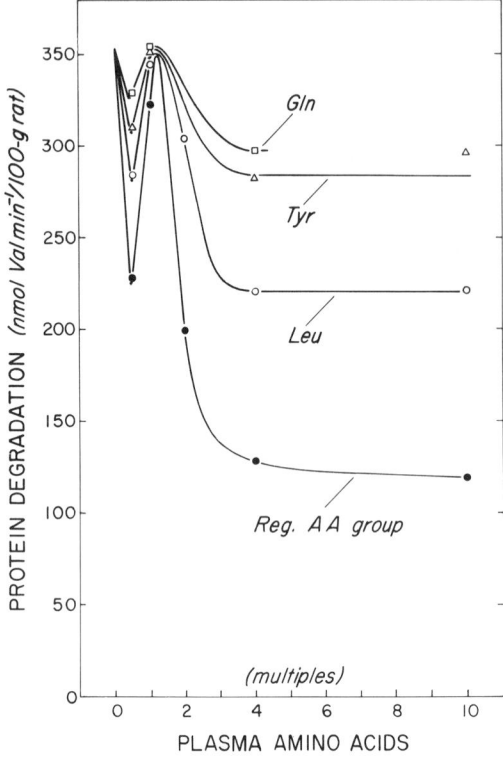

Fig. 8. *Proteolytic dose responses to regulatory amino acids. Livers from normal fed rats were perfused in the single-pass mode with various regulatory (Reg) amino acids at fractions/multiples of their concentrations in a 1× reference plasma mixture. The 1× regulatory mixture is composed of the following (µM): Leu, 204; Gln, 716; Tyr, 98; Pro, 437; Met, 60; His, 92; Trp, 93 [Mortimore et al., 1991; Venerando et al., 1994]. [From Mortimore et al., 1987, with permission of the publisher.]*

In nonfasted ad libitum fed rats, the L mode is commonly expressed late in the morning and at midday when most experiments are carried out. There are, however, exceptions to this, as noted by Venerando et al. [1994]. Typical proteolytic dose responses to leucine, glutamine, tyrosine, and the complete regulatory mixture (minus Phe) are depicted in Figure 8 for livers in the L mode. All reveal inhibition at low (0.5×) and high (4×) normal plasma concentrations and an intervening sharp (zonal) loss of effectiveness at about 1.25×. The zonal peak is directly related to the aforementioned coregulation by alanine and can be totally abolished by the addition of 0.5 mM (1×) alanine [Pösö and Mortimore, 1984; Mortimore et al., 1991]. Although 10- to 20-fold quantities of pyruvate/lactate or equimolar additions of glutamate or aspartate are effective substitutes, alanine is the only amino acid capable of accelerating proteolysis to near-maximal rates when it is deleted from a 1× normal plasma mixture [Pösö and Mortimore, 1984; Mortimore et al., 1991]. Because coregulation is not affected when transamination is blocked with aminooxyacetate, the site of alanine action must lie upstream from its transamination [Mortimore et al., 1991].

With the perifused rat hepatocyte, Meijer and coworkers have shown that leucine acts synergistically with alanine (and with glutamate and aspartate) at high physiological concentrations [Leverve et al., 1987; Caro et al., 1989]. While this interesting effect resembles alanine coregulation, it differs from it in exhibiting broader concentration and specificity requirements [Pösö and Mortimore, 1984; Mortimore et al., 1991] and by the fact that it is expressed only in the starved state [Mortimore et al., 1991].

The finding that the three strongest inhibitors of the regulatory group feature remarkably similar dose–response curves when external concentrations are plotted as multiples/fractions of their respective normal plasma values indicates that they act through a similar regulatory mechanism. The only known exception is phenylalanine. In contrast to the effects of tyrosine (Fig. 8), phenylalanine fails to inhibit proteolysis at 0.5× but evokes a dramatic decrease to near-basal values between 0.5 and 1× [Kadowaki et al., 1992]; maximal effects of the two amino acids are the same [Kadowaki et al., 1992]. These findings plus the lack of a zonal loss of inhibition indicate that phenylalanine regulates autophagy/proteolysis through a mechanism different from that of tyrosine and more in keeping with leucine's action in muscle [Tischler et al., 1982]. Why this difference exists is not known. It could be vestigial in nature, or it might relate in a unique way to the unusually rapid hydroxylation of phenylalanine to tyrosine in liver [Shiman et al., 1982].

Specificity and site of control by leucine.
Past studies have strongly suggested that leucine mediates its inhibition in muscle through products of oxidation [Tischler et al., 1982; Mitch and Clark, 1984]. This notion, however, was never seriously considered in the hepatocyte because of the cell's extremely low transamination activity [reviewed in Pösö et al., 1982b]. This opened the possibility that leucine is recognized at a site earlier than its transamination. Impetus for this view was fueled by the finding that L-α-hydroxyisocaproate, the hydroxyl analog of L-leucine, closely mimics leucine's multiphasic (L mode) dose response [Mortimore et al., 1987], a result that would exclude leucyl-tRNA as a mediator of regulation. The discovery that isovaleryl-L-carnitine shares this property [Miotto et al., 1989] prompted inquiry into the nature of leucine recognition [Miotto et al., 1992]. A structural analog of leucine, isovalerylcarnitine, was found to react with leucine at the same low concentration sites and to interact with alanine in the expression of coregulation [Miotto et al., 1992]. Recognition was stereospecific [Miotto et al., 1992] and could not be duplicated by either isovalerate or L-carnitine alone, but only when they were covalently linked [Miotto et al., 1989].

From an examination of other leucine derivatives, it is possible to construct the following set of structural requirements for low concentration recognition. First, a side chain closely matching the isovaleryl moiety is required [Miotto et al., 1992]. Second, the presence of a reactive group at or near the α-carbon appears to be necessary. This is so since the α-keto and α-chloro derivatives, which will not bind, are ineffective except at high physiological concentrations [Mortimore et al., 1987; Miotto et al., 1992], while substitutions with reactive OH, NH_2, or a quaternary amine, as in isovalerylcarnitine, are fully bioreactive [Mortimore et al., 1987; Miotto et al., 1992]. Finally, there is no specific requirement for the carboxyl group [Wert et al., 1992].

As for the location of the site of recognition, accumulating evidence strongly points to the plasma membrane. Much of it is based on the strong correlation that exists between responses and external concentrations. This is illustrated in Figures 8 and 9, where the sharp inflections of the dose–response curves for leucine, glutamine, and tyrosine correspond closely with their respective normal plasma concentrations but not, in the case of glutamine, with internal pools, which are comparatively stable. A second example brings the question into clearer focus. Isovalerylcarnitine is rapidly metabolized by the hepatocyte [Miotto et al., 1992], and, as a consequence, its intracellular levels

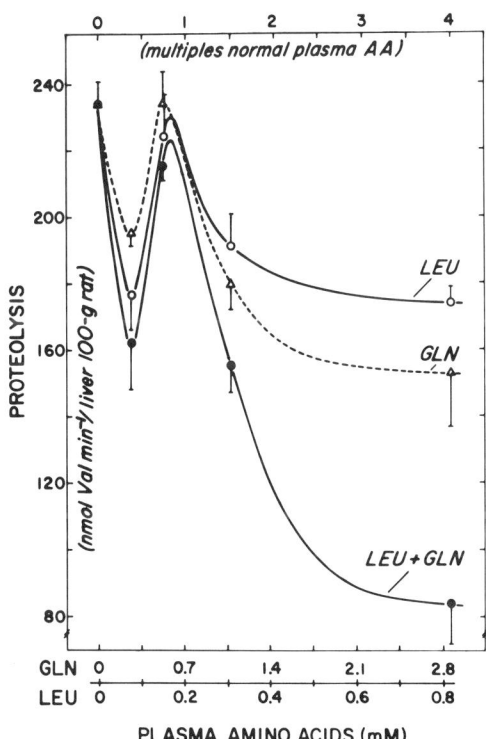

Fig. 9. *Proteolytic dose responses to leucine, glutamine, and leucine + glutamine. The experiments were carried out as in Figure 8, except that the livers were obtained from synchronously fed, 24-hour starved rats. The molar values for leucine and glutamine on the abscissa (bottom) correspond to fractions/multiples of their concentrations in the reference plasma mixture (top) mentioned in Figure 8. The values are means ± S.E. of 3–33 experiments, normalized to 100 g of initial body weight. [From Miotto et al., 1992, with permission of the publisher.]*

equilibrate at about ~10% of external values. By contrast, leucine is rapidly transported by system L, is metabolically stable, and its internal and external concentrations are approximately equal. Since both agents share the same recognition site, the most reasonable location for it is the plasma membrane [Miotto et al., 1992]; other possibilities, of course, may be considered.

Further support for a plasma membrane locus was obtained by using the multiple antigen peptide MAP [Tam, 1989] to form an ~1,900 kDa molecule with eight residues of leucine attached to its N termini (G Miotto, R Venerando, O Marin, N Siliprandi, GE Mortimore, manuscript in preparation). Its synthesis was prompted from evidence that the carboxyl group of leucine is not required for biological activity [Wert et al., 1992]. Leu_8-MAP proved to be as effective as leucine in suppressing proteolysis in the isolated hepatocyte, and macroautophagy was decreased in parallel with proteolysis; Ile_8-MAP was inactive. Because it was not transported into the cytosolic space, its inhibition could only have been initiated at the plasma membrane. Indeed, subsequent studies with an iodinatable, azide derivative of Leu-MAP have shown the existence of a M_r ~340 protein complex at the plasma membrane that specifically photoreacts with an apparent affinity of 10^{-5}–10^{-3} M (GE Mortimore, JJ Wert Jr, G Miotto, R Venerando, M Kadowaki, Biochem Biophys Res Commun, in press).

Concerted effects of regulatory amino acids in proteolytic control. The putative plasma membrane locus of leucine recognition has important implications concerning the location of other amino acid recognition sites. Because leucine, glutamine, and tyrosine individually and the regulatory amino acids as a group all evoke multiphasic responses at the same fractions/multiples of their molar concentrations in normal plasma (Figs. 8, 9), it is probable that they act in a concerted manner through sites in proximity to each other. Inhibitory effects of leucine, glutamine, and tyrosine have been shown to be additive in livers from ad libitum fed rats [Kadowaki et al., 1992]; in synchronously fed rats, a minimal combination of only two amino acids, leucine and glutamine, will evoke multiphasic responses equivalent to the complete regulatory group (Fig. 9).

As a coregulator, alanine is specifically required for expression of proteolytic inhibition by leucine, glutamine, and the regulatory group at normal plasma concentrations [Pösö and Mortimore, 1984; Mortimore et al., 1991]. Thus, in the presence of alanine, the zonal peak is lost and individual dose responses to leucine and glutamine become directly proportional over the range of regulation [Miotto et al., 1992]. These findings suggest that recognition sites for glutamine as well as for leucine are directly accessible to plasma amino acids and that proteolytic inhibition is determined by reversible reactions between the sites and the regulatory amino acids.

Support for this contention was obtained by the use of conventional Michaelis-Menten kinetics in plotting the relationship between proteolytic inhibition (V) and fraction/multiples of normal amino acid concentrations (S). In Figure 10, highly linear VS over V plots were obtained from which values of about 0.5× were computed for the *relative* apparent K_m of the complete amino acid mixture and of leucine plus alanine. In molar terms, this indicates that apparent K_m values for regulatory amino acids are one-half of their normal plasma concentrations. Thus the agreement between leucine and the complete amino acid mixture is probably not fortuitous, but an essential feature of a complex mechanism that is closely regulated in a concerted way by more than one amino acid. By use of the above probes, it may be possible to identify the recognition site of leucine and determine its cellular location and relationship to the putative complex.

Hormonal Regulation of Macro- and Microautophagy

Insulin. Proteolytic responses to insulin as well as to fetal calf serum and other growth-promoting factors in mammalian cells are uniformly inhibitory [reviewed by Ballard and Gunn, 1982; Mortimore and Pösö, 1987] and,

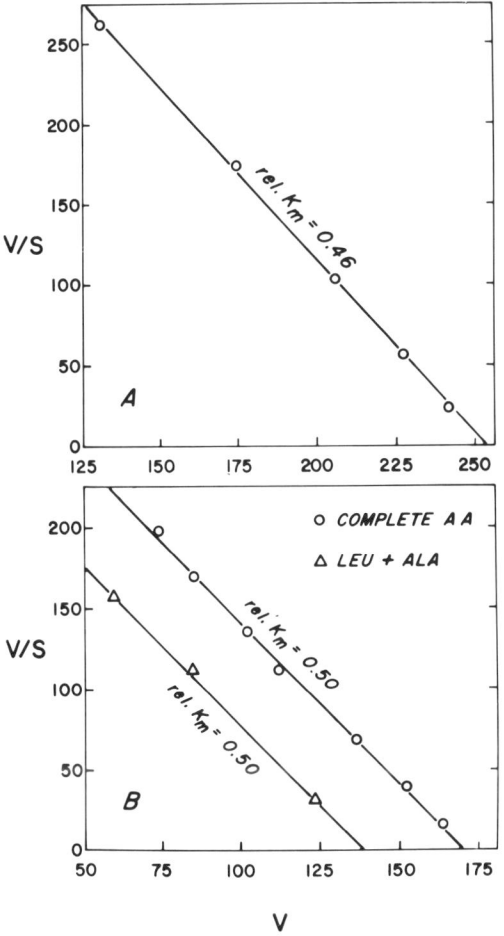

Fig. 10. *V/S versus V plots of proteolytic dose responses to the standard complete amino acid mixture and to leucine + alanine. V represents proteolytic inhibition, expressed as nmoles of valine min^{-1} per liver (100 g of initial body weight); S denotes fractions/multiples of amino acids in the medium. In this plot V_{max} is conveniently shown as the V-intercept. The apparent K_m values are also relative (rel.) in that they are based on fractions/multiples of the molar values in the standard plasma mixture. **A:** Responses of livers from normal fed rats to the complete amino acid mixture. **B:** Responses of livers from synchronously fed, 24-hour starved rats to the same mixture and to leucine + alanine. [From Miotto et al., 1992, with permission of the publisher.]*

in many instances, accompanied by a suppression of autophagy [Jefferson et al., 1974; Neely et al., 1977; Mortimore et al., 1987]. Virtually nothing is known of the molecular mechanism by which insulin mediates its effects. In cultured kidney cells, Tsao et al. [1990] tested the possibility that the autophagic inhibition is the result of Na^+-H^+ antiporter stimulation, an effect associated with cytoplasmic growth. No enhancement of antiporter activity was found, however, despite the fact that protein degradation was strongly suppressed.

It is evident from the perfusion experiments of Figure 11 that insulin elicits two responses in liver, one that is amino acid concentration dependent and closely linked to amino acid control and a second that is concentration in-

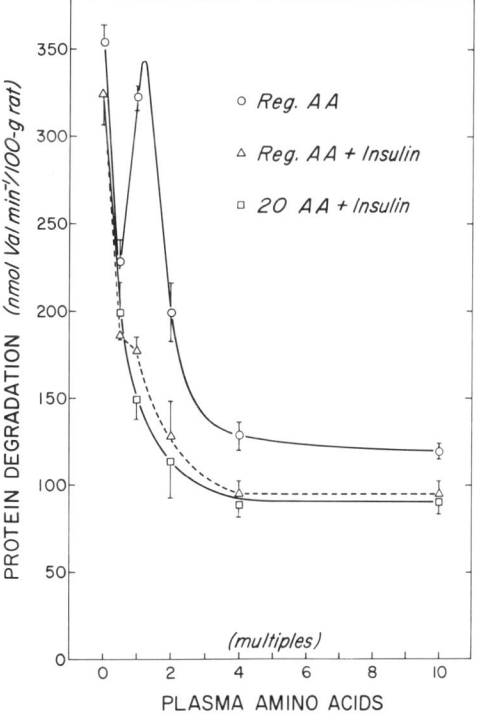

Fig. 11. *Effects of 10^{-9} M insulin on proteolytic dose responses to regulatory (Reg. AA) and complete plasma amino acid mixtures. In other respects the experiments were the same as those in Figure 8. [From Mortimore et al., 1987, with permission of the publisher.]*

dependent. In the first, insulin selectively abolishes the zonal loss of inhibitory effectiveness at normal amino acid concentrations without affecting its low concentration inhibition, thus mimicking the coregulatory effect of alanine. The resulting dose response curve is typical of the L mode of amino acid regulation [Mortimore et al., 1991]. Possibly through a similar mechanism, insulin has been shown to evoke a rapid switchover from H to L in the basic mode of regulation [Mortimore et al., 1991].

The second effect of insulin represents a small but significant decrease in proteolytic rates that has the same absolute value in the presence and absence of added amino acids. A similar response to insulin was observed with RNA degradation [Lardeux and Mortimore, 1987]. Because the effects on protein and RNA can be obtained in the virtual absence of macroautophagy (10× plasma amino acids), they are presumed to reflect reductions in microautophagic sequestration. Although small in comparison with the first effect, its magnitude nevertheless is sufficient to account for approximately half of the net loss of resident proteins during starvation. But of more immediate interest is the distinct possibility that it belongs to a class of autophagy below which is under direct hormonal control by glucagon.

Glucagon and β-Agonists. Glucagon is widely known as an inducer of hepatic macroautophagy [Ashford and Porter, 1962; Deter et al., 1967; Rosa, 1971; Schworer and Mortimore, 1979] and long-lived protein degradation [Woodside et al., 1974; Schworer and Mortimore, 1979; Mortimore et al., 1987; Lardeux and Mortimore, 1987] and can evoke responses similar to those seen after stringent amino acid deletion [Schworer and Mortimore, 1979]. Effects have also been observed in the isolated hepatocyte [Hopgood et al., 1980], but for unknown reasons their magnitude is small. Results in the perfused liver suggest that it acts by blocking the low concentration inhibition exerted by regulatory amino acids, thus accelerating macroautophagy and proteolysis maximally at normal plasma concentrations (Fig. 12). Inhibition, though, is ultimately achieved

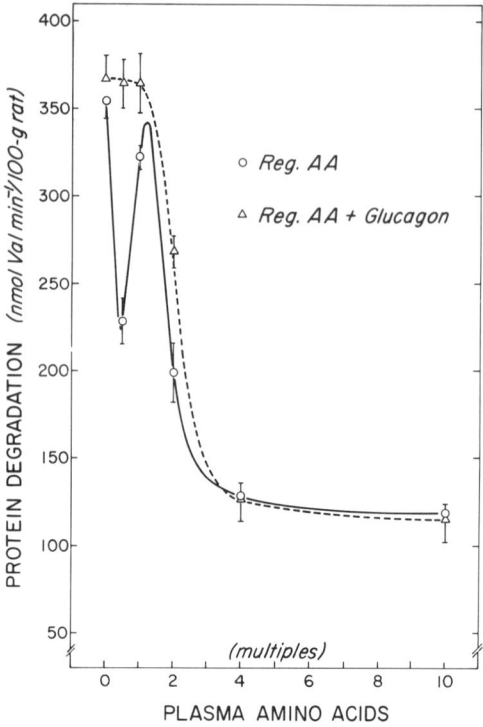

Fig. 12. *Effect of 8×10^{-9} M glucagon on the proteolytic dose response to regulatory (Reg. AA) amino acids. The conditions otherwise were the same as those in Figure 8. [From Mortimore et al., 1987, with permission of the publisher.]*

at higher levels through a second mechanism that completely overrides the glucagon enhancement. The resulting dose–response curve (Fig. 12) is strikingly reminiscent of the H mode that has been noted in livers of synchronously fed rats perfused 18 hours after food intake [Mortimore et al., 1991].

The response to glucagon appears to be mediated via cyclic AMP, since comparable effects have been elicited with epinephrine [Rosa, 1971; Woodside et al., 1974; Hopgood et al., 1980] and cyclic AMP itself [Rosa, 1971; Hopgood et al., 1980]. The α-adrenergic pathway, however, does not seem to be utilized inasmuch as vasopressin and phenylephrine fail to inhibit proteolysis in rat liver perfusion experiments (GE Mortimore, unpublished findings). Curiously, in muscle, cyclic AMP and β-agonists suppress rather than stimulate proteolysis. In rat heart, for example, glucagon and isoproterenol inhibit macroautophagy and proteolysis [Chua et al., 1978; Dämmrich and Pfeifer, 1981]; similar effects have been reported in skeletal muscle [Garber et al., 1976; Li and Jefferson, 1977].

How cyclic AMP enhances autophagy is not well understood, although it is clear that protein synthesis is not required [Woodside et al., 1974; Hopgood et al., 1980] and it is not mediated by the depletion of glucogenic amino acids as was once believed [Schworer and Mortimore, 1979]. Current evidence invokes two mechanisms. The first is related to the above observation that cyclic AMP switches the mode of amino acid control from L to H, possibly at site(s) of amino acid recognition. The second derives from a recent finding (Fig. 13) that glucagon and cyclic AMP can stimulate autophagy independently of amino acids (GE Mortimore, M Kadowaki, KK Khurana, JJ Wert Jr, manuscript in preparation). The effect, which in magnitude equals the maximal deprivation response, is consistently elicited in the presence and absence of 10× plasma amino acids in livers of synchronously fed rats perfused 18 hours after food intake. It is inhibited by chloroquine (Fig. 13) and disappears after short-term starvation.

Of interest in the above experiments are the autophagic changes induced by glucagon at 10× plasma concentrations. Type A lysosomes, resembling glycogenosomes [Amherdt et al., 1974], increase in both size and number, and a unique single-walled autophagic vacuole is a prominent feature. Small increases in macroautophagy are also seen. These findings are potentially important because they demonstrate that basal/microautophagy (and possibly macroautophagy as well) is subject to direct hormonal regulation. In addition, they point to the existence of an independent pathway involving phosphorylation/dephosphorylation that can bypass the amino acid control mechanism. The extent to which insulin is capable of counterbalancing the stimulation has not been determined, although there is little doubt that

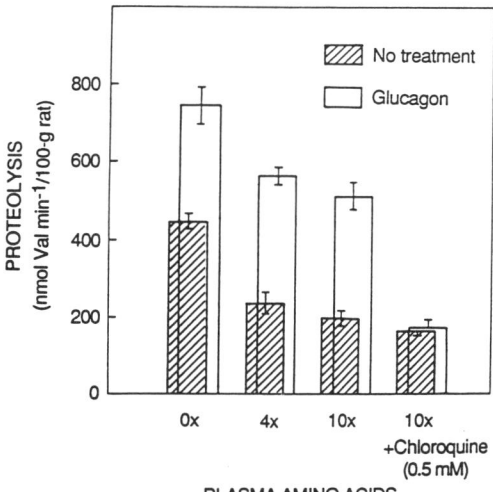

Fig. 13. *Amino acid–independent responses to glucagon in livers of synchronously fed rats, perfused in the single-pass mode. The experiments were carried out 18 hours after the start of feeding as described earlier [Mortimore et al., 1991]; glucagon (8×10^{-9} M) was infused in the absence and presence of 4× and 10× complete amino acid mixtures. The coadministration of chloroquine (0.5 mM) eliminated the stimulation. (From Mortimore GE, Kadowaki M, Khurana KK, Wert JJ Jr, manuscript in preparation.)*

it will express inhibition (see Fig. 11). The microautophagic stimulation could be mediated through a regulatory phosphoprotein attached to forming membranes of the nascent vacuoles. Since microautophagic elements can be separated on gradients [Surmacz et al., 1983a,b], eventual identification of the putative site would appear to be feasible.

ACKNOWLEDGMENTS

Studies from the laboratory of G.E.M. were supported by USPHS grant DK-21624.

REFERENCES

Ahlberg J, Berkenstam A, Henell F, Glaumann H (1985): Degradation of short and long lived proteins in isolated rat liver lysosomes. Effects of pH, temperature, and proteolytic inhibitors. J Biol Chem 260:5847–5854.

Amenta JS, Brocher SC (1981): Mechanisms of protein turnover in cultured cells. Life Sci 28:1195–1208.

Amherdt M, Harris V, Renold AE, Orci L, Unger RH (1974): Hepatic autophagy in uncontrolled experimental diabetes and its relationships to insulin and glucagon. J Clin Invest 54:188–193.

Aplin A, Jasionowski T, Tuttle DL, Lenk SE, Dunn WA Jr (1992): Cytoskeletal elements are required for the formation and maturation of autophagic vacuoles. J Cell Physiol 152:458–466.

Ashford TP, Porter KR (1962): Cytoplasmic components in hepatic cell lysosomes. J Cell Biol 12:198–202.

Auteri JS, Okada A, Bochaki V, Dice JF (1983): Regulation of intracellular degradation in IMR-09 human diploid fibroblasts. J Cell Physiol 115:167–174.

Belavoine S, Feldmann G, Lardeux B (1990): Rates of RNA degradation in isolated hepatocytes. Effects of amino acids and inhibitors of lysosomal function. Eur J Biochem 189:617–623.

Ballard FJ, Gunn JM (1982): Nutritional and hormonal effects on intracellular protein catabolism. Nutr Rev 40:33–42.

Blouin A, Bolender RP, Weibel EW (1977): Distribution of organelles and membranes between hepatocytes and nonhepatocytes in the rat liver parenchyma. J Cell Biol 72:441–455.

Buse MG, Reid SS (1975): A possible regulator of protein turnover in muscle. J Clin Invest 56:1250–1261.

Canut H, Alibert G, Boudet AM (1985): Hydrolysis of intracellular proteins in vacuoles isolated from *Acer pseudoplatanus* L. cells. Plant Physiol 79:1090–1093.

Caro LHP, Leverve XM, Plomp PJAM, Meijer AJ (1989): A combination of intracellular leucine with either glutamate or aspartate inhibits autophagic proteolysis in isolated rat hepatocytes. Eur J Biochem 181:717–720.

Chua BHL, Kao RL, Rannels DE, Morgan HE (1979): Inhibition of protein degradation by anoxia and ischemia in perfused rat hearts. J Biol Chem 254:6617–6623.

Chua BHL, Watkins CA, Siehl DL, Morgan HE (1978): Effects of epinephrine and glucagon on protein turnover in perfused rat heart. Fed Proc 37:1333A.

Ciechanover A, Finley D, Varshavsky A (1984): The ubiquitin-dependence of selective protein degradation demonstrated in the mammalian cell cycle mutant ts85. Cell 37:57–66.

Dämmerich J, Pfeifer U (1981): Acute effects of isoproterenol on cellular autophagy: Inhibition in myocardium but stimulation in liver parenchyma. Virchows Arch B 38:209–218.

de Duve C, Wattiaux R (1966): Functions of lysosomes. Annu Rev Physiol 28:435–492.

Deter RL, Baudhuin P, de Duve C (1967): Participation of lysosomes in cellular autophagy induced in rat liver by glucagon. J Cell Biol 35:c11–c15.

de Waal EJ, Vreeling-Sindelárová H, Schellens JPM, Houtkooper JM, James J (1986): Quantitative changes in the lysosomal vacuolar system of the rat hepatocytes during short-term starvation. A morphometric analysis with special reference to macro- and microautophagy. Cell Tissue Res 243:641–648.

Dunn WA Jr (1990a): Studies on the mechanisms of autophagy: Formation of autophagic vacuoles. J Cell Biol 110:1923–1933.

Dunn WA Jr (1990b): Studies on the mechanisms of autophagy: Maturation of the Autophagic vacuoles. J Cell Biol 110:1935–1945.

Epstein D, Elias-Bishko S, Hershko A (1975): Requirement for protein synthesis in the regulation of protein breakdown in cultured hepatoma cells. Biochemistry 14:5199–5204.

Fulks RM, Li JB, Goldberg AL (1975): Effects of insulin, glucose, and amino acids on protein turnover in rat diaphragm. J Biol Chem 250:290–298.

Garber AJ, Karl IE, Kipnis DM (1976): Alanine and glutamine synthesis and release from skeletal muscle. J Biol Chem 251:851–857.

Gordon PB, Høyvik H, Seglen PO (1992): Prelysosomal and lysosomal connections between autophagy and endocytosis. Biochem J 283:361–369.

Gordon PB, Seglen PO (1988): Prelysosomal convergence of autophagic and endocytic pathways. Biochem Biophys Res Commun 151:40–47.

Gropper R, Brandt RA, Elias S, Bearer CF, Mayer A, Schwartz AL, Ciechanover A (1991): The ubiquitin-activating enzyme, E1, is required for stress-induced lysosomal degradation of cellular proteins. J Biol Chem 266:3602–3610.

Henell F, Glaumann H (1984): Effect of leupeptin on the autophagic vacuolar system of rat hepatocytes. Lab Invest 51:46–56.

Heydrick SJ, Lardeux BR, Mortimore GE (1991): Uptake and degradation of cytoplasmic RNA by hepatic lysosomes: Quantitative relationship to RNA turnover. J Biol Chem 266:8790–8796.

Hirsimäki Y, Hirsimäki P (1984): Vinblastine-induced autophagocytosis: The effect of disorganization of microfilaments by cytochalasin B. Exp Mol Pathol 40:61–69.

Hopgood MF, Clark MG, Ballard FJ (1980): Protein degradation in hepatocyte monolayers: Effects of glucagon, adenosine 3′:5′-cyclic monophosphate, and insulin. Biochem J 186:71–79.

Hutson NJ, Mortimore GE (1982): Suppression of cytoplasmic uptake by lysosomes as the mechanism of protein regain in livers of starved–refed mice. J Biol Chem 257:9548–9554.

Jefferson LS, Rannels DE, Munger BL, Morgan HE (1974): Insulin in the regulation of protein turnover in heart and skeletal muscle. Fed Proc 33:1098–1104.

Kadowaki M, Pösö AR, Mortimore GE (1992): Parallel control of hepatic proteolysis by phenylalanine and phenylpyruvate through independent inhibitory sites at the plasma membrane. J Biol Chem 267:22060–22065.

Kadowaki M, Venerando R, Miotto G, Mortimore GE

(1994): De novo formation of macroautophagic vacuoles in hepatocytes permeabilized by *Staphylococcus aureus* α-toxin: Inhibition by nonhydrolyzable GTP analogs. J Biol Chem 269:3703–3710.

Khairallah EA, Mortimore GE (1976): Assessment of protein turnover in perfused rat liver: Evidence for amino acid compartmentation from differential labeling of free and tRNA-bound valine. J Biol Chem 251:1375–1384.

Knowles SE, Ballard FJ (1976): Selective control of the degradation of normal and aberrant proteins in Rueber H35 hepatoma cells. Biochem J 156:609–617.

Kominami E, Hashida S, Khairallah EA, Katunuma N (1983): Sequestration of cytoplasmic enzymes in an autophagic vacuole–lysosomal system induced by the injection of leupeptin. J Biol Chem 258:6093–6100.

Kopitz J, Kisen GØ, Gordon PB, Bohley P, Seglen PO (1990): Nonselective autophagy of cytosolic enzymes by isolated rat hepatocytes. J Cell Biol 111:941–953.

Kovács J, Fellinger E, Kárpáti AP, Kovács AL, László L, Réz G (1987): Morphometric evaluation of the turnover of autophagic vacuoles after treatment with Triton X-100 and vinblastine in murine pancreatic acinar and seminal vesicle epithelial cells. Virchows Arch B 53:183–190.

Kovács AL, Grinde B, Seglen PO (1981): Inhibition of autophagic vacuole formation and protein degradation by amino acids in isolated hepatocytes. Exp Cell Res 133:431–436.

Lardeux BR, Heydrick SJ, Mortimore GE (1987): RNA degradation in perfused rat liver as determined from the release of [^{14}C]cytidine. J Biol Chem 262:14507–14513.

Lardeux BR, Mortimore GE (1987): Amino acid and hormonal control of macromolecular turnover in perfused rat liver: Evidence for selective autophagy. J Biol Chem 262:14514–14519.

Lawrence BP, Brown WJ (1992): Autophagic vacuoles rapidly fuse with preexisting lysosomes in cultured hepatocytes. J Cell Sci 102:515–526.

Lenk SE, Dunn WA Jr, Trausch JS, Ciechanover A, Schwartz AL (1992): Ubiquitin-activating enzyme is associated with maturation of autophagic vacuoles. J Cell Biol 118:301–308.

Leverve XM, Caro LHP, Plomp PJAM, Meijer AJ (1987): Control of proteolysis in perifused rat hepatocytes. FEBS Lett 219:455–458.

Li JB, Jefferson LS (1977): Effect of isoproterenol on amino acid levels and protein turnover in skeletal muscle. Am J Physiol 232:E32–E37.

Marty F (1978): Cytochemical studies on GERL, provacuoles, and vacuoles in root meristematic cells of *Euphorbia*. Proc Natl Acad Sci USA 75:852–856.

Marzella L, Ahlberg J, Glaumann H (1980): In vitro uptake of particles by lysosomes. Exp Cell Res 129:460–465.

May LT, Anderson OR, Hogg JF (1982): Changes of cellular structure and subcellular enzymatic patterns during the activation of glyconeogenesis in *Tetrahymena pyriformis*. J Ultrastruct Res 81:271–289.

Miotto G, Venerando R, Khurana KK, Siliprandi N, Mortimore GE (1992): Control of hepatic proteolysis by leucine and isovaleryl-L-carnitine through a common locus: Evidence for a possible mechanism of recognition at the plasma membrane. J Biol Chem 267:22066–22072.

Miotto G, Venerando R, Siliprandi N (1989): Inhibitory action of isovaleryl-L-carnitine on proteolysis in perfused rat liver. Biochem Biophys Res Commun 158:797–802.

Mitch WE, Clark AS (1984): Specificity of the effects of leucine and its metabolites on protein degradation in skeletal muscle. Biochem J 222:570–586.

Mitchener JS, Shelburne JD, Bradford WD, Hawkins HK (1976): Cellular autophagocytosis induced by deprivation of serum and amino acids in HeLa cells. Am J Pathol 83:485–498.

Mortimore GE, Hutson NJ, Surmacz CA (1983): Quantitative correlation between macro- and microautophagy in mouse hepatocytes during starvation and refeeding. Proc Natl Acad Sci USA 80:2179–2183.

Mortimore GE, Khurana KK, Miotto G (1991): Amino acid control of proteolysis in perfused livers of synchronously fed rats: Mechanism and specificity of alanine co-regulation. J Biol Chem 266:1021–1028.

Mortimore GE, Lardeux BR, Adams CE (1988a): Regulation of microautophagy and basal protein turnover in rat liver: Effects of short-term starvation. J Biol Chem 263:2506–2512.

Mortimore GE, Mondon CE (1970): Inhibition by insulin of valine turnover in liver. J Biol Chem 245:2375–2383.

Mortimore GE, Neely AN, Cox JR, Guinivan RA (1973): Proteolysis in homogenates of perfused rat liver: Responses to insulin, glucagon and amino acids. Biochem Biophys Res Commun 54:89–95.

Mortimore GE, Pösö AR (1984): Lysosomal pathways in hepatic protein degradation: Regulatory role of amino acids. Fed Proc 43:1289–1294.

Mortimore GE, Pösö AR (1987): Intracellular protein catabolism and its control during nutrient deprivation and supply. Annu Rev Nutr 7:539–564.

Mortimore GE, Pösö AR, Kadowaki M, Wert JJ Jr (1987): Multiphasic control of hepatic protein degradation by regulatory amino acids: General features and hormonal modulation. J Biol Chem 262:16322–16327.

Mortimore GE, Schworer CM (1977): Induction of autophagy by amino acid deprivation in perfused rat liver. Nature 270:174–176.

Mortimore GE, Ward WF (1981): Internalization of cytoplasmic protein by hepatic lysosomes in basal and deprivation-induced proteolytic states. J Biol Chem 256:7659–7665.

Mortimore GE, Wert JJ Jr, Adams CE (1988b): Modulation of the amino acid control of hepatic protein degradation by caloric deprivation: Two modes of alanine co-regulation. J Biol Chem 263:19545–19551.

Mortimore GE, Woodside KH, Henry JE (1972): Compartmentation of free valine and its relation to protein turnover in perfused rat liver. J Biol Chem 247:2776–2784.

Neely AN, Cox JR, Fortney JA, Schworer CM, Mortimore GE (1977): Alterations of lysosomal size and density during rat liver perfusion: Suppression by amino acids and insulin. J Biol Chem 252:6948–6954.

Neely AN, Nelson PB, Mortimore GE (1974): Osmotic alterations of the lysosomal system during rat liver perfusion: Reversible suppression by insulin and amino acids. Biochim Biophys Acta 338:458–472.

Nishimura M, Beevers H (1979): Hydrolysis of protein in vacuoles isolated from higher plant tissue. Nature 277:412–413.

Novikoff AB, Shin WY (1978): Endoplasmic reticulum and autophagy in rat hepatocytes. Proc Natl Acad Sci USA 75:5039–5042.

Paavola LG (1978a): The corpus luteum of the guinea pig. II. Cytochemical studies on the Golgi complex, GERL, and lysosomes in luteal cells during maximal progesterone secretion. J Cell Biol 79:45–58.

Paavola LG (1978b): The corpus luteum of the guinea pig. III. Cytochemical studies on the Golgi complex and GERL during normal postpartum regression of luteal cells, emphasizing the origin of lysosomes and autophagic vacuoles. J Cell Biol 79:59–73.

Papadopoulos T, Pfeifer U (1986): Regression of rat liver autophagic vacuoles by locally applied cycloheximide. Lab Invest 54:100–107.

Pfeifer U (1973): Cellular autophagy and cell atrophy on the rat liver during long-term starvation. Virchows Arch B 12:195–211.

Pfeifer U (1978): Inhibition by insulin of the formation of autophagic vacuoles in rat liver. J Cell Biol 78:152–167.

Pfeifer U (1981): Morphological aspects of intracellular protein degradation: Autophagy. Acta Biol Med Ger 40:1619–1624.

Plomp PJAM, Gordon PB, Meijer AJ, Høyvik H, Seglen PO (1989): Energy dependence of different steps in the autophagic-lysosomal pathway. J Biol Chem 264:6699–6704.

Poole B, Wibo M (1973): Protein degradation in cultured cells. J Biol Chem 248:6221–6226.

Pösö AR, Mortimore GE (1984): Requirement for alanine in the amino acid control of deprivation-induced protein degradation in liver. Proc Natl Acad Sci USA 81:4270–4274.

Pösö AR, Schworer CM, Mortimore GE (1982a): Acceleration of proteolysis in perfused rat liver by deletion of glucogenic amino acids: Regulatory role of glutamine. Biochem Biophys Res Commun 107:1433–1439.

Pösö AR, Wert JJ Jr, Mortimore GE (1982b): Multifunctional control by amino acids of deprivation-induced proteolysis in liver: Role of leucine. J Biol Chem 257:12114–12120.

Rabkin R, Tsao T, Shi JD, Mortimore GE (1991): Amino acids regulate kidney cell protein breakdown. J Lab Clin Med 117:505–513.

Rannels DE, Kao R, Morgan HE (1975): Effect of insulin on protein turnover in heart muscle. J Biol Chem 250:1694–1701.

Rosa F (1971): Ultrastructural changes produced by glucagon, cyclic 3´:5´-AMP, and epinephrine on perfused livers. J Ultrastruct Res 34:205–213.

Schwartz AL, Brandt RA, Geuze HJ, Ciechanover A (1992): Stress-induced alteration in the autophagic pathway: Relationship to ubiquitin system. Am J Physiol 262:C1031–C1038.

Schworer CM, Mortimore GE (1979): Glucagon-induced autophagy and proteolysis in rat liver: Mediation by selective deprivation of intracellular amino acids. Proc Natl Acad Sci USA 76:3169–3173.

Schworer CM, Shiffer HA, Mortimore GE (1981): Quantitative relationship between autophagy and proteolysis during graded amino acid deprivation in perfused rat liver. J Biol Chem 256:7652–7658.

Seglen PO, Gordon PB (1982): 3-Methyladenine: Specific inhibitor of autophagic/lysosomal protein degradation in isolated rat hepatocytes. Proc Natl Acad Sci USA 79:1889–1892.

Seglen PO, Gordon PB, Poli A (1980): Amino acid inhibition of the autophagic/lysosomal pathway of protein degradation in isolated rat hepatocytes. Biochim Biophys Acta 693:103–118.

Seglen PO, Solheim AE (1978): Valine uptake and incorporation into protein in isolated rat hepatocytes. Eur J Biochem 85:15–25.

Shiman R, Mortimore GE, Schworer CM, Gray DW (1982): Regulation of phenylalanine hydroxylase activity by phenylalanine in vivo, in vitro, and in perfused rat liver. J Biol Chem 257:11213–11216.

Solheim AE, Seglen PO (1980): Subcellular distribution of proteolytically generated valine in isolated rat hepatocytes. Eur J Biochem 107:587–596.

Sommercorn JM, Swick RW (1881): Protein degradation in primary monolayer cultures of adult rat hepatocytes. J Biol Chem 256:4816–4821.

Surmacz CA, Pösö AR, Mortimore GE (1987): Regulation of lysosomal fusion during deprivation-induced autophagy in perfused liver. Biochem J 242:453–458.

Surmacz CA, Wert JJ Jr, Mortimore GE (1983a): Role of particle interaction on distribution of liver lysosomes in colloidal silica. Am J Physiol 245:C52–C60.

Surmacz CA, Wert JJ Jr, Mortimore GE (1983b): Metabolic alterations and distribution of rat liver lysosomes in colloidal silica. Am J Physiol 245:C61–C67.

Tam JP (1989): High-density multiple antigenic-peptide

system for preparation of antipeptide antibodies. Methods Enzymol 168:7–15.

Tischler ME, Desautels M, Goldberg AL (1982): Does leucine, leucyl-tRNA, or some metabolite of leucine regulate protein synthesis and degradation in rat skeletal and cardiac muscle? J Biol Chem 257:1613–1621.

Tooze J, Hollinshead M, Ludwig T, Howell K, Hoflack B, Kern H (1990): In exocrine pancreas, the basolateral pathway converges with the autophagic pathway immediately after the early endosome. J Cell Biol 111:329–345.

Tsao TC, Shi JD, Mortimore GE, Cragoe EJ, Rabkin R (1990): Modulation of kidney cell protein degradation by insulin. J Lab Clin Med 116:369–376.

Vandenburgh H, Kaufman S (1980): Protein degradation in embryonic skeletal muscle. Effects of medium, cell type, inhibitors, and passive stretch. J Biol Chem 255:5826–5833.

Venerando R, Miotto G, Kadowaki M, Siliprandi N, Mortimore GE (1994): Multiphasic control of proteolysis by leucine and alanine in the isolated rat hepatocyte. Am J Physiol 266:C455–C461.

Ward WF, Cox JR, Mortimore GE (1977): Lysosomal sequestration of intracellular protein as a regulatory step in hepatic proteolysis. J Biol Chem 252:6955–6961.

Wert JJ Jr, Miotto G, Kadowaki M, Mortimore GE (1992): 4-Amino-6-methylhept-2-enoic acid: A leucine analogue and potential probe for localizing sites of proteolytic control in the hepatocyte. Biochem Biophys Res Commun 186:1327–1332.

Wheatley DN (1984): Intracellular protein degradation: Basis of a self-regulating mechanism for the proteolysis of endogenous proteins. J Theor Biol 107:127–149.

Woodside KH, Mortimore GE (1972): Suppression of protein turnover by amino acids in the perfused rat liver. J Biol Chem 247:6474–6481.

Woodside KH, Ward WF, Mortimore GE (1974): Effects of glucagon on general protein degradation and synthesis in perfused rat liver. J Biol Chem 249:5458–5463.

Young VR, Munro HN (1978): N^t-methylhistidine (3-methylhistidine) and muscle protein turnover: An overview. Fed Proc 37:2291–2300.

ABOUT THE AUTHORS

GLENN E. MORTIMORE is Professor of Molecular and Cellular Physiology at the College of Medicine of The Pennsylvania State University at Hershey, Pennsylvania, where he conducts research in cellular and metabolic regulation and teaches medical and graduate courses. After receiving a B.S. from Oregon State University he entered the University of Oregon School of Medicine in Portland, earning an M.D. in 1952. This was followed by clinical studies at Emory and Duke Universities and by postdoctoral research at the University of California in San Francisco with Peter Forsham and at the National Institutes of Health in Bethesda in the laboratory of DeWitt Stetten. As a Senior Scientist at the National Institutes of Health, Dr. Mortimore pioneered methods and approaches for investigating the autophagic turnover of macromolecules in the hepatocyte and its regulation by insulin, glucagon, and specific regulatory amino acids. He has held Visiting Professorships at the University of Connecticut and the University of Pennsylvania and was a Fellow of the National Foundation. He is an Associate Editor of *Metabolism* and has served on the Editorial Board of the *Journal of Biological Chemistry*.

MOTONI KADOWAKI is Associate Professor of Applied Biochemistry at Niigata University in Niigata, Japan. After obtaining a Ph.D. in Applied Biochemistry at Tokyo University, Dr. Kadowaki carried out postdoctoral studies in the laboratory of Glenn Mortimore, first as a Fellow and later as a Research Associate.

Hepatic Endosomes Are a Major Physiological Locus of Insulin and Glucagon Degradation In Vivo

François Authier, Barry I. Posner, and John J.M. Bergeron

INTRODUCTION

Cellular proteolysis is a highly controlled process that takes place in virtually all compartments of cells. Polypeptide hormones brought into the cell by endocytosis are the target of proteases that are present in each compartment throughout the endocytic pathway (Fig. 1), namely, endopeptidases (or proteinases, which cleave peptide bonds internally in polypeptide chains), of the four classes (serine, cysteine, aspartic and metallo-endopeptidases) and exopeptidases (which cleave peptide bonds near the N and C termini of polypeptide chains) of the two classes (amino- and carboxy-peptidases). Thus, plasma membrane proteases that have been identified are mainly serine or metallo-endopeptidases, whereas those identified in lysosomes are mainly cysteine and aspartic endopeptidases or exopeptidases. Only three endopeptidases (cysteine, aspartic, and metallo-endopeptidases) have been identified associated with the endosomal apparatus.

The present review discusses the degradation of internalized proteins in the endosomal and lysosomal apparatus and emphasizes insulin and glucagon degradation pathways in endosomes that illustrate a new locus for internalized polypeptide degradation within the cells. Until recently, endosomes were viewed as sites of sorting and segregation, that is, a meeting place where pathways enter from endocytosis and the trans-Golgi network and exit to exocytosis/secretion and lysosomal degradation (Fig. 1). Therefore, it is generally believed that the major degradative pathway used by ligands taken up by the cell consists of delivering them to the lysosomal compartment for degradation by acid hydrolases. However, endosomes contain a number of lysosomal enzymes that appear to reside in this organelle (like cathepsin B, cathepsin D, and acid phosphatase) and several nonlysosomal proteinases (like endosomal acidic insulinase), indicating that endosomes also have hydrolytic capacity.

PROTEIN DEGRADATION IN LYSOSOMES

The lysosome is the major site of intracellular protein degradation. Containing approximately 20 proteolytic enzymes (Table I) as well as many hydrolases (such as acid phosphatase), lysosomes are well equipped to degrade polypeptides and macromolecules that are continually transported to these organelles from other regions of the cell [Barrett, 1977; Bond and Butler, 1987; Hasilik, 1992]. The concentration of lysosomal proteases (and lysosomes) is particularly high in liver, spleen, kidney, and macrophages.

The lysosomal proteases, called *cathepsins,* display similar properties from various species and cell types: a low molecular mass (20–42 kDa), a monomeric structure, an acid–pH optimum, a broad substrate specificity, and gener-

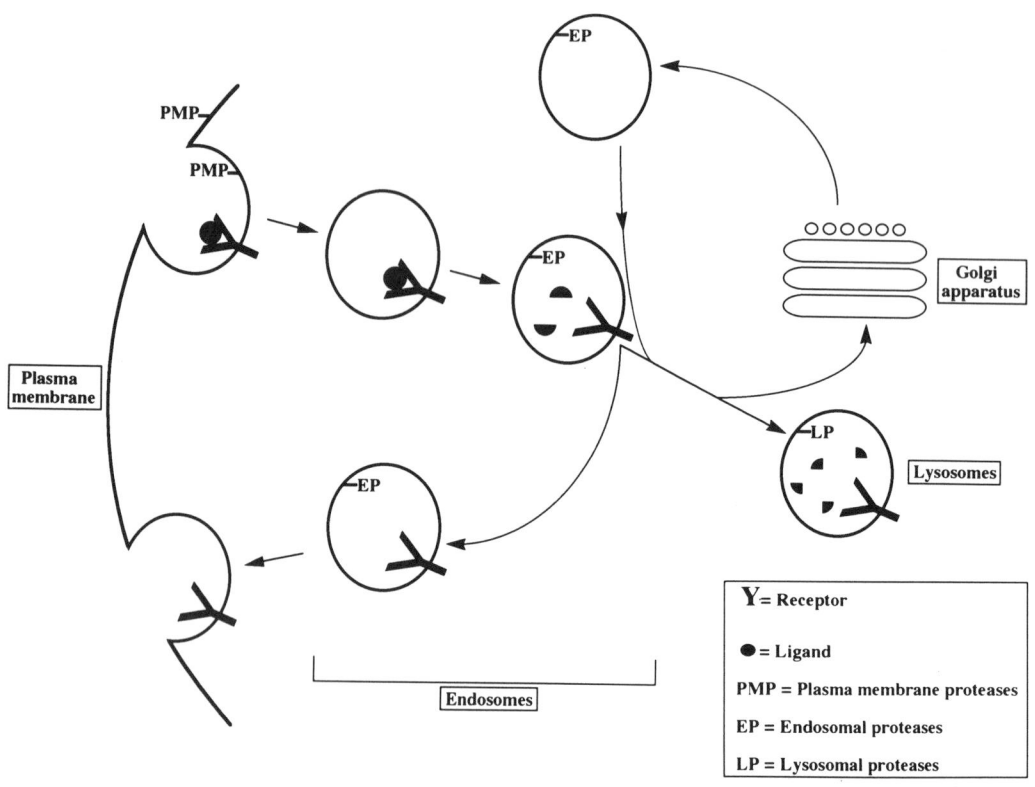

Fig. 1. *Model for the transport of proteases into compartments of the endocytic pathway.*

TABLE I. Proteolytic Activities of Lysosomes

	M_r (subunit)	Optimum pH	Class of protease
Endopeptidases			
Cathepsin B	25	5	Cysteine
Cathepsin L	24	5	Cysteine
Cathepsin H	28	5	Cysteine
Cathepsin M	30	5–7	Cysteine
Cathepsin N	20	3.5	Cysteine
Cathepsin S	25	3.5	Cysteine
Cathepsin T	35	6	Cysteine
Cathepsin D	42	3.5	Aspartic
Cathepsin E	100	2.5	Aspartic
Exopeptidases			
Cathepsin C	20–40	5–6	Aminopeptidase
Dipeptidyl aminopeptidase II	130	4.5–5.5	Aminopeptidase
Cathepsin III	ND	6	Aminopeptidase
Cathepsin I (or A)	100–650	5.5	Carboxypeptidase
Cathepsin IV (or B2)	50	5–6	Carboxypeptidase
Lysosomal carboxypeptidase C	25	5.5	Carboxypeptidase

TABLE II. Proteases Identified in the Endosomal Apparatus of Various Cell Types

Proteases	Cell type	Physiological substrates
Endosomal acidic insulinase	Hepatocyte	Insulin
Endosomal acidic glucagonase	Hepatocyte	Glucagon
Cathepsin D	Macrophage	Mannose-BSA
		PTH-(1-84)
		Ricin A-chain
	Oocyte	Vitellogenin
	Hepatocyte	Apolipoprotein-B100?
Cathepsin B	Fibroblast	EGF?
	HepG2	?
Carboxypeptidase B	Hepatocyte	EGF
	Fibroblast	EGF
Trypsin-like protease	Hepatocyte	EGF
Leucine aminopeptidase	HeLa	?

ally a glycosylated form. The best characterized cathepsins are cathepsins B, L, H, and D (Table I) [Barrett, 1977].

Acid phosphatase is transported to lysosomes as a transmembrane protein precursor, passing through the Golgi apparatus, the trans-Golgi network, the plasma membrane, and the endosomes on their way to the lysosomes. In some cells, several rounds of enzyme recycling between the plasma membrane and the endosome may occur [Braun et al., 1989]. Proteases, like cathepsins, are transported to endosomes and lysosomes by the cation independent (215 kDa) mannose-6-phosphate receptor (MPR) acting to sort cathepsin transport at the level of the trans-Golgi network. The MPR sorts the majority of lysosomal proteases and delivers them to endosomes, where dissociation occurs and the MPR returns to the trans-Golgi network while the enzymes enter the lysosomes [Kornfeld and Mellman, 1989].

PROTEIN DEGRADATION IN ENDOSOMES

Our knowledge of the structure, mechanisms, and function of the endosomal proteases is very limited compared with those located in the lysosomes. Cell-free studies with intact endosomes by Diment and Stahl [1985] and Pease et al. [1985] provided the first indication that endosomes might be proteolytically active against internalized ligands such as mannosylated bovine serum albumin (mannose-BSA) and insulin. Subsequent investigations have shown that other hormones (such as glucagon and parathyroid hormone), epidermal growth factor (EGF), toxins (such as ricin toxin isolated from the plant *Ricin communis* and cholera toxin produced by the bacterium *Vibrio cholaerae*) and foreign antigens are cleaved by endosomal proteases (Table II). Endosomal proteolysis of internalized molecules has been observed in a number of cells and especially in hepatocytes, macrophages, and fibroblasts (Table II).

The responsible proteases have not been extensively characterized except one, the aspartyl protease cathepsin D (E.C. 3.4.23.5). This protease has been localized in endosomes isolated from rabbit alveolar macrophages using immunoblotting and isolated by affinity chromatography on pepstatin-agarose [Diment and Stahl, 1985; Diment et al., 1988]. Biosynthetic studies revealed that macrophage cathepsin D is first synthesized as an inactive membrane-associated precursor of 53 kDa. The precursor is processed to an active membrane-associated form of 46 kDa, which is then released into the lumen of endosomes and lysosomes [Diment et al., 1988]. The distribution of inactive and active cathepsin D to membranes and lumina, respectively, has been confirmed in early and late endosomes isolated from rat liver [Runquist and Havel, 1991]. In macrophages, mature cathepsin D is responsible for the endosomal proteolysis of mannose-BSA [Diment and Stahl, 1985] and parathyroid hormone PTH-(1-84) [Diment et al., 1989]. In the case of PTH-(1-84), the hormone is cleaved to fragments that include a bioactive peptide PTH-(1-34). Then, the degradation product is rapidly shuttled back to the plasma membrane and released into the extracellular medium without delivery to lysosomes [Diment et al., 1989]. In hepatic endosomes, cathepsin D does not seem to be involved in the degradation of apolipoprotein B-100 of endocytosed low density lipoproteins in vivo, presumably because the enzyme displays a low catalytic activity at the endosomal pH [Runquist and Havel, 1991]. Thus, the presence of endosomal proteases with different pH optima may regulate the specificity of proteolytic cleavage throughout the endocytic apparatus. Carboxypeptidase B-like and trypsin-like enzymes were observed to induce a limited proteolysis of EGF in endosomes of rat fibroblasts [Matrisian et al., 1984; Planck et al., 1984], human fibroblasts [Wiley et al., 1985], and rat liver [Renfrew and Hubbard, 1991]. Three forms of EGF processed at their C termini at positions Leu^{47}-Lys^{48}, Lys^{48}-Trp^{49}, and Leu^{52}-Arg^{53} were generated sequentially in hepatic endosomes [Renfrew and Hubbard, 1991]. They corresponded to the three distinct acidic forms of EGF identified by isoelectric focusing that were generated in a sequential fashion within endosomes of rat fibroblasts [Planck et al., 1984]. Despite the fact that the three forms remained trichloroacetic acid (TCA) precipitable, only the proteolytic product EGF-(1-52) bound identically to EGF receptors as intact EGF-(1-53). Endosomes do not contain the full complement of active proteases capable of completely degrading EGF. However, after its entry into lysosomes, it is subsequently degraded to TCA-soluble peptides and amino acids by one or more leupeptin-inhibitable proteases [Wiley et al., 1985; Renfrew and Hubbard, 1991]. The cysteine protease cathepsin B (E.C. 3.4.22.1) was also detected in endosomes of various mammalian cultured cells using fluorogenic substrate [Roederer et al., 1987; Bowser and Murphy, 1990] and by morphological studies [Guagliardi et al., 1990]. However, its physiological substrates have not been firmly established. Leucine aminopeptidase activity has also been detected in endosomes of HeLa cells [Ajioka and Kaplan, 1987]. Proteolytic activities responsible for foreign protein antigen processing within endosomes of immune cells have not been investigated. Furthermore, endosomal proteases and/or reductases involved in toxin degradation and/or reduction have not yet been characterized.

METABOLIC FATE OF INSULIN AND GLUCAGON IN LIVER PARENCHYMA

The major hormones regulating glucose homeostasis are insulin and glucagon. The islets of Langerhans in the pancreas are anatomically positioned to deliver these hormones to the portal circulation of the liver, and the liver via hepatic glycogen in turn represents the major target for the physiological actions of these hormones on blood glucose homeostasis. Remarkably, the liver extracts in one pass greater than 40% of the insulin [Jaspan et al., 1981] and glucagon [Hagopian and Tager, 1987] presented via the portal circulation from the secretion of the islets (Fig. 2).

Internalization and Degradation of Insulin and Glucagon in Hepatic Endosomes

Insulin. Electron microscopy radioautography was instrumental in visualizing the sites of initial accumulation and uptake of insulin in hepatic parenchyma (Fig. 3). Insulin receptors were deduced to be randomly distributed over the hepatic cell surface. They were rapidly internalized with associated insulin into an intracellular compartment now designated as the *endosomal apparatus* [Bergeron et al., 1985]. In rat liver, circulating lipoproteins accumulate in the same compartment, hence the earlier designation of *lipoprotein-filled vesicles* [Bergeron et al., 1979; Posner et al., 1980; Khan et al.,

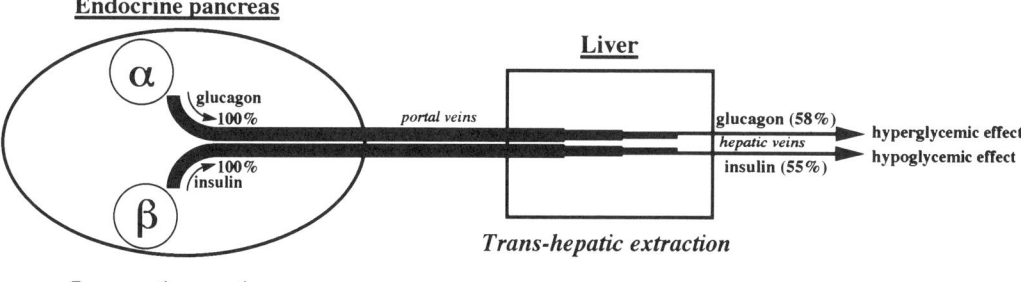

Fig. 2. *Scheme of the trans-hepatic extraction of the pancreatic hormones insulin and glucagon. After secretion by pancreatic α-cells for glucagon and β-cells for insulin, polypeptide hormones are delivered to the liver via the portal vein. In one pass, approximately 45% of insulin [Jaspan et al., 1981] and 42% of glucagon [Hagopian and Tager, 1987] are extracted by liver parenchyma.*

1982]. Remarkably, even in the early studies on internalization little uptake (less than 5%) was found in secondary lysosomes as defined by acid phosphatase cytochemistry [Bergeron et al., 1979]. Subcellular fractionation confirmed and extended the morphological studies. We have demonstrated that insulin internalization was receptor mediated and that both insulin and its receptor accumulated in nonlysosomal subcellular fractions originally designated as *Golgi fractions, low-density vesicles,* or *lipoprotein-filled vesicles* [Posner et al., 1980, 1982; Khan et al., 1982]. By the application of the diaminobenzidine shift protocol of Courtoy et al. [1984] it was conclusively demonstrated that these structures were non-Golgi [Kay et al., 1984].

A comparison of receptor-mediated prolactin and insulin uptake into liver endocytic components demonstrated a similar pathway for both ligand-receptor complexes. The accumulation of prolactin occurred over a longer time course and to a greater extent than that of insulin [Khan et al., 1985; Bergeron et al., 1986, 1988]. This paradox was resolved when the acidotropic agent chloroquine was administered to rats. Chloroquine was shown to accumulate in endosomal components (as well as lysosomes) and was accompanied by a marked increase in the insulin content within isolated endosomal vesicles while only minimally affecting prolactin accumulation therein (Fig. 4) [Posner et al., 1982; Khan et al., 1985]. Taken together with earlier observations that intact as well as degraded insulin was found in endosomes, the hypothesis that the liver endosome was the major site of degradation of internalized insulin was elaborated [Posner et al., 1980]. This was supported by subsequent studies of Pease et al. [1985, 1987], Smith et al. [1989], Hamel et al. [1988, 1991], and Backer et al. [1990].

The hypothesis was tested by two approaches. In the first, insulin fragments were extracted from liver endosomes and evaluated by reverse-phase high-performance liquid chromatography (HPLC) with the fragmentation sequence concluded to be that outlined in Figure 5 [Hamel et al., 1988]. This corresponded to that previously deduced for insulin degradation in vivo.

The second approach employed a cell-free system in which isolated intact endosomes were shown to degrade selectively previously internalized insulin [Doherty et al., 1990]. Endosomal fractions, isolated shortly (5 minutes) after intravenous injection of ^{125}I-insulin into rats, were incubated in isotonic medium buffered to pH 3–8. In these conditions, the endosomes remained intact and maintained proteolytic capacity intraluminally. Although detectable at pH 7.0, the insulin-degrading activity was maxi-

125I - Insulin Internalization

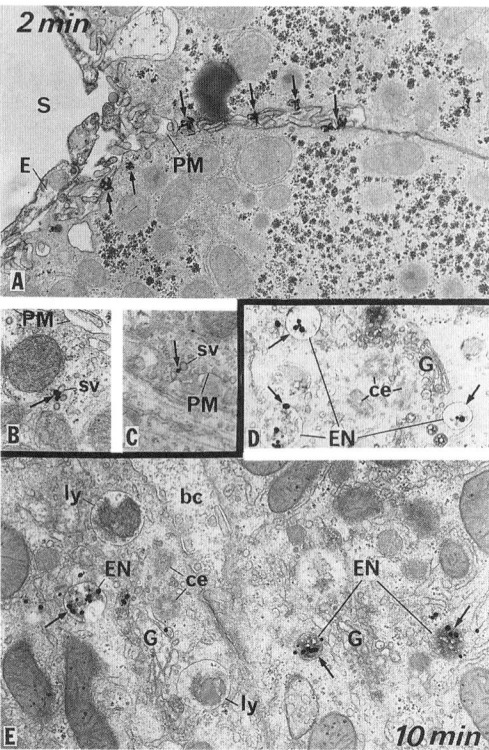

Fig. 3. *The visualization by electron microscopy radioautography of ^{125}I-insulin-binding to receptor on the hepatic cell surface at 2 minutes after insulin injection (A) and its internalization into small endocytic components (B) followed by internalization at 10 minutes into components of the endosomal apparatus near the Golgi apparatus (D,E). The arrows in A indicate filamentous silver grains from ^{125}I-insulin disintegrations and in B, C, D, and E the arrows indicate fine silver grains as developed by the solution physical technique [Kopriwa, 1975]. S, liver sinusoid; E, endothelium; sv, small vesicles; EN, endosomes, ce, centrioles, G, Golgi apparatus, and PM, plasma membrane. The data are taken from Bergeron et al. [1977, 1979].*

mal at pH 5.5, demonstrating that an acidic pH is required for maximal proteolytic velocity. Proteolytic fragments of native ^{125}I-insulin and ^{125}I-tyrosine were generated by this acidic insulinase activity, indicating that proteolysis to component amino acids occurred within the endosomes. The specificity of this proteolytic process and its relationship with ATP-dependent endosomal acidification were also demonstrated [Doherty et al., 1990]. Utilizing endosomes preloaded with ^{125}I-insulin, ^{125}I-prolactin, and ^{125}I-EGF, we demonstrated that, whereas the generation of TCA-soluble radioactive products from insulin occurred rapidly, with a half-life of 6 minutes, neither internalized EGF nor prolactin was converted into TCA-soluble radioactive products even after long incubation (Fig. 6). Requirements for the effect of ATP on the pH-dependent insulin degradation were similar to those for ATP-driven acidification, as measured by the endosomal accumulation of acridine orange. Thus, ATP-dependent endosomal degradation of insulin was markedly inhibited in the presence of ionophores (e.g., nigericin, monensin, and valinomycin), weak base (e.g., chloroquine), and proton pump inhibitors (e.g., N-ethylmaleimide [NEM] and dicyclohexylcarbodiimide), suggesting that the low endosomal pH generated by the vacuolar H^+-ATPase is required for ligand processing. Use has been made of various inhibitors to attempt to classify the type of proteolytic activity. Even in the presence of ATP, degradation of insulin was significantly inhibited by metal-chelating agents (e.g., 1,10-phenanthroline), thiol-blocking agents (e.g., p-chloromercuribenzoic acid [pCMB]), and bacitracin. However, insulin degradation in intact endosomes was insensitive to ethylenediamine-tetraacetic acid (EDTA), serine protease inhibitors (e.g., phenylmethanesulfonyl fluoride [PMSF]), the cathepsin D inhibitor pepstatin, and the lysosomal protease inhibitor leupeptin. The data also indicated that inhibiting the dissociation of insulin from its receptor in endosomes reduced insulin degradation. Hence, free intraluminal insulin was concluded to be the physiological substrate for endosomal acidic insulinase(s).

Insulin degradation products extracts from rat hepatic endosomes following the injection of ^{125}I-insulin suggested three proteolytic cleavages in the B chain (Tyr^{16}-Leu^{17}, Gly^{23}-Phe^{24}, and Phe^{24}-Phe^{25}) occurred prior to cleavage of the A chain [Hamel et al., 1988; Clot et al.,

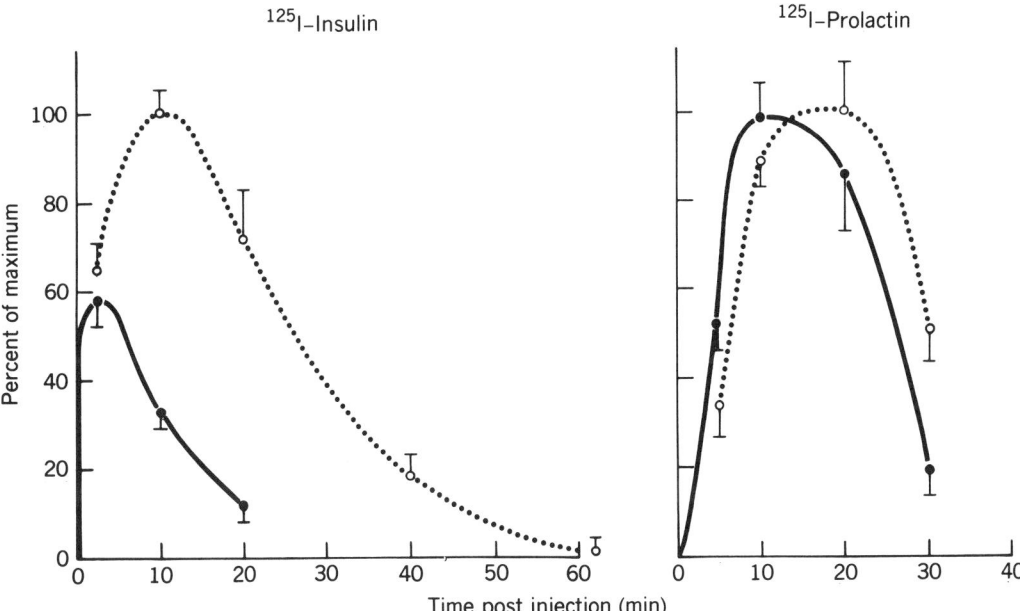

Fig. 4. Effect of chloroquine treatment on the time course of ^{125}I-insulin and ^{125}I-prolactin uptake into the Golgi intermediate fraction. At various times following the injection of ^{125}I-insulin or ^{125}I-prolactin into control (solid lines) or chloroquine-treated (dotted lines) animals, the Golgi intermediate fraction was prepared and its content of radiolabeled ligand determined. Results are expressed as a percent of the radioactive content of the Golgi intermediate fraction at the peak time of uptake. Chloroquine treatment increases selectively the accumulation and time course of ^{125}I-insulin into endosomal components of the Golgi light (not shown) and Golgi intermediate fractions. By contrast, chloroquine shows no similar effect on ^{125}I-prolactin uptake. The data are taken from Bergeron et al. [1988].

1990]. Further studies showed that the degradation products generated from ^{125}I-insulin degraded in cell-free endosomes were qualitatively similar to those extracted from freshly prepared endosomal fractions. The products generated in vitro contain a major cleavage at the Tyr^{16}-Leu^{17} bonds and two minor cleavages at the Gly^{23}-Phe^{24} and Phe^{24}-Phe^{25} bonds [Clot et al., 1990], demonstrating that cell-free endosomes are capable of generating metabolic products similar to those identified in vivo [Hamel et al., 1988].

Data supporting the involvement of a prelysosomal (endosomal) compartment in the degradation of internalized insulin by Fao cells and the importance of the acidity in the proteolytic process have also been presented by Backer et al. [1990]. In Fao cells, rapid intracellular degradation, as assessed by TCA precipitation, was incompatible with the time required to deliver ligand to lysosomes as judged by the internalization of insulin-like growth factor II (IGF-II). This confirmed that insulin processing occurred prelysosomally, probably within the endosomal apparatus. The striking similarity between the kinetics of dissociation and degradation of internalized insulin also suggested that these two endosomal processes are tightly coupled. However, the ordering of these two events was not established. To investigate the possibility that the endosomal degradation of insulin was cell-type specific, Backer et al. [1990] examined the processing of insulin in CHO cells transfected with

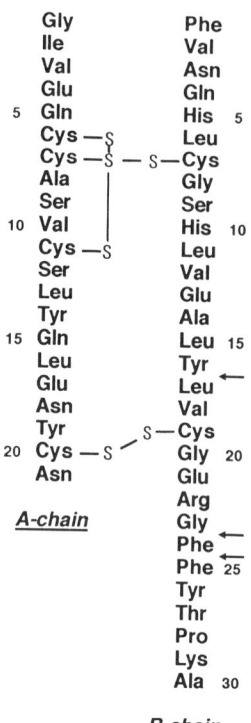

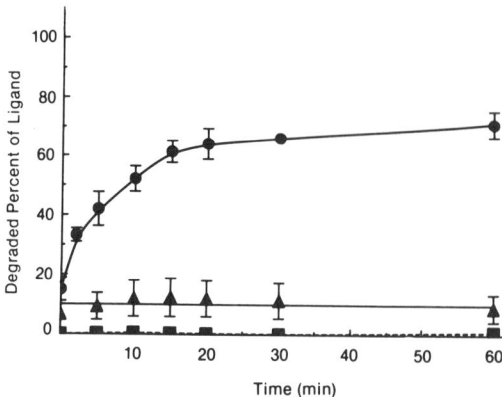

Fig. 6. *Selective degradation of internalized ^{125}I-insulin within isolated intact endosomes. Intact endosomes containing internalized ^{125}I-insulin (●), ^{125}I-EGF (▲), or ^{125}I-prolactin (■) were incubated in a cell-free system for the indicated times. The integrity of each ligand was assessed by precipitability of radiolabel in cold 10% TCA. The data are taken from Doherty et al. [1990].*

Fig. 5. *Primary structure of the A- and B-chain regions of insulin. Arrows indicate the sites of cleavage of insulin generated in hepatic endosomes [Hamel et al., 1988; Clot et al., 1990].*

the human insulin receptor. In contrast to Fao cells, CHO cells were unable to degrade insulin rapidly. Rather, the kinetics of degradation were identical to those for IGF-II, with the time course now being consistent with the delivery to lysosomes. This result suggests that endosomal acidic insulinase may be a cell-type specific enzyme and restricted to cells highly enriched in insulin receptors.

Recently, an endosomal insulinase has been partially purified using an affinity purification protocol at acidic pH [Authier et al., 1994]. The endosomal insulinase differed from insulin-degrading enzyme (IDE) (a neutral thiol-metalloprotease; see below) by virtue of 1) its acid pH optimum for insulin binding and degradation; 2) its lack of cross-reactivity with specific monoclonal and polyclonal IDE antibodies; 3) its inability to be linked covalently to iodinated insulin at neutral pH; and 4) its insensitivity to some inhibitors of IDE [Authier et al., 1994].

Glucagon. Upon interaction with the hepatocyte, glucagon undergoes receptor-mediated endocytosis and follows the same intracellular pathway as insulin, although quantitative differences in receptor mobility and the rate of internalization have been observed [Barazzone et al., 1980; De Diego et al., 1991]. As for insulin, this event has been morphologically and biochemically documented. Studies with isolated hepatocytes have shown that, with time, cell-associated glucagon becomes less dissociable by acid treatment, suggesting internalization of cell surface ligand [Horwitz and Gurd, 1988]. Using electron microscopy, internalized ligand was shown to associate preferentially with intracellular tubulovesicular components [Watanabe et al., 1984]. Finally, hepatic subcellular fractionation studies identified the endosomal compartment as a major site of accumulation of internalized glucagon [Authier et al., 1990, 1992; Authier and Desbuquois, 1991].

The observation of TCA-soluble radioactive products of glucagon in endosomal fractions indicated that ^{125}I-iodoglucagon was taken up by the liver and progressively degraded within this compartment. The injection of chloroquine into rats increased the accumulation of internalized glucagon in endosomes due to at least two factors: 1) the ability of chloroquine to elevate endosomal pH with a subsequent reduction in the dissociation of glucagon–receptor complexes and hence consequent lowering of endosomal glucagonase activity and 2) the capacity of chloroquine to inhibit (presumably by direct interaction) the endosomal glucagon-degrading activities. These findings indicated that, as with insulin, an acidic pH is required for the complete degradation of glucagon in the endosomal compartment. Contrasting with the high recovery of internalized glucagon in endocytic vesicles, little of this hormone was observed in lysosomes by cell fractionation studies [Authier et al., 1990]. As with insulin, chloroquine treatment only minimally augmented the lysosomal accumulation of glucagon, effecting a major accumulation of glucagon within endosomes [Authier et al., 1990]. Hence, the role (if any) of lysosomes in the degradation of glucagon must remain an open question. The use of mono-^{125}I-iodoglucagon to study the endosomal processing of glucagon allowed a probe of the metabolic fate of different regions of the molecule. Reverse-phase HPLC analysis of endosomal extracts from chloroquine-treated rats identified four degradation products. Radiosequence analysis revealed three cleavage sites in the glucagon sequence, namely, Ser^2-Gln^3, Thr^5-Phe^6, and Phe^6-Thr^7 bonds, and an undefined cleavage located C-terminal to Tyr^{13} (Fig. 7).

The endosomal degradation of glucagon was further characterized using a cell-free system similar to that for insulin [Authier and Desbuquois, 1991]. It was shown that glucagon degradation in hepatic endosomes was maximal at pH 4.0 and functionally linked to ATP-dependent endosomal acidification. As with insulin, ATP stimulated endosomal glucagonase activity at neutral pH by promot-

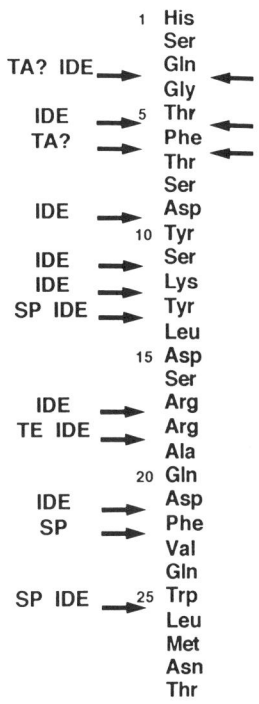

Fig. 7. *Comparison of the specificity of the endosomal glucagonase to that of other previously reported glucagon-degrading activities. The arrows at the right of the glucagon sequence show cleavage sites identified in hepatic endosomes. The arrows with no question mark at the left of the glucagon sequence (IDE, SP, and TE) show cleavage sites generated upon incubation of glucagon with the partially purified proteases. The cleavage bonds with a question mark (e.g., TA?) refer to potential cleavage sites deduced by specificity studies using substrates other than glucagon. TA, tripeptidyl-aminopeptidase; IDE, insulin-degrading enzyme; SP, serine-protease; TE, thiol-endopeptidase.*

ing endosomal acidification as judged by acridine orange uptake [Authier and Desbuquois, 1991]. The ability of ATP to stimulate glucagon degradation in endosomes did not occur if the isotonic medium was depleted of Cl$^-$ as expected from the electrogenic properties of the endosomal proton pump [Fuchs et al., 1989]. The effect of ATP was also suppressed by weak bases (e.g., chloroquine and dansyl cadaverine), by proton ionophore (e.g., monensin), and by inhibitor of the vacuolar-type H$^+$-ATPase (e.g., NEM). In contrast, ATP was fully effective in the presence of vanadate and ouabain, two inhibitors of Na$^+$,K$^+$-ATPase, suggesting a specific involvement of the vacuolar proton pump (Fig. 8). Dissociation of the glucagon–receptor complex was required for glucagon degradation, indicating that free intraluminal glucagon was the physiological substrate for the endosomal protease(s). These findings suggested the involvement of soluble protease(s) in the degradative process. Presumably, dissociation was enhanced by the presence in endosomes of low-affinity glucagon receptor complexes uncoupled from GTP-binding proteins and thus displaying rapid dissociation [Authier et al., 1992]. Radiosequence analysis of the degradation products generated by the incubation of endosomal fractions containing ^{125}I-glucagon and purified by reverse-phase HPLC revealed two cleavage sites affecting the Thr^5-Phe^6 and Phe^6-Thr^7 bonds (Fig. 7). Thus, isolated endosomes incubated in vitro in a cell-free system generated the same degradation products as those found in endosomes in vivo.

The sensitivity of the endosomal glucagon-degrading process to selected protease inhibitors and weak bases is compatible with the involvement of at least two acidic proteases, one of which is a thiol-endopeptidase [Authier and Desbuquois, 1991]. Thus, glucagon degradation was partially inhibited by thiol-blocking agents (e.g., pCMB), bacitracin, acidotropic agents (e.g., chloroquine and dansyl cadaverine), and chelator (e.g., 1,10-phenanthroline), but was unaffected by methylamine, ammonium chloride, and serine protease inhibitors (e.g., PMSF and benzamidine). However, none of these protease inhibitors were capable of fully inhibiting glucagon proteolysis within endosomes, suggesting that several proteases from various classes are involved in endosomal glucagon degradation. Although insulin, internalized within liver cells, is degraded in the same endocytic structures as glucagon, the metabolic fate of these peptides differs in several respects. First, insulin is internalized to endosomes more rapidly than glucagon and is also more rapidly cleared from these structures. Second, in cell-free endosomes, the pH for

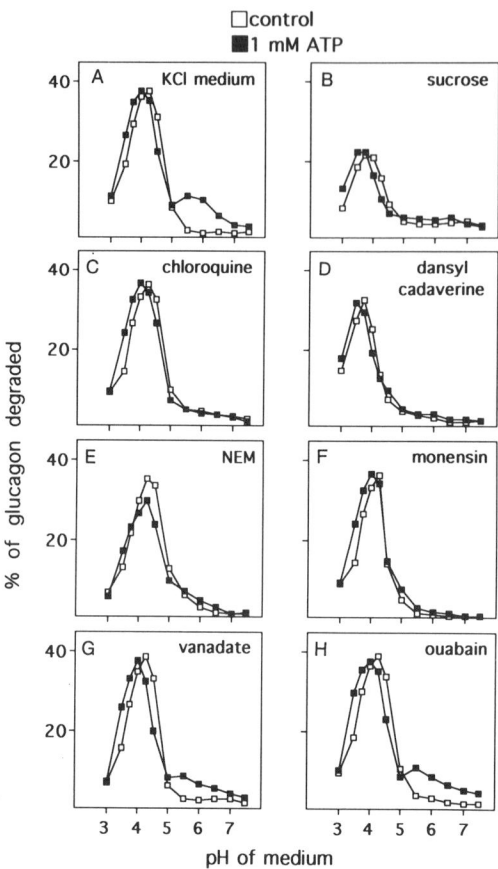

Fig. 8. *Effect of ATP on the pH dependence of endosomal ^{125}I-iodoglucagon degradation in the presence of various drugs. Endosomal fractions were isolated 20 minutes after injection of labeled glucagon. Samples were suspended in 0.15 M KCl (**A,C–H**) or 0.25 M sucrose (**B**) containing 25 mM citrate-phosphate at the indicated pH. For C–H, one of the following drugs were included at the indicated concentrations: C, chloroquine, 20 mM; D, dansyl cadaverine, 5 mM; E, N-ethylmaleimide, 1 mM; F, monensin, 10 μM; G, vanadate, 0.1 mM; H, ouabain, 1 mM. These suspensions were incubated for 5 minutes at 30°C in the absence (□) or presence (■) of 1 mM ATP and 5 mM Mg^{2+} ions as either chloride (KCl medium) or sulfate (sucrose medium). The percentage of ^{125}I-iodoglucagon degraded during incubation was estimated by TCA precipitation. ^{125}I-iodoglucagon degradation occurs mainly in the pH interval 3–5, with a sharp maximal at pH 4. Addition of ATP (1 mM) to the cell-free system broadens the pH range at which degradation of ^{125}I-iodoglucagon occurs, stimulating proteolysis by 5–10-fold at pH above 5.5. The data are taken from Authier and Desbuquois [1991].*

maximal insulin degradation is 5.5, whereas that for maximal glucagon degradation is 4.0. Third, the effect of ATP at neutral pH on the degradation of glucagon is less than that of insulin. Although these differences might reflect different pH-dependence dissociation of ligand–receptor complexes, it appears likely that distinct enzymes are involved in the degradation of insulin and glucagon within endosomes.

Insulin and Glucagon-Degrading Activities in Plasma Membranes

A large number of studies have proposed the plasma membrane and/or the cytosol as a physiological locus for insulin and glucagon degradation. Thus, isolated liver plasma membrane fractions have been found to degrade insulin [Duckworth, 1979; Yokono et al., 1979], but the lack of characterization of the enzyme(s) involved has prevented evaluation of the physiological relevance of these findings.

Other studies have suggested the plasma membrane as a major site of glucagon degradation (Fig. 9) [Pohl et al., 1972]. Incubation of rat or canine hepatocytes (or derived plasma membranes) with mono^{125}I-iodoglucagon isomers resulted in the identification of three degradation products that arose from proteolytic cleavages at the peptide bonds Gln^3-Gly^4 (glucagon-[4-29]), Phe^6-Thr^7 (glucagon-[7-29]), and Tyr^{13}-Leu^{14} (glucagon-[1-13]) [Hagopian and Tager, 1984; Sheetz and Tager, 1988a]. Cleavage at these bonds does not reflect the specificity of any known proteases. Nevertheless, deletion of three and six amino acids from the N terminus of glucagon (glucagon-[4-29] and -[7-29]) could involve the action of a tripeptidyl aminopeptidase that has been isolated from a microsomal extract of rat liver [Bålöw et al., 1983]. This neutral extralysosomal protease could act sequentially to effect cleavage at peptide bonds Gln^3-Gly^4 and Phe^6-Thr^7. Glucagon-(4-29) release into the cell incubation medium was unaffected by various acidotropic agents (e.g., chloroquine, dansyl cadaverine, and procaine), indicating that hormone processing probably occurred on the hepatocyte surface [Hagopian and Tager, 1987]. Evidence suggests that glucagon-(4-29) is produced in vivo. Thus, a peptide having the size and immunological properties expected of glucagon-(4-29) was demonstrated in the plasma of both man and dog. Furthermore, the concentration of glucagon-(4-29) in the posthepatic circulation exceeded that in the portal circulation. In contrast, the absence of a gradient for glucagon and glucagon-(4-29) across the canine kidney has suggested that the liver is the major (if not the sole) site for the production of this hormone fragment [Hagopian and Tager, 1987].

A glucagon receptor-linked protease in plasma membranes appears to be responsible for cleavages at internal and carboxy regions of the glucagon molecule and for the generation of two major fragments, glucagon-(1-13) and -(14-29), the former remaining tightly membrane bound [Sheetz and Tager, 1988a,b]. This is compatible with the observation of Balage et al. [1986] that about 20% of ^{125}I-glucagon bound to hepatic plasma membranes was of decreased apparent molecular weight as assessed by gel filtration. Formation of this fragment was inhibited by PMSF but not by NEM, aprotinin, bacitracin, or a variety of glucagon-related peptides, indicating that the relevant enzymatic activity was apparently glucagon-specific and resulted from a serine protease. Formation of this fragment also depended on the prior association of the hormone with high-affinity glucagon binding sites, its production being sensitive to GTP, which is known, via interaction with GTP-binding proteins, to promote dissociation of glucagon from its receptor. The partially purified enzyme appeared to have a different molecular size than endoprotease 24.11 and IDE. It showed a broad neutral pH optimum (pH 7–9) and was sensitive to the presence of salt. Several minor degradation products generated by this partially purified protease have been identified, including glucagon-(1-10), -(14-25), and -(23-29).

Rat liver plasma membranes contain another enzymatic system that is a receptor-independent mechanism and that processes glucagon

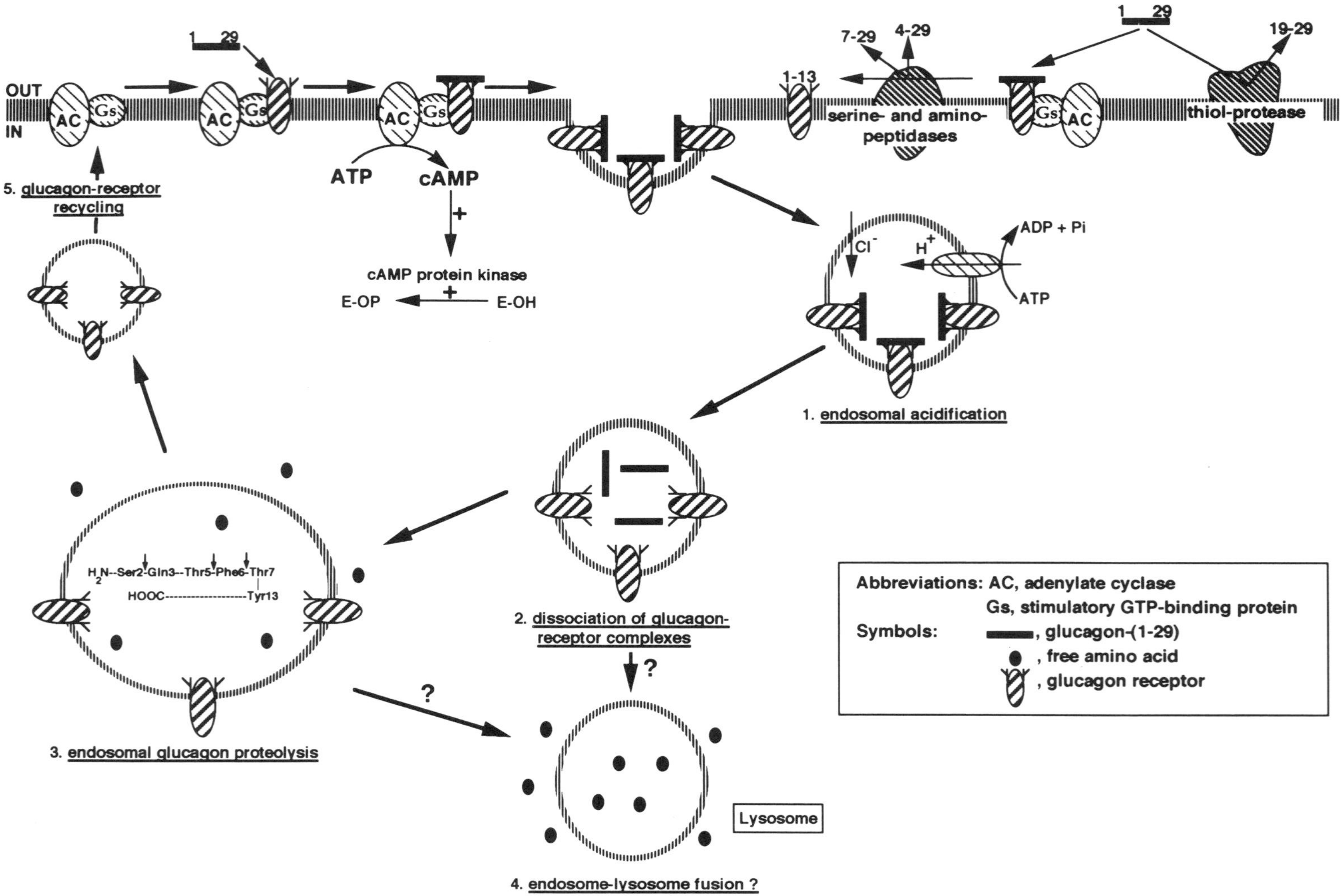

Abbreviations: AC, adenylate cyclase
Gs, stimulatory GTP-binding protein
Symbols: ▬, glucagon-(1-29)
●, free amino acid
, glucagon receptor

1. endosomal acidification

2. dissociation of glucagon-receptor complexes

3. endosomal glucagon proteolysis

H_2N-Ser2-Gln3-Thr5-Phe6-Thr7
HOOC----------Tyr13

4. endosome-lysosome fusion ?

Lysosome

5. glucagon-receptor recycling

OUT
IN

cAMP protein kinase
E-OP ← E-OH
ATP → cAMP +

thiol-protease
19-29
1-29
4-29
7-29
1-13
serine- and amino-peptidases

ADP + Pi
ATP
H^+
Cl^-

to glucagon-(19-29) [Blache et al., 1990]. It is of interest that this 19-29 fragment has been shown to inhibit the (Ca^{2+}-Mg^{2+})-ATPase pump of hepatic plasma membranes [Mallat et al., 1987]. Another aminopeptidase type enzyme, sensitive to amastatin and bestatin, degraded glucagon-(19-29) very rapidly following its production. Since glucagon-(19-29) is undetectable in plasma, and because it displays a very short half-life, its generation probably occurs largely within target tissues, especially the liver. The protease responsible is likely to be a thiol-metalloendopeptidase, since the production of glucagon-(19-29) was inhibited by the chelating agent 1,10-phenanthroline and by thiol-blocking agents such as NEM and pCMB. However, IDE, another thiol-metalloendopeptidase (see below), was described as an enzyme that degrades glucagon [Duckworth and Kitabchi, 1974] with an apparent K_m (ca. 6 µM) in the range of that reported for the protease responsible for glucagon-(19-29) production (12 µM). Although the major portion of IDE activity in hepatic tissue is cytosolic, plasma membrane preparations contain IDE activity (likely due to nonspecific adsorption) with characteristics similar to the cytosolic enzyme and seems to catalyse cleavage in the glucagon molecule at the Arg^{18}-Ala^{19} bond, making its involvement in the generation of the glucagon-(19-29) questionable [Baskin et al., 1975]. Using a fluorescence assay, Baskin et al. [1975] have proposed that the peptide bonds cleaved during the digestion of glucagon with highly purified IDE were Ser^{11}-Lys^{12}, Tyr^{13}-Leu^{14}, Arg^{18}-Ala^{19}, Trp^{25}-Leu^{26}, and either Thr^{5}-Phe^{6} or Asp^{21}-Phe^{22} and either Asp^{9}-Tyr^{10} or Lys^{12}-Tyr^{13}. Contrary to this report, Rose et al. [1988] has proposed that cleavage of glucagon by IDE occurs at only one region, namely, the double-basic doublet sequence Arg^{17}-Arg^{18}, with the release of glucagon-(1-17) and -(18-29).

Insulin and Glucagon-Degrading Activities in the Cytosol

IDE. The major enzyme has been designated as E.C. 3.4.22.11 and is known as *insulin-degrading enzyme* (IDE), *insulin protease* (IP), or insulinase. By conventional chromatography, a 50,000-fold purification has been achieved [Shii et al., 1986]. The properties of a red blood cell IDE (Table III) include a molecular weight of 300,000 by gel filtration; a single subunit of 110 kDa after SDS-PAGE under reducing conditions; a pH optimum of 7.0; an isoelectric point (pI) of 5.2; and inhibition of enzymatic activity by thiol-blocking agents such as NEM, pCMB, and *p*-hydroxymercuribenzoic acid (pHMB); other sulfhydryl reagents such as leupeptin and iodoacetate being inactive) and by metal chelating agents (e.g., EDTA and 1,10-phenanthroline) and bacitracin. Another IDE preparation from the same source has been reported by Kirschner and Goldberg [1983] with a 40,000-fold purification. The characteristics displayed by this purified enzyme suggest strongly that it is the same IDE as described by Shii et al. [1986]. However, there were two notable differences: a much higher pH optimum (8.6 instead of 7.0) and a much higher affinity

Fig. 9. *Proposed model for glucagon proteolysis in the hepatocyte. Glucagon metabolism begins at the cell surface. At the plasma membrane level, glucagon is proteolysed into glucagon-(19-29) by its direct interaction with a thiol-endopeptidase activity. After binding to its receptor, three other major degradation products are generated: first glucagon-(1-13) by a serine-protease activity, and lastly glucagon-(4-29) and -(7-29) by a potential tripeptidyl-aminopeptidase activity. The 4-29 and 7-29 fragments are released into the hepatic circulation, whereas the 1-13 fragment remains associated with high-affinity binding sites. Only glucagon-(19-29) is known to display a biological activity, namely, the modulation of the plasma membrane calcium pump. After internalization of the glucagon–receptor complex, glucagon is degraded in the endosomes. Endosomal acidity, maintained by an ATP-dependent proton pump, facilitates glucagon dissociation and subsequent degradation of free intraluminal hormone. The peptide bonds Ser^{2}-Gln^{3}, Thr^{5}-Phe^{6}, and Phe^{6}-Thr^{7} are cleaved, at least in part, by a thiol-endopeptidase activity. Afterwards the glucagon–receptor complex is recycled to the cell surface to associate again with the transducing system at the plasma membrane level.*

TABLE III. Similarities Between Mammalian and Submammalian IDEs

	Mammalian	*Drosophila*	*Neurospora crassa*	*Escherichia coli*
Subunit M_r (kDa)	110	110	120	110
Apparent sedimentation coefficient (S)	9.2	7.2	ND	5.3
Optimal pH	6.5–8.5	7–8	7.5	7–7.4
Isoelectric point	5.3	5.3	ND	ND
Inhibition by				
Sulfhydryl reagents	+	+	+	−
Metal-chelating agents	+	+	+	+
Bacitracin	+	+	+	ND
Metal activator	Zn^{2+} Mn^{2+}	Zn^{2+}	Mn^{2+}	$Mn^{2+}Zn^{2+}Co^{2+}$
Metal endogenously associated	Zn^{2+} Mn^{2+}	ND	ND	Zn^{2+}
K_m for insulin (µM)	0.029–0.13	0.1	ND	ND
Affinity labeling with				
^{125}I-insulin	+	+	+	+
^{125}I-TGF-α	+	+	ND	ND
^{125}I-EGF	−	+	ND	ND
Substrate specificity				
Insulin	+	+	+	+
IGF-I	+	−	ND	+
IGF-II	+	+	ND	+
TGF-α	+	+	ND	ND
EGF	−	−	ND	ND
Immunological cross reactivity				
Antihuman IDE	+	−	ND	ND
Anti-*Drosophila* IDE	−	+	−	ND
Antihuman EGF receptor	−	+	ND	ND

of the enzyme for separate A and B chains than for intact insulin. It has been proposed that the discrepancies between these two preparations were due to the insulin concentration used by Kirschner and Goldberg [1983], which was far above the K_m for insulin degradation by IDE [Duckworth, 1988]. Easily detectable by a cross-linking procedure (using ^{125}I-insulin and disuccinimidyl suberate as cross-linker), IDE has been identified in many tissue extracts, especially those of rat muscle, liver, kidney, or brain with relative concentrations being liver = brain > muscle > kidney [Shii et al., 1985]. This enzyme accounts for most of the insulinase activity in extracts of fat cells, fibroblasts, placenta, and pancreas [Duckworth, 1988]. The partially purified IDE from muscle, liver, brain, and fibroblast of various species display properties similar to those described for the erythrocyte IDE.

Complementary DNAs coding for human [Affholter et al., 1988, 1990b] and rat [Baumeister et al., 1993] IDE have been isolated. The deduced amino acid sequences exhibit 95% amino acid identity and do not contain consensus sequences for any of the known classes of proteases (i.e., metallo, cysteine, aspartic, or serine). However, they do show about 26% amino acid homology to an *Escherichia coli* protease (called *protease III* or *Pi*) [Finch et al., 1986], which can cleave insulin and is present in the periplasmic space, and about 48% homology to an enzyme from *Drosophila melanogaster* [Kuo et al., 1990]. Three regions of high homology are readily visible, sharing 57%–80% sequence identity. Blot analysis of total RNA from human lymphocytes and HepG2 cells indicated that the human IDE is translated from the 3.1 and 3.5 kb transcripts [Affholter et al., 1990b]. Rat IDE mRNA levels were recently examined by Northern blot analysis by Baumeister et al. [1993]. Two transcripts of 3.7 and 5.5 kb were identified in kidney, liver, adrenal, heart, brain, spleen, and lung.

In rat testis IDE transcripts of 3.7, 4.1, and 6.1 kb were observed. The relationship between the functional maturation of this tissue and the expression of the IDE gene was evaluated by Northern blotting at different stages of testis development. A significant increase in mRNA level occurred between days 14 and 28, indicating developmentally regulated expression of the IDE gene. The tissue and developmental distribution of rat IDE has also been studied by Kuo et al. [1993]. Their results indicate that IDE transcripts were present in all of the tissues analyzed, with a predominant expression in testis, tongue, and brain. The sizes of the major transcripts of rat IDE were 3.4 and 6.3 kb in all tissues analyzed, except testis, in which the major transcripts were shifted in size to 3.8 and 6.7 kb. Finally, during rat development from 6 to 7 days of age to adulthood, rat IDE mRNA levels increased in brain, testis, and tongue, decreased in muscle and skin, and did not significantly change in other tissues [Kuo et al., 1993].

The substrate specificity of this protease is not restricted to insulin but extends to some other biologically active peptides. In addition to insulin, IGF-I and -II are degraded by IDE with the following order of affinities: insulin > IGF-II > IGF-I [Affholter et al., 1990a; Misbin and Almira, 1989]. More recently, it has been shown that atrial natriuretic peptide (ANP) [Müller et al., 1991, 1992] and transforming growth factor-α (TGF-α) [Garcia et al., 1989a] are also high-affinity substrates (K_d = 1–60 nM). There are reports on the cleavage of glucagon [Duckworth and Kitabchi, 1974; Rose et al., 1988] and oxidatively damaged hemoglobin [Fagan et al., 1986; Fagan and Waxman, 1991], but the affinity of IDE for these substrates is relatively low. Proinsulin and EGF are poor substrates for this enzyme, although inhibitory effects of these ligands on insulin degradation have been described [Duckworth, 1988]. Mammalian IDEs can be labeled by cross-linking protocols with iodinated substrates with the following preferential affinity: insulin = ANP = bombyxin-II >> IGF-I and –II >> relaxin [Ding et al., 1992]. Not all the ligands that can be cross-linked to IDE are obligatory substrates, since substrate binding and substrate hydrolysis are distinguishable events in the course of IDE action. Cross-linking experiments with ^{125}I-insulin and ^{125}I-ANP have shown that sulfhydryl-modifying inhibitors (pCMB and pHMB) and bacitracin inhibit IDE activity at the level of substrate binding [Müller et al., 1992; Ogawa et al., 1992]. However, metal ion chelating inhibitors (1,10-phenanthroline and EDTA) do not affect substrate binding, but prevent substrate hydrolysis by removing divalent cations required for the hydrolysis step.

IDE cleavage sites in the insulin molecule reveal a peptide bond specificity that is unlike that of other known proteases. Two cleavages in the A chain and seven (four major and three minor) in the B chain of insulin have been identified [Duckworth et al., 1988]. The cleavage sites in the A chain are A13-A14 (Leu-Tyr) and A14-A15 (Tyr-Gln). The B-chain cleavages occur at B9-B10 (Ser-His), B13-B14 (Glu-Ala), B16-B17 (Tyr-Leu), and B25-B26 (Phe-Tyr) for the major sites and B10-B11 (His-Leu), B14-B15 (Ala-Leu), and B24-B25 (Phe-Phe) for the minor sites. Among the degradation products is one composed of a cleaved A chain and an intact B chain, which displays several insulin-like activities such as hypoglycemic effect, stimulation of glucose oxidation, and insulin receptor kinase activity at a low biological potency compared with that of insulin itself (1/40th to 1/60th) [Yonezawa et al., 1989]. When viewing the three-dimensional structure of insulin [Hodgkin, 1972], all the sites of cleavage appear to be in close proximity to leucine and tyrosine (hydrophobic residues) adjacent to α-helical turns. This same Leu-Tyr bond is cleaved in the glucagon molecule by IDE [Baskin et al., 1975], and the reported cleavage sites in the ANP molecule are not inconsistent with such a preference of IDE since three cleavage sites have a hydrophobic residue on their carboxyl side [Müller et al., 1992]. Therefore, the preferred cleavage sites for this protease appear to be N-terminal side to hydrophobic residues, but no enzymatic specificity for a particular amino acid sequence has been firmly established. This has raised the suggestion that this enzyme could recognize secondary structures (the molecule itself) rather than a particular amino acid sequence. It is interesting to note that the region implicated in the binding of insulin to its receptor is also responsible for its high-affinity interaction with IDE, suggesting that the structural determinants required for the binding of insulin to its receptor and to IDE are different [Affholter et al., 1990a].

As an approach to examine the physiological significance of these degradation products in vivo, Duckworth's group has compared the degradation products of insulin generated by intact hepatocyte and by purified IDE. The similarity of the HPLC patterns and the identical cleavage sites produced support the potential importance of the IDE in the physiological degradation of insulin [Duckworth et al., 1988].

The factors controlling IDE actions are poorly understood. However, at least two mechanisms of regulation have been proposed. Since IDE is a metal-requiring enzyme, divalent cations provide the first control of activity. Using neutron activation analysis, the metal content of partially purified IDE from rat skeletal muscle, rat liver, and human placenta has been determined [Ebrahim et al., 1991]. Results indicate that zinc and manganese are endogenously associated with the enzyme, with approximately 10 times more zinc than manganese. A second way to control the biological activity of IDE is through its endogenous inhibitor, which has been recently purified from the cytosol of rat liver [Ogawa et al., 1992]. Its characteristics include a molecular weight of 14 kDa on SDS-PAGE, partial association with IDE in the cytosolic fraction, heat stability, and inhibition of insulin degradation by IDE in a competitive binding manner.

Submammalian species have been suggested to contain insulin-like hormones [Meneses and Ortiz, 1975] and insulin-binding homologs of the insulin receptor [Petruzelli et al., 1986] and to express insulin-degrading activities. A 110 kDa IDE from *D. melanogaster* (dIDE) having properties very similar to those of its mammalian homolog has been purified and cloned [Garcia et al., 1987, 1988, 1989b; Kuo et al., 1990]. Overexpression of *Drosophila* IDE in COS monkey kidney cells led to five- to sevenfold increases in the rate of degradation of extracellular insulin, suggesting that IDE can regulate cellular insulin degradation [Kuo et al., 1991]. Analysis of the deduced amino acid sequence of dIDE indicated that this protease is evolutionarily conserved from bacteria to *Drosophila* to man, sharing 27% amino acid identity with bacterial protease III and 48% identity with the human IDE. Northern blot analysis indicated that the dIDE is translated from a 3.6 kb transcript similar in size to one of the two human IDE transcripts. Furthermore, a comparison of the molecular weight, inhibitor profiles, catalytic properties, and substrate specificities shared by the two enzymes (Table III) showed that structural and functional characteristics of IDE are highly conserved through evolution. Duckworth et al. [1989] have shown that dIDE cleaves the A chain at identical sites (A13-A14 and A14-A15) to that of rat IDE and cleaves the B chain at four sites (B10-B11, B14-B15, B16-B17, and B25-B26), corresponding to four of the seven cleavages observed with the rat IDE. These findings demonstrate a high degree of conservation for the specificity of cleavage bonds. Although ^{125}I-insulin, ^{125}I-TGF-α, and ^{125}I-EGF can be cross-linked to dIDE and despite the fact that insulin, TGF-α, and EGF inhibit insulin and TGF-α degradation by IDE or insulin and TGF-α binding to IDE, only insulin and TGF-α are equal substrates for this enzyme [Garcia et al., 1988, 1989a]. IGF-II is a good competitor for insulin degradation, whereas IGF-I, insulin A and B chains, and glucagon compete poorly. Despite the similarities between human IDE and dIDE in terms of structural and functional homology, polyclonal antibodies produced to dIDE and monoclonal antibodies produced to human IDE did not show cross-reactivity. However, anti-human EGF receptor antiserum recognized dIDE, but failed to recognize mammalian IDEs [Garcia et al., 1987].

In an attempt to define a possible role for IDE in cell differentiation, Rosner's group has studied the regulation of dIDE expression during development [Stoppelli et al., 1988]. Using a ^{125}I-TGF-α cross-linking assay, immunoblotting with antibodies to dIDE, and measurements of insulin degradation, they found that dIDE was expressed at 10% the level in the embryo compared with the adult organism. The level of dIDE expression increased gradually from embryo to larva through pupa to adult [Stoppelli et al., 1988]. These results are consistent with a role for the IDE in differentiation, and the pattern of its expression suggests a specific role for this enzyme in the later stages of development.

McKenzie et al. [1988] have suggested that the filamentous fungus *Neurospora crassa* possesses a functional insulin-induced signal transduction pathway. It has been recently established that an insulinase is present in *N. crassa* (Table III). This insulinase activity of *N. crassa* resembles IDE from other sources in two major ways: sensitivity to classic IDE inhibitors and similarities between HPLC elution profiles of degradation products from ^{125}I-insulin. However, it differs from previously reported IDEs in at least three respects: 1) a molecular weight (ca. 120 kDa) that is a little higher than that found for classic IDE (110 kDa); 2) a requirement for manganese but not zinc, whereas mammalian and insect IDEs are activated by both metal cations; and 3) an unusual low sensitivity to sulfhydryl inhibitors [Kole et al., 1992]. This recent report of insulinase activity in microbial cells extends the phylogenetic scope of this protease family, but sequence data are required to evaluate fully its membership in the IDE family.

Another IDE found in nonmammalian cells is protease III or protease Pi present in the periplasmic space of *E. coli* [Cheng and Zipser, 1979]. The purified enzyme displays similar properties to mammalian IDEs (i.e., molecular weight, ability to degrade insulin, and amino acid sequence identity) (Table III). It cleaves intact insulin and oxidized insulin B chain at Tyr^{16}-Leu^{17} and Phe^{25}-Tyr^{26} bonds, two of the major cleavage sites of the mammalian IDEs. However, it differs from mammalian IDEs in its insensitivity to sulfhydryl reagents (as pHMB and NEM) and its substrate specificity. Thus, the bacterial enzyme degrades equally IGF-I and IGF-II and can be cross-linked to ligands with the following affinity order: IGF-II > insulin = IGF-I >> relaxin >> bombyxin [Ding et al., 1992].

Most metallo-endopeptidases contain zinc as an essential metal ion for their catalytic activity [Jiang and Bond, 1992]. The coordination of the zinc atom in the active site has been established. Based on studies with thermolysin (the prototypic zinc metallo-endopeptidase), it has been proposed that the active site sequence is -His-Glu-Xaa-Xaa-His- (or HEXXH), a sequence that has been identified in many zinc-dependent metallo-endopeptidases and confirmed by mutagenesis studies to be the active site of endopeptidase 24.11 (an enzyme also called *enkephalinase*) [Devault et al., 1988] and angiotensin-converting-enzyme [Wei et al., 1991]. In this HXXEX sequence, the two histidines coordinate the binding of the essential zinc atom, and the glutamate acts as a general base in catalysis. Based on the lack of this consensus active site sequence [found only in carboxypeptidases and members of the IDE family], a new superfamily of metallo-endopeptidases including human IDE, dIDE, *E. coli* protease III, and the mitochondrial matrix processing peptidases from the rat, the yeast *Saccharomyces cerevisiae,* and the ascomycete *N. crassa* has been defined by Rawlings and Barrett [1991]. On the basis of an alignment of the sequences of these proteins, they have postulated that His^{88}, Glu^{169}, and His^{283} may represent the three zinc-binding ligands in these proteinases. However, using a sequence alignment of only the three confirmed members of the IDE family, Becker and Roth [1992] have proposed an alternative zinc-binding site sequence HXXEH (His^{88}, Glu^{91}, and His^{92}) corresponding to an inversion of the consensus sequence seen in classic metallo-endopeptidases. Mutation of the two histidines (His^{88} and His^{92}) in *E. coli* protease III results in a lack of proteolytic activity toward insulin substrates and a strong decrease in zinc signal of histidine mutants, consistent with this region being the active site in these proteins [Becker and Roth, 1992]. This active-site sequence is also found in the processing-enhancing protein subunit of the mitochondrial processing enzyme. Glutamate-169 has been proposed to be the third zinc-binding residue in proteinase III [Becker and Roth, 1993].

The absence of a functional insulin signal transduction pathway in *E. coli* raises the question of the physiological role for the bacterial IDE. In these bacteria, this enzyme has been proposed to be responsible for the rapid elimination of abnormal polypeptides—hence its broad specificity [Cheng and Zipser, 1979]. Moreover, the question of when and how IDE interacts with insulin in mammalian cells where it is endocytosed via a receptor-mediated pathway still remains unclear. However, it is possible that IDE plays other roles in the cell. Thus, Kayalar and Wong [1989] demonstrated that metalloprotease inhibitors that inhibit IDE activity block the differentiation of L_6 myoblasts, indicating that IDE may have a role in cell differentiation. Furthermore, the reports by Stoppelli et al. [1988], Baumeister et al. [1993], and Kuo et al. [1993] implicate *Drosophila* IDE and rat IDE in growth and development. Thus, various functions for IDE have been suggested, while its involvement in insulin degradation remains uncertain.

Protein disulfide isomerase. Protein disulfide isomerase (PDI) or glutathione-insulin transhydrogenase was proposed to be involved in the cellular metabolism of insulin [Duckworth, 1988]. This enzyme has been purified from bovine, rat, and human liver and placenta. A decrease in the transcription of the PDI gene in rat liver has been shown during streptozotocin-induced diabetes with normalization after insulin treatment [Nieto et al., 1990]. PDI has been shown to reduce interchain disulfide bonds in intact insulin. This led to the sequential theory of insulin degradation of Varandani et al. [1972] in which two enzymes act in concert to degrade insulin with disulfide bridge reduction preceding proteolysis of the individual chains. In fact, early reports on insulin degradation by rat liver cytosol relegated PDI to a nonexistent or minor role in this proteolytic process. The studies of Shii and Roth [1986] in HepG2 cells used monoclonal antibodies to IDE and antibodies to PDI (as a negative control) to demonstrate significant inhibition of cellular insulin metabolism by the former but none by the latter. Insulin degradation in the cytosol of rat liver, kidney, and skeletal muscle was reduced by up to 90% after immunoprecipitation of IDE [Akiyama et al., 1990]. Finally, many studies have isolated from cell extracts degradation products of insulin with intact disulfide bonds [Duckworth, 1988]. Recently, Wroblewski et al. [1992] reevaluated the possible role of PDI by measuring the IDE and PDI contribution to the degradation of human insulin by hepatic cytosol from human, monkey, and rat. The findings confirm previous reports on rodent liver that indicated that the contribution of PDI to insulin degradation by rat hepatic cytosol was undetectable or very low. IDE was concluded to be the sole enzyme detected in cytosol by Western blot analysis. In contrast, the same procedure indicates the presence of both PDI and IDE in the cytosolic fraction of human and monkey liver, suggesting that PDI may play a role in the in vitro insulin degradation in human and nonrodent liver. Furthermore, unlabeled human insulin is degraded by human liver cytosol into three major metabolic products, two of which contain intact A and B chains. The prevention of the in vitro metabolism of human insulin after immunoprecipitation of PDI in monkey and human liver cytosol indicates that IDE in these species appears incapable of initiating significant degradation of human insulin without prior disulfide bond reduction. These results are consistent with the involvement of both PDI and IDE in the in vitro degradation of insulin by hepatic cytosol from these species, but their involvement in the in vivo degradation of the hormone remains unclear.

Physiological Relevance of Insulin Degradation by IDE and PDI

Several studies support the role for cytosolic IDE in the catabolism of insulin, including 1) antibodies directed against IDE when introduced into the cytosol of a hepatoma cell line led to the inhibition of 18%–54% of insulin degradation [Shii et al., 1986]; 2) the degradation products of insulin from intact cells are nearly identical with those seen with purified IDE [Assoian and Tager, 1981; Duckworth et al., 1988] and 3) overexpression of IDE has been found to increase the rate of insulin degradation in intact cells [Kuo et al., 1991].

A potentially disturbing observation is that IDE was undetectable in hepatic endosomes as evaluated by immunoblotting, immunoprecipitation, or chemical cross-linking procedures, whereas its presence in cytosolic fractions was easily detected [Authier et al., 1994]. Furthermore, a small but detectable presence of IDE in particulate nuclear (N) and large granule (ML) fractions was observed by differential centrifugation. By analytical centrifugation, IDE cosedimented with the organelle containing the peroxisomal marker catalase and 3-ketoacyl-CoA thiolase. Finally, highly purified peroxisomes were observed to be enriched in IDE. Since all cloned cDNA of IDE (human, rat, and *Drosophila*) reveal a deduced classic peroxisomal targeting sequence -A/SKL at their C termini (Fig. 10), this may account for the peroxisomal location of IDE. Indeed, this tripeptide has been shown to be sufficient for protein import into peroxisomes in yeasts, plants, insects, and mammalian cells [Gould et al., 1989, 1990]. Furthermore, the chemical addition of the SKL motif onto nonperoxisomal proteins leads to direct these entities into peroxisomes as evaluated by immunofluorescence following microinjection into mammalian cells [Gould et al., 1989; Walton et al., 1992] or when such proteins are introduced into permeabilized cells [Wendland and Subramani, 1993]. Since endocytosed insulin in peroxisomes has not been observed [reviewed by Bergeron et al., 1985], the physiological substrate(s) in vivo of the peroxisomal-associated IDE are polypeptides other than insulin [Authier et al., 1994]. Then, it may be predicted that IDE is intraperoxisomal, providing an even more complicated scenario for insulin presentation than is even effected by its previously suggested cytosolic location.

The situation with PDI is as complicated. All eukaryotic proteins with PDI activity are found in the lumen of the endoplasmic reticulum (ER) [Freedman, 1989], which is the sole site of disulfide bond formation documented as of yet [Hwang et al., 1992]. Furthermore, all identified proteins with PDI activity that have been cloned and sequenced reveal an ER retention motif at the C termini, i.e., -KDEL [Munro and Pelham, 1987]. The source of cytosolic PDI involved in insulin degradation is unknown but may reflect a consequence of the homogenization protocols. Hence, and especially since PDI activity appears to be required for IDE activity in monkey and human liver cytosolic fractions, then the predicted segregation of these two enzyme in disparate organelles, i.e., the peroxisome and the lumen of the ER away from endocytic components, renders interpretation

Fig. 10. *Carboxy-terminal sequences of rat, human, and* Drosophila *insulin-degrading enzymes.*

of a physiological role for these enzymes in insulin or glucagon degradation less than straightforward.

CONCLUSION

The proton pump and the Na^+,K^+-ATPase in endosomal membranes maintain a moderately acidic lumen pH, which favors the dissociation of various ligands from their receptors. Lysosomes lack the Na^+,K^+-ATPase and are therefore more acidic. Thus, the fluctuating pH and the limited number of proteases within endosomes provide a selective and suitable processing environment in comparison to lysosomes.

The biological importance of endosomal proteolysis is suggested by the low number of proteases and by the cleavage of specific proteins within this intracellular compartment. One potential role of endosomal proteolysis of insulin and EGF may be to modulate tyrosine kinase activities by processing ligands for termination of insulin and EGF receptor signal transduction in target cells [Kay et al., 1986; Authier et al., 1994]. Another role may be to generate bioactive peptides derived from native ligands so that they can be externalized and serve functions. The proteolytic processing involved in lymphoid cell class II antigen presentation [Guagliardi et al., 1990] and limited proteolysis of PTH-(1-84) to a smaller bioactive form (PTH-[1-34]) [Diment et al., 1989] have been proposed to involve such a mechanism.

However, it is not clear how the proteases are retained in endosomes after delivery. Interesting questions concern the specificity of the endosomal degradative process, the extent of endosomal proteolysis in different cell types, and the point(s) in the endocytic pathway where proteases are delivered and activated. Purification of these endosomal proteases, elucidation of their primary structure by cDNA cloning, and development of antibody probes should enable elucidation of the physiological significance of proteolytic processes within endosomes.

ACKNOWLEDGMENTS

This work was supported by a grant from the Medical Research Council of Canada to B.I.P. and J.J.M.B. F.A. is the recipient of a postdoctoral fellowship from the Medical Research Council of Canada.

REFERENCES

Affholter JA, Cascieri MA, Bayne ML, Brange J, Casaretto M, Roth RA (1990a): Identification of residues in the insulin molecule important for binding to insulin-degrading enzyme. Biochemistry 29: 7727–7733.

Affholter JA, Fried VA, Roth RA (1988): Human insulin-degrading enzyme shares structural and functional homologies with *E. coli* protease III. Science 242: 1415–1418.

Affholter JA, Hsieh C-L, Francke U, Roth RA (1990b): Insulin-degrading enzyme: Stable expression of the human complementary DNA, characterization of its protein product, and chromosomal mapping of the human and mouse genes. Mol Endocrinol 4:1125–1135.

Ajioka RS, Kaplan J (1987): Characterization of endocytic compartments using the horse-radish peroxidase-diaminobenzidine density shift technique. J Cell Biol 104:77–85.

Akiyama H, Yokono K, Shii K, Ogawa W, Taniguchi H, Baba S, Kasuga M (1990): Natural regulatory mechanisms of insulin degradation by insulin degrading enzyme. Biochem Biophys Res Commun 170:1325–1330.

Assoian RK, Tager HS (1981): [^{125}I]Iodotyrosyl insulin: Semisynthesis receptor binding and cell mediated degradation of a B chain labeled insulin. J Biol Chem 256:4042–4049.

Authier F, Desbuquois B (1991): Degradation of glucagon in isolated liver endosomes. Biochem J 280: 211–218.

Authier F, Desbuquois B, De Galle B (1992): Ligand-mediated internalization of glucagon receptors in intact rat liver. Endocrinology 131:447–457.

Authier F, Janicot M, Lederer F, Desbuquois B (1990): Fate of injected glucagon taken up by rat liver in vivo. Biochem J 272:703–712.

Authier F, Rachubinski RA, Posner BI, Bergeron JJM (1994): Endosomal proteolysis of insulin by an acidic thiol-metalloprotease unrelated to insulin degrading enzyme. J Biol Chem 269:3010–3016.

Backer JM, Kahn CR, White MF (1990): The dissociation and degradation of internalized insulin occur in the endosomes of rat hepatoma cells. J Biol Chem 265:14828–14835.

Balage M, Grizard J, Grizard G (1986): Binding and degradation of ^{125}I-glucagon by highly purified rat liver plasma membranes. Biochim Biophys Acta 884: 101–108.

Bålöw R-M, Ragnarsson U, Zetterqvist Ö (1983): Tripeptidyl aminopeptidase in the extralysosomal fraction of rat liver. J Biol Chem 258:11622–11628.

Barazzone P, Gorden P, Carpentier J-L, Orci L (1980): Binding, internalization, and lysosomal association of ^{125}I-glucagon in isolated rat hepatocytes: A quantitative electron microscope autoradiographic study. J Clin Invest 66:1081–1093.

Barrett AJ (1977): Research Monographs in Cell and Tissue Physiology: Proteinases in Mammalian Cells and Tissues. North-Holland: Elsevier Biomedical Press.

Baskin FK, Duckworth WC, Kitabchi AE (1975): Sites of cleavage of glucagon by insulin-glucagon protease. Biochem Biophys Res Commun 67:163–169.

Baumeister H, Müller D, Rehbein M, Richter D (1993): The rat insulin-degrading enzyme: Molecular cloning and characterization of tissue-specific transcripts. FEBS Lett 317:250–254.

Becker AB, Roth RA (1992): An unusual active site identified in a family of zinc metalloendopeptidases. Proc Natl Acad Sci USA 89:3835–3839.

Becker AB, Roth RA (1993): Identification of glutamate-169 as the third zinc-binding residue in proteinase III, a member of the family of insulin-degrading enzymes. Biochem J 292:137–142.

Bergeron JJM, Cruz J, Khan MN, Posner BI (1985): Uptake of insulin and other ligands into receptor-rich endocytic components of target cells: The endosomal apparatus. Annu Rev Physiol 47:383–403.

Bergeron JJM, Kay DG, Lai WH, Doherty II JJ, Smith CE, Khan MN, Posner BI (1988): Functional characteristics of the endosomal apparatus of rat liver parenchyma. In Morré DJ, et al. (eds): Cell-Free Analysis of Membrane Traffic. New York: Alan R. Liss, pp 391–409.

Bergeron JJM, Levine G, Sikstrom R, O'Shaughnessy D, Kopriwa B, Nadler NJ, Posner BI (1977): Polypeptide hormone binding sites in vivo: Initial localization of ^{125}I-labeled insulin to hepatocyte plasmalemma as visualized by electron microscope radioautography. Proc Natl Acad Sci USA 74:5051–5055.

Bergeron JJM, Searle N, Khan MN, Posner BI (1986): Differential and analytical subfractionation of rat liver components internalizing insulin and prolactin. Biochemistry 25:1756–1764.

Bergeron JJM, Sikstrom R, Hand AR, Posner BI (1979): Binding and uptake of ^{125}I-insulin into rat liver hepatocytes and endothelium: An in vivo radioautographic study. J Cell Biol 80:427–443.

Blache P, Kervran A, Dufour M, Martinez J, Le-Nguyen D, Lotersztajn S, Pavoine C, Pecker F, Bataille D (1990): Glucagon-(19-29), a Ca^{2+} pump inhibitory peptide, is processed from glucagon in the rat liver plasma membrane by a thiol endopeptidase. J Biol Chem 265:21514–21519.

Bond JS, Butler PE (1987): Intracellular proteases. Annu Rev Biochem 56:333–364.

Bowser R, Murphy RF (1990): Kinetics of hydrolysis of endocytosed substrates by mammalian cultured cells: Early introduction of lysosomal enzymes into the endocytic pathway. J Cell Physiol 143:110–117.

Braun M, Waheed A, Von Figura K (1989): Lysosomal acid phosphatase is transported to lysosomes via the cell surface. EMBO J 8:3633–3640.

Cheng Y-SE, Zipser D (1979): Purification and characterization of protease III from *Escherichia coli*. J Biol Chem 254:4698–4706.

Clot J-P, Janicot M, Fouque F, Desbuquois B, Haumont P-Y, Lederer F (1990): Characterization of insulin degradation products generated in liver endosomes: In vivo and in vitro studies. Mol Cell Endocrinol 72:175–185.

Courtoy PJ, Quintart J, Baudhuin P (1984): Shift of equilibrium density induced by 3,3´-diaminobenzidine cytochemistry: A new procedure for the analysis and purification of peroxidase-containing organelles. J Cell Biol 98:870–876.

De Diego JG, Gorden P, Carpentier J-L (1991): The relationship of ligand receptor mobility to internalization of polypeptide hormones and growth factors. Endocrinology 128:2136–2140.

Devault A, Nault C, Zollinger M, Fournie-Zaluski M-C, Roques BP, Crine P, and Boileau G (1988): Expression of neutral endopeptidase (Enkephalinase) in heterologous COS-1 cells. J Biol Chem 263:4033–4040.

Diment S, Leech MS, Stahl PD (1988): Cathepsin D is membrane-associated in macrophage endosomes. J Biol Chem 263:6901–6907.

Diment S, Martin KJ, Stahl PD (1989): Cleavage of parathyroid hormone in macrophage endosomes illustrates a novel pathway for intracellular processing of proteins. J Biol Chem 264:13403–13406.

Diment S, Stahl PD (1985): Macrophage endosomes contain proteases which degrade endocytosed protein ligands. J Biol Chem 260:15311–15317.

Ding L, Becker AB, Suzuki A, Roth RA (1992): Comparison of the enzymatic and biochemical properties of human insulin-degrading enzyme and *Escherichia coli* protease III. J Biol Chem 267:2414–2420.

Doherty II J-J, Kay DG, Lai WH, Posner BI, Bergeron JJM (1990): Selective degradation of insulin within rat liver endosomes. J Cell Biol 110:35–42.

Duckworth WC (1979): Insulin degradation by liver cell membranes. Endocrinology 104:1758–1764.

Duckworth WC (1988): Insulin degradation: Mechanisms, products, and significance. Endocrine Rev 9:319–345.

Duckworth WC, Garcia JV, Liepnieks JJ, Hamel FG, Hermodson MA, Frank BH, Rosner MR (1989): *Drosophila* insulin degrading enzyme and rat skeletal muscle insulin protease cleave insulin at similar sites. Biochemistry 28:2471–2477.

Duckworth WC, Hamel FG, Peavy DE, Liepnieks JJ,

Ryan MP, Hermodson MA, Frank BH (1988): Degradation products of insulin generated by hepatocytes and by insulin protease. J Biol Chem 263:1826–1833.

Duckworth WC, Kitabchi AE (1974): Insulin and glucagon degradation by the same enzyme. Diabetes 23:536–543.

Ebrahim A, Hamel FG, Bennett RG, Duckworth WC (1991): Identification of the metal associated with the insulin degrading enzyme. Biochem Biophys Res Commun 181:1398–1406.

Fagan JM, Waxman L (1991): Purification of a protease in red blood cells that degrades oxidatively damaged haemoglobin. Biochem J 277:779–786.

Fagan JM, Waxman L, Goldberg AL (1986): Red blood cells contain a pathway for the degradation of oxidant-damaged hemoglobin that does require ATP or ubiquitin. J Biol Chem 261:5705–5713.

Finch PW, Wilson RE, Brown K, Hickson ID, Emmerson PT (1986): Complete nucleotide sequence of the *Escherichia coli ptr* gene encoding protease III. Nucleic Acids Res 14:7695–7703.

Freedman RB (1989): Protein disulfide isomerase: Multiple roles in the modification of nascent secretory proteins. Cell 57:1069–1072.

Fuchs R, Mâle P, Mellman I (1989): Acidification and ion permeabilities of highly purified rat liver endosomes. J Biol Chem 264:2212–2220.

Garcia JV, Fenton BW, Rosner MR (1988): Isolation and characterization of an insulin-degrading enzyme from *Drosophila melanogaster*. Biochemistry 27:4237–4244.

Garcia JV, Gehm BD, Rosner MR (1989a): An evolutionarily conserved enzyme degrades transforming growth factor-alpha as well as insulin. J Cell Biol 109:1301–1307.

Garcia JV, Stoppelli MP, Decker SJ, Rosner MR (1989b): An insulin epidermal growth factor-binding protein from *Drosophila* has insulin-degrading activity. J Cell Biol 108:177–182.

Garcia JV, Stoppelli MP, Thompson KL, Decker SJ, Rosner MR (1987): Characterization of a *Drosophila* protein that binds both epidermal growth factor and insulin-related growth factors. J Cell Biol 105:449–456.

Gould SJ, Keller GA, Hosken N, Wilkinson J, Subramani S (1989): A conserved tripeptide sorts proteins to peroxisomes. J Cell Biol 108:1657–1664.

Gould SJ, Keller GA, Schneider M, Howell SH, Garrard LJ, Goodman JM, Distel B, Tabak H, Subramani S (1990): Peroxisomal protein import is conserved between yeast, plants, insects and mammals. EMBO J 9:85–90.

Guagliardi LE, Koppelman B, Blum JS, Mark MS, Cresswell P, Brodsky FM (1990): Colocalization of molecules involved in antigen processing and presentation in an early endocytic compartment. Nature 343:133–139.

Hagopian WA, Tager HS (1984): Receptor binding and cell-mediated metabolism of ^{125}I-monoiodoglucagon by isolated canine hepatocytes. J Biol Chem 259:8986–8993.

Hagopian WA, Tager HS (1987): Hepatic glucagon metabolism: Correlation of hormone processing by isolated canine-hepatocytes with glucagon metabolism in man and in the dog. J Clin Invest 79:409–417.

Hamel FG, Mahoney MJ, Duckworth WC (1991): Degradation of intraendosomal insulin by insulin-degrading enzyme without acidification. Diabetes 40:436–443.

Hamel FG, Posner BI, Bergeron JJM, Frank BH, Duckworth WC (1988): Isolation of insulin degradation products from endosomes derived from intact rat liver. J Biol Chem 263:6703–6708.

Hasilik A (1992): The early and late processing of lysosomal enzymes: Proteolysis and compartmentation. Experientia 48:130–157.

Hodgkin DC (1972): The structure of insulin. Diabetes 21:1131–1150.

Horwitz EM, Gurd RS (1988): Quantitative analysis of internalization of glucagon by isolated hepatocytes. Arch Biochem Biophys 267:758–769.

Hwang C, Sinskey AJ, Lodish HF (1992): Oxidized redox state of glutathione in the endoplasmic reticulum. Science 257:1496–1502.

Jaspan JB, Polonsky KS, Lewis M, Pensler J, Pugh W, Moossa AR, Rubenstein AH (1981): Hepatic metabolism of glucagon in the dog: Contribution of the liver to overall metabolic disposal of glucagon. Am J Physiol 240:233–244.

Jiang W, Bond JS (1992): Families of metalloendopeptidases and their relationship. FEBS Lett 312:110–114.

Kay DG, Khan MN, Posner BI, Bergeron JJM (1984): In vivo uptake of insulin into hepatic Golgi fractions: Application of the diaminobenzidine-shift protocol. Biochem Biophys Res Commun 123:1144–1148.

Kay DG, Lai WH, Uchihashi M, Khan MN, Posner BI, Bergeron JJM (1986): Epidermal growth factor receptor kinase translocation and activation in vivo. J Biol Chem 261:8473–8480.

Kayalar C, Wong WT (1989): Metalloendoprotease inhibitors which block the differentiation of L_6 myoblasts inhibit insulin degradation by the endogenous insulin-degrading enzyme. J Biol Chem 264:8928–8934.

Khan MN, Posner BI, Khan RJ, Bergeron JJM (1982): Internalization of insulin into rat liver Golgi elements: Evidence for vesicle heterogeneity and the path of intracellular processing. J Biol Chem 257:5969–5976.

Khan MN, Savoie S, Khan RJ, Bergeron JJM, Posner BI (1985): Insulin and insulin receptor uptake into rat liver: Evidence for site of chloroquine action on receptor recycling. Diabetes 34:1025–1030.

Kirschner RJ, Goldberg AL (1983): A high molecular weight metalloendoprotease from the cytosol of mammalian cells. J Biol Chem 258:967–976.

Kole HK, Smith DR, Lenard J (1992): Characterization and partial purification of an insulinase from *Neurospora crassa*. Arch Biochem Biophys 297:199–204.

Kopriwa BM (1975): A comparison of various procedures for fine grain development in electron microscopic radioautography. Histochemistry 44:201–224.

Kornfeld S, Mellman I (1989): The biogenesis of lysosomes. Annu Rev Cell Biol 5:483–525.

Kuo W-L, Gehm BD, Rosner MR (1990): Cloning and expression of the cDNA for a *Drosophila* insulin-degrading enzyme. Mol Endocrinol 4:1580–1591.

Kuo W-L, Gehm BD, Rosner MR (1991): Regulation of insulin degradation: Expression of an evolutionarily conserved insulin-degrading enzyme increases degradation via an intracellular pathway. Mol Endocrinol 5:1467–1476.

Kuo W-L, Montag AG, Rosner MR (1993): Insulin-degrading enzyme is differentially expressed and developmentally regulated in various rat tissues. Endocrinology 132:604–611.

Mallat A, Pavoine C, Dufour M, Lotersztajn S, Bataille D, Pecker F (1987): A glucagon fragment is responsible for the inhibition of the liver Ca^{2+} pump by glucagon. Nature 325:620–622.

Matrisian LM, Planck SR, Magun BE (1984): Intracellular processing of epidermal growth factor. I. Acidification of ^{125}I-epidermal growth factor in intracellular organelles. J Biol Chem 259:3047–3052.

McKenzie MA, Fawell SE, Cha M, Lenard J (1988): Effects of mammalian insulin on metabolism, growth, and morphology of a wall-less strain of *Neurospora crassa*. Endocrinology 122:511–517.

Meneses P, Ortiz MD (1975): A protein extract from *Drosophila* with insulin-like activity. Comp Biochem Physiol 51A:483–485.

Misbin RI, Almira EC (1989): Degradation of insulin and insulin-like growth factors by enzyme purified from human erythrocytes: Comparison of degradation products observed with A14- and B26-[^{125}I]monoiodoinsulin. Diabetes 38:152–158.

Müller D, Baumeister H, Buck F, Richter D (1991): Atrial natriuretic peptide (ANP) is a high-affinity substrate for rat insulin-degrading enzyme. Eur J Biochem 202:285–292.

Müller D, Schulze C, Baumeister H, Buck F, Richter D (1992): Rat insulin-degrading enzyme: Cleavage pattern of the natriuretic peptide hormones ANP, BNP, and CNP revealed by HPLC and mass spectrometry. Biochemistry 31:11138–11143.

Munro S, Pelham HRB (1987): A C-terminal signal prevents secretion of luminal ER proteins. Cell 48:899–907.

Nieto A, Mira E, Castano JG (1990): Transcriptional regulation of rat liver protein disulpide-isomerase gene by insulin and in diabetes. Biochem J 267:317–323.

Ogawa W, Shii K, Yonezawa K, Baba S, Yokono K (1992): Affinity purification of insulin-degrading enzyme and its endogenous inhibitor from rat liver. J Biol Chem 267:1310–1316.

Pease RJ, Smith GD, Peters TJ (1985): Degradation of endocytosed insulin in rat liver is mediated by low-density vesicles. Biochem J 228:137–146.

Pease RJ, Smith GD, Peters TJ (1987): Characterization of insulin degradation by rat-liver low density vesicles. Eur J Biochem 164:251–257.

Petruzzelli L, Herrera R, Garcia-Arenas R, Fernandez R, Birnbaum MJ, Rosen OM (1986): Isolation of a *Drosophila* genomic sequence homologous to the kinase domain of the human insulin receptor and detection of the phosphorylated *Drosophila* receptor with an anti-peptide antibody. Proc Natl Acad Sci USA 83:4710–4714.

Planck SR, Finch JS, Magun BE (1984): Intracellular processing of epidermal growth factor. II. Intracellular cleavage of the COOH-terminal region of ^{125}I-epidermal growth factor. J Biol Chem 259:3053–3057.

Pohl SL, Krans HMJ, Birnbaumer L, Rodbell M (1972): Inactivation of glucagon by plasma membranes of rat liver. J Biol Chem 247:2295–2301.

Posner BI, Patel BA, Bergeron JJM (1982): Effect of chloroquine on the internalization of ^{125}I-insulin into subcellular fractions of rat liver: Evidence for an effect of chloroquine on Golgi elements. J Biol Chem 257:5789–5799.

Posner BI, Patel B, Verma AK, Bergeron JJM (1980): Uptake of insulin by plasmalemma and Golgi subcellular fractions of rat liver. J Biol Chem 255:735–741.

Rawlings ND, Barrett AJ (1991): Homologues of insulinase, a new superfamily of metalloendopeptidases. Biochem J 275:389–391.

Renfrew CA, Hubbard AL (1991): Sequential processing of epidermal growth factor in early and late endosomes of rat liver. J Biol Chem 266:4348–4356.

Roederer M, Bowser R, Murphy RF (1987): Kinetics and temperature dependence of exposure of endocytosed material to proteolytic enzymes and low pH; evidence for a maturation model for the formation of lysosomes. J Cell Physiol 131:200–209.

Rose K, Savoy L-A, Muir AV, Davies JG, Offord RE, Turcatti G (1988): Insulin proteinase liberates from glucagon a fragment known to have enhanced activity against Ca^{2+}-Mg^{2+}–dependent ATPase. Biochem J 256:847–851.

Runquist EA, Havel RJ (1991): Acid hydrolases in early and late endosome fractions from rat liver. J Biol Chem 266:22557–22563.

Sheetz MJ, Tager HS (1988a): Receptor-linked proteolysis of membrane-bound glucagon yields a membrane-associated hormone fragment. J Biol Chem 263:8509–8514.

Sheetz MJ, Tager HS (1988b): Characterization of a glucagon receptor-linked protease from canine hepatic plasma membranes. J Biol Chem 263:19210–19217.

Shii K, Baba S, Yokono K, Roth RA (1985): Covalent

linkage of ^{125}I-insulin to a cytosolic insulin-degrading enzyme. J Biol Chem 260:6503–6506.

Shii K, Roth RA (1986): Inhibition of insulin degradation by hepatoma cells after microinjection of monoclonal antibodies to a specific cytosolic protease. Proc Natl Acad Sci USA 83:4147–4151.

Shii K, Yokono K, Baba S, Roth RA (1986): Purification and characterization of insulin-degrading enzyme from human erythrocytes. Diabetes 35:675–683.

Smith GD, Christensen JR, Rideout JM, Peters TJ (1989): Hepatic processing of insulin: Characterization of differential inhibition by weak bases. Eur J Biochem 181:287–294.

Stoppelli MP, Garcia JV, Decker SJ, Rosner MR (1988): Developmental regulation of an insulin-degrading enzyme from *Drosophila melanogaster*. Proc Natl Acad Sci USA 85:3469–3473.

Varandani PT, Shroyer LA, Nafz MA (1972): Sequential degradation of insulin by rat liver homogenates. Proc Natl Acad Sci USA 69:1681–1684.

Walton P, Gould SJ, Feramisco JR, Subramani S (1992): Transport of microinjected proteins into peroxisomes of mammalian cells: inability of Zellweger cell lines to import proteins with the SKL tripeptide peroxisomal targeting signal. Mol Cell Biol 12:531–641.

Watanabe J, Kanamura S, Asada-Kubota M, Kanai K, Ika M (1984): Receptor-mediated endocytosis of glucagon in isolated mouse hepatocytes. Anat Rec 210:557–567.

Wei L, Alhenc-Gelas F, Corvol P, Clauser E (1991): The two homologous domains of human angiotensin I-converting enzyme are both catalytically active. J Biol Chem 266:9002–9008.

Wendland M, Subramani S (1993): Cytosol-dependent peroxisomal protein import in a permeabilized cell system. J Cell Biol 120:675–685.

Wiley HS, VanNostrand W, McKinley DN, Cunningham DD (1985): Intracellular processing of epidermal growth factor and its effect on ligand-receptor interactions. J Biol Chem 260:5290–5295.

Wroblewski VJ, Masnyk M, Khambatta SS, Becker GW (1992): Mechanisms involved in degradation of human insulin by cytosolic fractions of human, monkey and rat liver. Diabetes 41:539–547.

Yokono K, Imamura Y, Sakai H, Baba S (1979): Insulin-degrading activity of plasma membranes from rat skeletal muscle: Its isolation, characterization, and biologic significance. Diabetes 28:810–817.

Yonezawa K, Yokono K, Shii K, Hari J, Yaso S, Sakamoto T, Kawase Y, Adiyama H, Taketomi S, Baba S (1989): Biological properties of an initial degradation product of insulin by insulin-degrading enzyme. Endocrinology 124:496–504.

ABOUT THE AUTHORS

FRANÇOIS AUTHIER is a Postdoctoral Fellow in the Department of Anatomy and Cell Biology at McGill University, Montreal, Canada. He received a Master's degree in biochemistry at Orsay University in France and a doctorate in Pharmacy at Châtenay Malabry University in France. He pursued doctoral research in biochemistry as a fellowship of ministry of research at Necker Hospital for Sick Children in Paris in the division of Prof. Raphaël Rappaport, where he earned a Ph.D in 1992 for work on endocytosis of glucagon and its receptor in rat liver. Since 1992, he has worked with Dr. J.J.M. Bergeron on endosomal proteolysis of insulin in hepatocytes in the Department of Anatomy and Cell Biology at McGill University in Montreal, Canada, as a Postdoctoral Fellow of the Medical Research Council of Canada. Dr. Authier will shortly move to this new permanent position as a researcher in the "Institut National de la Santé et de la Recherche Médicale" at Necker Hospital for Sick Children in Paris.

BARRY I. POSNER is Professor of Medicine and Director of the Polypeptide Laboratory and the Endocrine Training Program at McGill University and the Royal Victoria Hospital in Montreal, Canada. After receiving his M.D. from the University of Manitoba in 1961 he pursued graduate research at the Massachusetts Institute of Technology (supervisor: B. Magasanik) and the New England Medical Center (supervisors: E. Astwood and M. Rabin) where he worked on the molecular mechanisms of cardiac hypertrophy. Subsequent postdoctoral work was in the Laboratory of Biochemistry at the National Heart and Lung Institute of the NIH under the supervision of Drs. E. Stadtman and M. Flavin, where he focused on the study of enzyme reaction mechanisms, especially those involving pyridoxal phosphate. He was recruited to McGill in 1970, where he began work on receptors for insulin, growth hormone, and prolactin. In collaboration with Dr. J.M. Bergeron at McGill he explored the in vivo localization of receptors and the internalization of hormone–receptor complexes. Other work has involved the study of peptide hormone receptors in the CNS, the production and function of insulin-like growth factors and their binding proteins, and the development of a novel class of phosphotyrosine phosphatase inhibitors (the peroxovanadium compounds) as insulin-mimetic agents for use in diabetes mellitus. Dr. Posner is a Fellow of the Royal College of Physicians and the Royal Society of Canada. He is an editor of *Endocrinology* and the *Journal of Laboratory and Clinical Medicine*.

JOHN J.M. BERGERON is Professor of Anatomy and Cell Biology and cross-appointed to the Departments of Biochemistry and Medicine at McGill University in Montreal, Canada, where he teaches undergraduate, graduate, and medical courses dealing with topics in biochemistry, cell biology, and histology. After receiving a B.Sc. Honours in Biochemistry at McGill University, he pursued doctoral research as a Rhodes Scholar at Oxford University in the laboratory of Charles Pasfernak and obtained a D.Phil. for studies on the biosynthesis of phospholipids of cell membranes during the cell cycle. This was followed by postdoctoral research at The Rockefeller University, New York, where he worked with Drs. George Palade and Phil Siekevitz on the Golgi apparatus. Since 1975 Dr. Bergeron has closely collaborated with Dr. Barry Posner at McGill on the physiological significance of hormone receptor internalization into the endosomal apparatus. Dr. Bergeron's research interests also include studies on the mechanism of protein transport in the endoplasmic reticulum, and Golgi apparatus, as well as the molecular mechanisms by which these organelles and the endosomal apparatus sort the molecular traffic that they regulate. Dr. Bergeron is an Associate Editor of *Biochemistry* and *Cell Biology*.

Proteolysis in the Yeast Vacuole

Elizabeth W. Jones and Deborah G. Murdock

INTRODUCTION

The yeast vacuole shares features with lysosomes. The vacuole of *Saccharomyces cerevisiae* contains an ensemble of hydrolases, including endo- and exoproteases, ribonuclease(s), α-mannosidase (α-MS), trehalase, and alkaline phosphatase, and it is an acidic compartment with a proton gradient generated and maintained by a V-type ATPase [Wiemken et al., 1979; Anraku, 1987; Londesborough and Varimo, 1984; Mittenbühler and Holzer, 1988]. In addition, however, the vacuole serves as a repository for polyphosphate, amino acids, ions, and other small molecules [Wiemken and Durr, 1974; Wiemken et al., 1979; Anraku 1987].

Vacuolar Proteases

There are six lumenal proteases in the vacuole: two endoproteinases, proteinases A and B (PrA, PrB); two carboxypeptidases, carboxypeptidases Y and S (CpY, CpS); and two aminopeptidases, aminopeptidases I and Co (ApI, ApCo) [Lenney et al., 1974; Wiemken et al., 1979; Achstetter et al., 1982]. Dipeptidylaminopeptidase B (DPAP-B) is an integral membrane protein of the vacuole [Rendueles et al., 1981; Roberts et al., 1989]. In Table I are summarized for these enzymes the catalytic mechanism, physical properties, encoding gene, cellular function(s), and whether or not they are delivered to the vacuole by way of the secretory pathway. Their biochemical properties have been thoroughly summarized [Rendueles and Wolf, 1988; Jones, 1991a]. The biggest discrepancy between recent information and previous reports concerns ApI. Although it was reported to be a large glycoprotein [Metz and Röhm, 1976], a recent study indicates that it is not a mannoprotein [Klionsky et al., 1992].

Proteases within the vacuole are known to catalyze specific, limited cleavages and to participate in more global proteolytic orgies. Which mode of proteolysis predominates is in part a function of the environment in which a cell finds itself and in part a function of its recent nutritional history.

Proteasomes

Large multisubunit protease(s) are present in the nucleus and cytosol of eukaryotic cells [for reviews, see Orlowski, 1990; Goldberg, 1992; Rechsteiner et al., 1993]. Most commonly there are two proteases, a 20S and a 26S proteasome; the two have subunits in common. Work in yeast has centered on the 20S proteasome [Achstetter et al., 1984; Tanaka et al., 1988; Kleinschmidt et al., 1988]. The complex catalyzes hydrolysis of bonds on the carboxyl side of neutral/hydrophobic residues, basic residues, and acidic amino acids, referred to as the *chymotrypsin-like*, *trypsin-like*, and *peptidyl-glutamyl peptide–hydrolyzing* activities, respectively. A high-molecular-mass proteasome of only two subunits is found in the archaebacterium *Thermoplasma acidophilum*, although it only possesses chymotrypsin-like activity. The two subunits, α and β, show sequence similarity to subsets of the 20S proteasome subunits [Zwickl et al., 1992].

To date 10 genes known to encode or thought to encode (based on sequence similarity) sub-

TABLE I. Proteases of the Yeast Vacuole

Protease	Type	Mol. mass (kDa)	Glycosylation	Secretory pathway	Gene	Function*
PrA	Aspartic	42	N-linked	Yes	PEP4 (PRA1, PHO9)	Hydrolase precursor processing, protein degradation
PrB	Serine (subtilisin family)	31	O-linked	Yes	PRB1	Hydrolase precursor processing; protein degradation
CpY	Serine	61	N-linked	Yes	PRC1	Peptide metabolism
CpS	Metallo (Zn^{2+})	74, 77**	N-linked	Yes	CPS1 (DUT1)	Peptide metabolism
ApI	Metallo (Zn^{2+})	50†	None	No	LAP4 (APE1)	Peptide metabolism
ApCo	Metallo (Co^{2+})	110	—	—	—	
DPAP-B	Serine	120		Yes	DPA2(DPP2)	

*Inferred from the study of mutants.
**Differ by one glycosyl chain.
†Monomer; the oligomerization state of the active enzyme is unknown. It was reported to be 12 by the group that purified it as a glycoprotein.

units of the yeast proteasome have been cloned and sequenced. In Table II are presented salient features of the data available for these subunits. Subunits have been assigned to the α or β class by sequence similarity; *PRS3* encodes a protein very similar to HC5 and RC5 [Lee et al., 1992], which fall outside the α,β designation [Zwickl et al., 1992]. All of the genes except *PRS13* prove to be essential upon disruption (see references in Table II). Viable mutants are available that carry mutations in *PRE1*, *PRE2*, or *PRE4*. (There are viable *pre3* mutants as well; these, like the *pre4*-1 mutant, are deficient in peptidylglutamyl peptide–hydrolyzing activity [Hilt ct al., 1993]). Studies with the mutants have implicated the corresponding β-subunits in degradation of ubiquitinated proteins.

Proteolysis occurs at all times, whether the cells are growing vegetatively, being subjected to minor or major changes in nutrition and/or stress, or undergoing changes in life cycle. This review will be organized according to these life states. It focuses on vacuolar proteolysis, but information on proteasomal participation is included when available.

PROTEOLYSIS IN CELLS DURING NORMAL VEGETATIVE GROWTH

Yeast cells growing exponentially in a glucose-based minimal medium degrade protein at a rate of 0.5%–1% per hour, 2% per hour if ethanol based [Halvorson, 1958b; Betz and Weiser, 1976; Lopez and Gancedo, 1979]. For vegetative cells about 35% of the turnover of short-lived proteins and 45% of the turnover of long-lived proteins are eliminated in cells that have reduced levels of PrA and PrB [Teichert et al., 1989]. A number of proteolytic processes and proteolyses have been identified that take place constitutively and depend on the vacuolar proteases. These include proteolytic processing of vacuolar hydrolase precursors, degradation of by-products of these processing events, degradation of abnormal and/or misfolded proteins of the secretory pathway, turnover of specific integral membrane proteins, and degradation of some examples of unassembled polypeptide components of oligomeric assemblies. In addition, vacuolar proteases catalyze hydrolysis of peptides generated internally and/or taken up from the medium.

Precursor Processing

All of the hydrolases of the vacuolar lumen that have been studied to date are synthesized as precursors that undergo proteolytic processing en route or upon delivery to the vacuole [see Jones, 1991b]. Known cleavages are summarized in Table III and Figure 1.

PrA. Three proteolytic cleavages occur during the post-translational maturation of Pep4p, the PrA precursor (Fig. 1). After removal of the signal sequence by signal peptidase [Klionsky et al., 1988], the product polypeptide undergoes autocatalytic removal of an N-terminal peptide [Rupp et al., 1991; van den Hazel et al., 1992; Woolford et al., 1993]. We presume the cleavage occurs in the vacuole. The final cleavage that generates the mature N terminus is catalyzed by PrB [Jones et al., 1989; van den Hazel et al., 1992; Hirsch et al., 1992b; Woolford et al., 1993]. The final intermediate, which is enzymatically active, has not been detected kinetically; it accumulates in a *prb1*Δ mutant.

PrB. Four proteolytic cleavages occur during the post-translational maturation of Prb1p, the PrB precursor. After removal of the signal sequence by signal peptidase (V. Nebes, personal communication), the product polypeptide undergoes autocatalytic scission of a large (260aa) N-terminal peptide in the endoplasmic reticulum [Moehle et al., 1989; Mechler et al., 1988]. If exit of the N-terminal peptide from the endoplasmic reticulum is prevented in a *sec18* mutant, this normally short-lived peptide is stable (A. Bachhawat, personal communication). It is likewise stabilized in a *pep4* mutant [Hirsch et al., 1992a]. We infer, therefore, that the N-terminal peptide is degraded in the vacuole. The third cleavage, which converts the 40-kDa product of autocatalysis to a 37-kDa intermediate by removal of a C-terminal peptide, is catalyzed by PrA [Moehle et al., 1989].

TABLE II. Yeast Proteasome Subunits and Functions*

Gene	Class of subunit	Homologs human/rat	Function essential	In vivo (genetically)	Method of identification — Purified subunit of 20S proteasome	Method of identification — Sequence similarity	In vivo function
PRS1	α	HC8/RC8	Yes		YC1		
PRS2 (SCL1)	α	IOTA/	Yes	(SCL1)	YC7-α, Y8		
PRS7**	α	HC3/RC3	Yes		Y7		
PRS13**	α	HC9/RC9	No		Y13		
PUP2	α	ZETA/	Yes			YC1, YC7-α Y7, Y13	
PRE1	β	/RC7-1	Yes	PRE1			Chymotrypsin-like activity; protein degradation—short-lived proteins via N-end rule pathway, ubiquitinated proteins, canavanine proteins
PRG1 (PRE2)	β	RING10/	Yes	PRG1 (PRE2)			Chymotrypsin-like activity; protein degradation—short-lived proteins via N-end rule pathway, ubiquitinated proteins, canavanine proteins
PRE4	β		Yes	PRE4			Peptidylglutamyl peptide–hydrolyzing activity; indirect evidence for involvement in degradation of ubiquitinated and canavanine proteins
PUP1	β	HC5/RC5	Yes				
PRS3	HC5/RC5	HC5/RC5	Yes			Several HC5/RC5	

*PRS1, PRS2 [Fujiwara et al., 1990]] SCL1 [Balzi et al., 1989]; PRS7 (Y7), PRS13 (Y13) [Emori et al., 1991]; PUP2 [Georgatsou et al., 1992]; PRE1 [Heinemeyer et al., 1991]; PRG1 [Friedman et al., 1992]; PRE2 [Heinemeyer et al., 1992]; PRE4 [Hilt et al., 1993; Haffter and Fox, 1991]; PRS3 [Lee et al., 1992].
**Named according to the convention established by the Tanaka laboratory [seeFujiwara et al., 1990; Lee et al., 1992].

TABLE III. Proteolytic Cleavages in Processing of Vacuolar Hydrolase Precursors

Hydrolase	Abbrev.	Gene	Cleavage of precursor by	Cleavage(s) required for activity
Proteinase A	PrA	*PEP4*	self (*proPrA*); PrB; SP	Self only, SP
Proteinase B	PrB	*PRB1*	self (*proPrB post1,2*); PrA; PrB or PrB *post1*; SP	*proPrB post1,2*; PrA or PrB; SP
Carboxypeptidase Y	CpY	*PRC1*	PrA; PrB; SP	PrA or PrB; SP
Carboxypeptidase S	CpS	*CPS1*	PrB	None
Aminopeptidase I	ApI	*LAP4*	PrB	PrB
Aminopeptidase Co	ApCo	—	—	—
Alkaline phosphatase	AlP	*PHO8*	PrA*, PrB	PrA* or PrB
Ribonuclease(s)	RNase(s)	—	PrA*	PrA*
Acidic Trehalase			PrA*	PrA*
α-Manosidase	αMS	*AMS1*	PrA*	None
Dipeptidylamino-peptidase B	DPAPB	*DAP2*	None	None

prb1 mutant has not been tested. If the *prb1* mutant does not process the precursor, then these entries will be PrB alone. See text for justification and details.

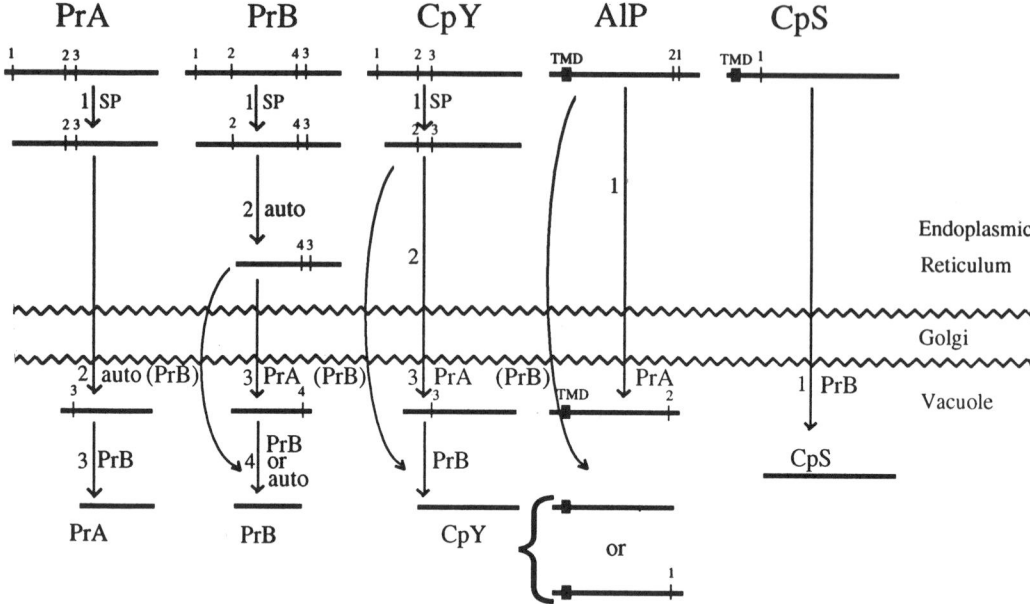

Fig. 1. *Maturation pathway for selected vacuolar hydrolases. For each enzyme the top line depicts the initial translation product; sites of proteolytic cleavage are numbered in the order in which they occur. Above the line the numbered reaction steps correspond to the numbered sites of cleavage; the enzyme activity identified adjacent to the reaction step catalyzes the cleavage in the compartment in which the enzyme is placed. SP, signal peptidase; auto, autocatalytic; TMD, transmembrane domain; PrA, proteinase A; PrB, proteinase B; CpY, carboxypeptidase Y; CpS, carboxypeptidase S; AlP, alkaline phosphatase. PrB is thought to catalyze three cleavages that bypass the requirement for PrA, giving rise to the phenomenon of phenotypic lag [Zubenko et al., 1982; Jones et al., 1989; Mechler et al., 1987]. The three suspected cleavages are signified by (PrB) adjacent to curved reaction arrows. The locations of the PrA- and PrB-catalyzed cleavages at the C-terminal end of AlP are uncertain because the prb1Δ mutant has not been tested for AlP activity. Not to scale.*

Since active PrA is only found in the vacuole, this cleavage must occur there. We infer that the fourth and final cleavage, which converts the 37-kDa species to the mature 31-kDa enzyme, results from autocatalysis [Moehle et al., 1989; Nebes and Jones, 1991], although we infer that PrB can catalyze this reaction in trans, even if the third cleavage has not occurred, and that PrB catalysis of this step is responsible for the phenomenon called *phenotypic lag*: delayed expression of the CpY⁻ phenotype resulting from the *pep4* mutation [Zubenko et al., 1982]. All four of these cleavages seem to be required for production of enzymatically active PrB.

CpY. Three proteolytic cleavages occur during maturation of Prc1p, the CpY precursor. After signal peptidase cleavage [Blachly-Dyson and Stevens, 1987}, PrA catalyzes removal of an N-terminal peptide in the vacuole [Hasilik and Tanner, 1978; Hemmings et al., 1981; Stevens et al., 1982]. A final cleavage catalyzed by PrB [Mechler et al., 1987] generates the mature N terminus. This PrB-catalyzed reaction can bypass the PrA-catalyzed cleavage and is presumed to account for phenotypic lag in expression of the CpY⁻ phenotype of *pep4* mutants [Zubenko et al., 1982].

CpS. Only one proteolytic cleavage occurs during maturation of Cps1p, the CpS precursor [Spormann et al., 1992]. Cps1p travels to the vacuole as a type II glycoprotein. Upon delivery of Cps1p to the vacuole, PrB catalyzes a reaction, presumed to be near the inside face of the vacuolar membrane, that releases CpS as a soluble hydrolase [Spormann et al., 1992]. This cleavage is not required for enzyme activity, however [Spormann et al., 1992]. Absence of cleavage in the *pep4* mutant reflects dependence of PrB production on PrA.

ApI. ApI does not enter the vacuole through the secretory pathway. Lap4p, the ApI precursor, is synthesized as a cytoplasmic protein that is then imported into the vacuole [Klionsky et al., 1992]. Within the vacuole, Lap4p is cleaved by PrB, a reaction that is required for enzyme activity [Klionsky et al., 1992]. Whether the cleavage severs a small N- or C-terminal peptide is not known. Initially processing of Lap4p was thought to be PrA dependent, since activity is not produced in the *pep4* mutant [Trumbly and Bradley, 1983]. However, lack of activation of the ApI zymogen in the absence of PrA is now known to arise through the dependence of production of active PrB on PrA [Klionsky et al., 1992].

AlP. Alkaline phosphatase, an integral membrane protein, undergoes at least one proteolytic cleavage [Klionsky and Emr, 1989], a cleavage that is required for activity [Jones et al., 1982]. Pho8p travels to the vacuole as a type II glycoprotein. It is processed at the C terminus in the vacuole [Klionsky and Emr, 1989]. The cleavage does not occur in the *pep4* mutant [Jones et al., 1982]. This presumed PrA-catalyzed cleavage can be by-passed by PrB-catalyzed cleavage, since production of AlP shows phenotypic lag in *pep4* mutant spores [Zubenko et al., 1982]. Whether the absence of activity in the *pep4* mutant reflects the absence of a PrA-catalyzed cleavage or dependence of PrB production on PrA is unknown, since the *prb1* mutant has yet to be tested for AlP activity.

Vacuolar (acidic) trehalase is delivered to the vacuole via the secretory pathway [Harris and Cotter, 1988; Mittenbühler and Holzer, 1991]. Neither it nor vacuolar RNAse(s) are produced in a *pep4* mutant [Jones et al., 1982; Harris and Cotter, 1987]. Nothing more has been reported on their syntheses.

Delivery of α-mannosidase to the vacuole is very slow and does not proceed via the secretory pathway but shares features with delivery of Lap4p [Yoshihisa and Anraku, 1990]. Once in the vacuole Ams1p, the α-MS precursor, undergoes a cleavage reaction that does not occur in the *pep4* mutant [Yoshihisa and Anraku, 1990]. Because the *pep4* mutant is not deficient for α-MS activity [Jones et al., 1982], we presume Ams1p to be enzymatically active.

Degradation of "By-Product" Peptides and Mutant or Misfolded Peptides in the Secretory Pathway

As mentioned above, the *pro* peptide released in the endoplasmic reticulum (ER) from the PrB precursor appears to be degraded in the vacuole. Whether the *pro* peptides released as a result of autocatalysis in the ER from Kex2p [Wilcox and Fuller, 1991] or from the α-factor precursor by Kex2 protease catalyzed cleavage in the Golgi [Julius et al., 1983, 1984; Redding et al., 1991] are also degraded in the vacuole has not been determined. A number of "by-product" peptides are released from hydrolase precursors in the vacuole itself, viz., the "*pro*" regions of Prc1p, Pep4p, Lap4p, Pho8p, and Ams1p, and the "*post*" peptides released from Prb1p. We presume these to be degraded in the vacuole through action of vacuolar proteases.

Abnormal forms of the PrA precursor are ordinarily highly unstable, but are stabilized in the *sec18* mutant at restrictive temperature [Klionsky et al., 1988], raising the possibility that they are degraded in the vacuole. A mutant CpY precursor that is normally unstable is stabilized by incubation of a *sec18* mutant at the restrictive temperature, suggesting that degradation may be vacuolar. In contradiction of this observation, introduction of the *pra1-1* and *prb1-1* mutations (*pra1-1* is a mutation in the *PEP4* gene) into the strain did not stabilize the mutant precursor [Teichert et al., 1989]. This latter observation may be very misleading, however, since everything we know about the dependence of PrB and CpY production on PrA activity indicates that the *pra1-1* mutation must not eliminate all PrA activity in vivo [see

Woolford et al., 1993], since the *pra*1-1 mutant produces active PrB and CpY [Mechler and Wolf, 1981]. Thus, in the *pra*1-1 *prb*1-1 strain, degradation of the mutant precursor may be due to residual PrA activity. Abnormal forms of the PrB precursor were not stabilized by an insertion mutation in the *PEP*4 gene and a deletion of the wild type *PRB*1 gene [Nebes and Jones, 1991; V. Nebes, personal communication]. The effect of the *sec*18 mutation was not tested.

Turnover of Membrane Proteins

The Kex2 protease, which cleaves to the C-terminal side of paired basic residues, is located in the Golgi [Redding et al., 1991]. In wild-type cells at 30° the half-life of Kex2 protease is 80 minutes. Increased growth temperatures and increased levels of expression of *KEX*2 each leads to a decrease in half-life of the protein (at 37° and 150×, the half-life is 7 minutes; at 30° and 150×, it is 22 minutes). Turnover is virtually eliminated by an insertion mutation in the *PEP*4 gene (at 150× and 30°, the half-life is >180 minutes) [Wilcox et al., 1992]. Thus Kex2 protease must travel to and be degraded in the vacuole.

The **a**-factor receptor, an integral plasma membrane protein encoded by the *STE*2 gene, is taken up by endocytosis constitutively in a ligand-independent but clathrin-dependent manner. The receptor is unstable in wild-type cells, showing a half-life around 20 minutes. In a *pep*4Δ background, the receptor is greatly stabilized, having a half-life much greater than 2 hours [Davis et al., 1993; Tan et al., 1993].

The α-factor receptor, an integral plasma membrane protein encoded by the *STE*2 gene, was also inferred to undergo endocytosis, albeit ligand-triggered, since treatment of **a** cells with α-factor led to loss of surface α-factor binding sites [Jenness and Spatrick, 1986] and the α-factor pheromone is internalized and degraded through a pathway that involves a vesicular compartment [Chvatchko et al., 1986; Singer and Riezman, 1990]. Recent results indicate that the α-factor receptor is internalized and delivered to the vacuolar membrane in the absence of ligand; the rate of internalization increases severalfold if α-factor is present. Once delivered to the vacuole the α-factor receptor is degraded in a *PEP*4-dependent manner, once again indicating vacuolar degradation [Jenness et al., 1993].

Ste6p, a polytopic plasma membrane protein required for export of the **a**-factor mating pheromone, has a half-life of about 30 minutes. Turnover of the protein requires internalization of the protein by endocytosis; degradation is *PEP*4 dependent. Thus all three mating factor–related, polytopic, plasma membrane proteins undergo endocytosis and vacuolar degradation [Berkower and Michaelis, 1993].

In contrast, two other polytopic plasma membrane proteins, Gap1p, the general amino acid permease, and Pma1p, the plasma membrane ATPase, are ordinarily stable [Berkower and Michaelis, 1993]. However, mutant forms of the ATPase are degraded prior to delivery to the plasma membrane in a *PEP*4-dependent manner [Chang and Fink, 1993].

Several sugar permeases in yeast undergo catabolite inactivation if glucose is added to the medium [Robertson and Halvorson, 1957; Gorts, 1969; Matern and Holzer, 1977]. It will be of interest to see whether turnover of these proteins requires delivery to the vacuole and the action of vacuolar proteases.

Taken together these results place plasma membrane proteins into two, and possibly three classes. The mating factor–related proteins are internalized and degraded in the vacuole in a constitutive mode, that, for at least one protein, can be stimulated by ligand. A second class of proteins seems not to be internalized or degraded. Whether there is a third class, normally stable, whose internalization and degradation is triggered by glucose, remains to be seen.

Overproduction of the HMG1 isozyme of HMG CoA reductase, an integral membrane protein of the endoplasmic reticulum, results in formation of *karmellae*, a proliferation of stacked membranes that nearly surround the yeast nucleus [Wright et al., 1988]. The membranes turn over, and swirls of karmellae are observed within vacuoles, implicating vacuolar proteases in turnover of the HMG1 isozyme

[Wright et al., 1988]. By contrast, overproduction of the HMG2 isozyme does not result in karmellae. HMG2 in inherently unstable, and degradation is not blocked in the *sec*18 mutant at the restrictive temperature. It is thus presumed to be degraded in the ER [Hampton and Rine, 1993].

Peptide Metabolism

Wild-type cells can transport di-, tri-, and oligopeptides into the cell and use them as sources of amino acids [Becker and Naider, 1980; Island et al., 1991]. For a dipeptide like Cbz-Gly-Leu, with a blocked α-amino group, cleavage by CpY and/or CpS is required if cells are to be able to use the peptide as a leucine and nitrogen source [Wolf and Ehmann, 1981]. *pep*4 and *prc*1 mutants (CpY$^-$ CpS$^+$) remain able to grow well on Cbz-Gly-Leu; *dut*1 mutants (CpY$^+$ CpS$^-$) grow less well [Jones, 1977; Wolf and Ehmann, 1981; E. Jones, unpublished observations]. Whether this reflects intrinsic activity of each enzyme toward the dipeptide or the fact that the level of CpS but not CpY is greatly increased when Cbz-Gly-Leu is the only nitrogen source [Wolf and Ehmann, 1978] is not known. As expected from in vitro experiments [Wolf and Weiser, 1977], cleavage by CpY is required for mobilization of the leucine in Cbz-Phe-Leu, since CpY-deficient mutants cannot use the peptide as a source of leucine (E. Jones, unpublished data).

That other peptides might be cleaved by these peptidases, including those generated by endoproteolytic activities, can be inferred from the observation that the *prb*1 mutation is epistatic to the *prc*1 mutation in starvation-induced proteolysis [Zubenko and Jones, 1981]. Presumably a similar relationship occurs during vegetative growth, although no studies have been reported.

It is certainly the case that the α-factor tridecapeptide, after being taken up by endocytosis, is hydrolyzed within the vacuole, for cleavage does not occur in a *pep*4 strain. This degradation would appear to be a normal component of the adaptation and recovery mechanism, for *pep*4 mutants are slowed in recovery from GI arrest [Singer and Riezman, 1990].

Adjustments for Purposes of Stoichiometry

The α and β-subunits of fatty acid synthetase, a cytosolic enzyme, are normally present in a 1:1 ratio within cells. If the ratio of the two subunits is altered, typically by deleting one of the two encoding genes, *FAS*1 or *FAS*2, the remaining subunit is highly unstable. The β-subunit, but not the α-subunit, can be stabilized in vivo, particularly in stationary phase, by eliminating PrB activity [Schüller et al., 1992]. The subunit is taken up into the vacuole [Egner et al., 1993]. The α-subunit is partially stabilized in the *pre*1-1 mutant, implicating the proteasome in turnover of the α-subunit [Egner et al., 1993].

Ribosomal proteins in yeast also typically are present in fixed ratios one to the other. Overexpression of a ribosomal protein leads to rapid degradation of the excess protein. This degradation is not altered by elimination of the vacuolar endoproteinases by the *pep*4 mutation [Tsay et al., 1988].

These results taken together indicate that there is not a unitary mechanism that the cell uses to deal with excess protein. There is even evidence for more than one pathway of degradation responsible for maintenance of appropriate stoichiometries of polypeptides within proteins.

Ubiquitinated Proteins

Some yeast proteins, like the α$_2$-repressor, are very short lived, having a 5-minute half-life [Hochstrasser and Varshavsky, 1990]. The α$_2$-repressor is ubiquitinated by four ubiquitin-conjugating enzymes [Chen et al., 1993]. Degradation of the ubiquitinated protein ensues.

A well-defined set of β-galactosidase fusion proteins are substrates of the N-end rule pathway of protein degradation, which requires ubiquitination of the substrate proteins [Bachmair et al., 1986; Chau et al., 1989]. Some of the fusion proteins have very short half-lives, others longer half-lives, and some are fairly stable.

Ubiquitinated proteins are known to be degraded by the proteasome in yeast [Heinemeyer et al., 1991, 1993; Hilt et al., 1993]. However, under certain conditions and/or in some strains, ubiquitinated proteins can be found in the vacuole [Simeon et al., 1992].

Richter-Ruoff et al., [1992] reported that the very short-lived β-galactosidase fusion proteins are stabilized in the *pre1* and *pre2* mutants; the more stable fusion proteins were unaffected by the mutations. These results clearly implicate the proteasome in degradation of these short-lived fusion proteins via the N-end rule pathway. Whether naturally occurring short-lived proteins like α_2-repressor will prove to be degraded by the proteasome, as seems likely, or will be transferred to the vacuole awaits testing.

PROTEOLYSIS IN CELLS DURING STATIONARY PHASE

Resting cells show higher rates of protein turnover than growing cells [Halvorson, 1958a,b], and rates of protein degradation increase two- to threefold as cells enter stationary phase [Bakalkin et al., 1976]. The total proteolytic activity in cells increases as they enter stationary phase; the levels of the vacuolar proteases rise substantially, with PrB showing quite dramatic increases that result from increased transcription [Betz and Weiser, 1976; Hansen et al., 1977; Moehle and Jones, 1990]. Transcription of *UBI4*, which encodes polyubiquitin, also increases markedly as cells approach stationary phase [Finley et al., 1987]. These observations imply that vacuolar proteolysis and ubiquitin-triggered proteolysis both increase as cells enter stationary phase. That the two processes may be more than temporally related is suggested by the finding of ubiquitinated proteins in the vacuoles of stationary phase cells of a mutant deficient for PrB and PrA but not in wild-type cells [Simeon et al., 1992]. This result implicates the vacuole in turnover of ubiquitinated proteins in stationary phase cells. Whether the proteasome also participates in this stationary phase turnover is unknown, and whether specific proteins or a random sample of the cell's protein species are degraded is completely unknown at this time.

PROTEOLYSIS IN CELLS UNDERGOING METABOLIC CHANGE
Changes in Carbon Source

The levels of enzymes of carbon metabolism depend on the carbon source used for growth. As cells exhaust the glucose in a medium, they derepress synthesis of the respiratory components and enzymes they will need for the new metabolic state. These include enzymes of the tricarboxylic cycle, the glyoxylate pathway, synthetic pathways for reserve carbohydrates, gluconeogenic enzymes, and mitochondrial components [Polakis and Bartley, 1965; Polakis et al., 1965; Düntze et al., 1969; Perlman and Mahler, 1974; Wales et al., 1980; Francois et al., 1991].

If, on the other hand, glucose is restored to cells growing on acetate, some enzymes, including enzymes of gluconeogenesis, reserve carbohydrate biosynthesis, and mitochondrial components decrease more rapidly than expected from dilution [Chapman and Bartley, 1968; Ferguson et al., 1967; Francois et al., 1991; Gancedo, 1971; Gancedo and Schwerzmann, 1976; Witt et al., 1966]. Fructose bisphosphatase (FBPase), PEP carboxykinase (PEPCK), the cytoplasmic form of malate dehydrogenase (cMDH), and isocitrate lyase are rapidly inactivated. For FBPase and cMDH the enzyme protein is known to be degraded [Hägele et al., 1978; Neeff et al., 1978; Funayama et al., 1980]. The phenomenon was named *catabolite inactivation* by Holzer [1976]. Glucose also triggers catabolite inactivation of fermentative capacity for galactose and maltose; the effect is on transport of the sugars [Robertson and Halvorson, 1957; Gorts, 1969; Matern and Holzer, 1977]. Whether the transport proteins are degraded is not yet known. ApI, in the vacuole, undergoes catabolite inactivation; loss of the antigen accompanies inactivation [Frey and Röhm, 1979]. It is logical that vacuolar proteinases may be responsible, but mutants have not been tested.

Recently Chiang and Schekman [1991] demonstrated that degradation of fructose 1,6-bisphosphatase takes place in the vacuole. Following a short lag after addition of high glucose, the FBPase is transferred from the cytosol to the vacuole. Transfer to the vacuole apparently requires synthesis of protein(s) that pass through the Golgi stack in the secretory pathway, since either cycloheximide or incubation of *sec* mutants blocked in vesicular transfer from the ER to the Golgi or in exit from the Golgi at the restrictive temperature can prevent import of FBPase. The destination of the required protein(s) is presumed to be the vacuole, since late but not early *sec* mutants make the required protein(s). Degradation of transferred FBPase is greatly slowed in a *pep*4 mutant, implicating vacuolar proteases in the degradation and resolving earlier conflicting reports [Teichert et al., 1989; Funaguma et al., 1985; Schäfer et al., 1987].

Three examples of direct import of cytosolic proteins into vacuoles or lysosomes, including the regulated import of FBPase described above, have been reported. A family of cytosolic proteins bearing the sequences KFERQ are imported into and degraded within lysosomes of mammalian cells upon serum starvation [Dice et al., 1986; Chiang and Dice, 1988; Dice and Chiang, 1989]. This stress response requires an hsc70 protein [Chiang et al., 1989; Terkecky et al., 1992].

As described above, precursors to both α-MS and ApI (Yoshihisa and Anraku, 1990; Klionsky et al., 1992] are imported directly from the cytoplasm into vacuoles. There is an indication that these two enzymes may compete for the same import system [Klionsky et al., 1992]; neither has been tested for competition with FBPase import.

Starvation

Carbon starvation. If cells are transferred from a glucose ammonia medium to a medium lacking carbon, the rate of protein degradation increases and may reach 3%–4% per hour [Lopez and Gancedo, 1979; Takeshige et al., 1992]. The degradation depends on PrB activity [Takeshige et al., 1992]. The only specific inactivation and degradation that is known to be triggered by carbon starvation is that of NADP-dependent glutamate dehydrogenase [Mazon, 1978; Mazon and Hemmings, 1979]. The rates of loss of activity and of NADP-GDH subunit protein upon carbon starvation are unaffected by mutations in the structural genes for PrA, PrB, or CpY [Wolf and Ehmann, 1979; Hemmings et al., 1980; Mechler and Wolf, 1981]. Hence the responsible protease(s) are presumably not located in the vacuole.

Nitrogen Starvation

When cells are transferred from a rich or minimal glucose-based medium to a glucose-based medium that lacks a nitrogen source, the rate of protein degradation increases markedly, by as much as fivefold [Lopez and Gancedo, 1979; Wolf and Ehmann, 1979]. Cells deficient in PrB activity show a rate of degradation about 60% of the wild-type rate [Wolf and Ehmann, 1979], whereas mutants lacking CpY and CpS show apparently normal rates [Wolf and Ehmann, 1981]. PrA-deficient mutants have not been examined in this experimental paradigm.

Nitrogen starvation of cells deficient for PrA and PrB results in accumulation of considerable amounts of large ubiquitinated proteins in the vacuoles, whereas wild-type cells accumulate lesser amounts of smaller ubiquitinated proteins [Simeon et al., 1990]. These results once again implicate the vacuole, in addition to the proteasome [Heinemeyer et al., 1991], in degradation of ubiquitinated proteins as a normal part of this cell's repertoire.

If cells growing on an acetate-based medium are subsequently starved for nitrogen in the presence of acetate, protein is degraded at a rate of 2%–3% per hour [Betz and Weiser, 1976; Zubenko and Jones, 1981; Teichert et al., 1989]. Mutations in the structural gene for CpY (*PRC*1) decrease degradation by 10%–20% [Zubenko and Jones, 1981]. Mutations in the PrB structural gene (*PRB*1) reduce degradation by 40%–50% [Zubenko and Jones, 1981; Teichert et al., 1989]. The *prb*1 mutation is epistatic to the *prc*1 mutation, implying that PrB supplies the substrates on which CpY acts, despite the presence of PrA in the cells [Zubenko and Jones, 1981]. Mutations in the PrA structural gene, *PEP*4 (*PRA*1), reduce protein degradation by 30% [Mechler and Wolf, 1981], but more typically by 60%–75% [Zubenko and Jones, 1981; Teichert et al., 1989]. Strains that carry mutations in genes for both PrA and PrB have degradation rates about 15% of wild-type rates. Thus, the bulk of the massive amount of protein degradation that takes place in this medium [Betz and Weiser, 1976; Zubenko and Jones, 1981; Hopper et al., 1974; Klar and Halvorson, 1975] is catalyzed by vacuolar proteases.

The medium typically employed for these nitrogen starvations (in the presence of acetate) is one or more of the standard sporulation mediums. Wild-type cells of *MATa/MATα* genotype will sporulate in these mediums, whereas wild-type diploid cells homozygous for *MATa* or *MATα* and haploid cells will not. Two groups reported that these asporogenous cells degrade protein at the same rates as sporogenous *MATa/MATα* cells [Betz and Weiser, 1976; Zubenko and Jones, 1981]; a third group reported that **a/a** and α/α diploids degrade proteins at lower rates than their **a**/α counterparts [Hopper et al., 1974].

As might be expected from the great dependence of sporulation-associated protein degradation on PrA and PrB and the massive scale of the protein degradation during sporulation, the mutations that reduce PrA and/or PrB activity typically have profound effects on the ability of diploids to sporulate. Pleiotropic *pep*4 mutations eliminate sporulation [Zubenko and Jones, 1981]; the milder *pra*1-1 reduces sporulation by 80% [Teichert et al., 1989]; the *pep*4-625 mutation, which is not at all pleiotropic [Woolford et al., 1993], has little effect on sporulation (E. Jones, unpublished data). Mutations in *PRB*1 typically reduce the frequency of sporulation somewhat [Zubenko and Jones, 1981; Teichert et al., 1989], depending on the allele. However, the cytoplasmic matrix from which the spores are carved remains virtually intact in the *prb*1 homozygote, making microscopic detection of spores problematic; that haploidization has occurred is easily demonstrated, however [Zubenko and Jones, 1981]. The *pra*1 *prb*1 double mutant diploids, like *pep*4-3 homozygotes, are unable to sporulate [Teichert et al., 1989; Zubenko and Jones, 1981]. No one has reported any effect of vacuolar protease deficiency on ascospore germination. We are not aware of any studies of this kind, however.

Analog-Containing Proteins

Incorporation of amino acid analogs like canavanine into proteins triggers an increased rate of protein degradation. Initial rates of degradation of such proteins are comparable for wild-type and protease-deficient (PrA$^-$ PrB$^-$) cells. However, within 1 hour the rate falls off in the protease-deficient cells, to a rate about 65% of wild-type cells [Teichert et al., 1989]. Presumably, the remaining 65% is largely attributable to the action of the proteasome, for Wolf and colleagues have demonstrated that degradation of analog-containing proteins is impaired in the mutants with defective proteasome subunits [Heinemeyer et al., 1991, 1993; Hilt et al., 1993].

Autophagy

In the process of autophagy, organelles and cytosol are taken up into the lysosome for degradation. In mammalian cells, autophagy is performed under conditions of nutritional deprivation to recycle required nutrients, and, after a shift in nutritional conditions, to hydrolyze cellular components no longer needed. Autophagy has been seen in yeast in response to several environmental conditions. In wild-type *S. cerevisiae*, autophagy can be seen in response to high temperature and to acridine orange, conditions that cause extensive breakdown of cellular proteins [Matile, 1979]. Several additional instances of autophagy have become apparent through the use of proteinase deficient cells.

General autophagy. In cells deficient for PrA and PrB, autophagic bodies appear in response to starvation for a carbon, nitrogen, or sulfate source or for required amino acids [Takeshige et al., 1992]. These autophagic bodies appear as small spherical bodies enclosed by a thin membrane that surrounds portions of the cytoplasm. They have been seen to include ribosomes, mitochondria, membranes of rough endoplasmic reticulum, lipid granules, membrane vesicles, and glycogen granules. Vacuoles isolated from such cells were shown to contain the cytosolic enzyme glucose-6-phosphate dehydrogenase. These autophagic bodies, if made in proteinase-proficient cells, would appear to be quickly degraded under these conditions, as autophagy is nearly undetectable when vacuolar proteinase functions are intact.

The accumulation of the autophagic bodies in proteinase-deficient yeast appears to be due mainly to the lack of the *PRB1* gene product PrB [Takeshige et al., 1992]. Cells disrupted only for *PRB1* accumulated autophagic bodies to the same extent as the multiply deficient strains. In addition, autophagic bodies were induced in nitrogen-starved wild-type cells by the addition of phenylmethylsulfonyl fluoride (PMSF), which irreversibly inactivates serine proteases, including PrB.

Accumulation of autophagic bodies requires de novo synthesis of protein, since addition of cycloheximide at the time of nutrient deprivation completely inhibits autophagic body formation. Autophagic body formation is also sensitive to N-ethylmaleimide (NEM), a reagent shown to inhibit membrane fusion events [Malhotra et al., 1988; Wilson et al., 1989]. Formation of autophagic bodies is not, however, sensitive to a calcium ionophore, or to chloramphenicol, or bafilomycin, an inhibitor of vacuolar type H^+-ATPases, leading to the conclusion that acidification of the vacuole is not necessary for the accumulation of autophagic bodies.

Ubiquitin–protein conjugates have been seen to accumulate in vacuoles of proteinase-deficient yeast starved for nitrogen [Simeon et al., 1992]. Historically, ubiquitin has been well characterized in its role in nonlysosomal degradation in mammalian cells and also in yeast cells. More recently, however, ubiquitin has been found in autophagic vacuoles and lysosomes of hepatoma cells [Schwartz et al., 1988]. Furthermore, the ubiquitin-activating enzyme E1 has been shown to be required for stress-induced degradation of cellular proteins, a process that occurs most probably in autophagic vacuoles [Gropper et al., 1991], and E1 is also associated with maturation of autophagic vacuoles [Lenk et al., 1992]. The presence of ubiquitin in autophagic bodies in vacuoles suggests a role for ubiquitin in autophagy in yeast, also. The details and essentiality of this role remain to be elucidated.

Selective autophagy. In response to growth in methanol, the methylotrophic yeasts *Pichia pastoris* and *Hansenula polymorpha* synthesize large peroxisomes that can occupy over 50% of the total cellular volume [Veenhuis et al., 1978, 1979]. When glucose replaces methanol as the carbon source, these peroxisomes are degraded, as are the enzymes required for methanol metabolism, including alcohol oxidase. This degradation appears to be a case of selective autophagy in yeast, as other organelles and cytoplasmic components are absent from vacuoles containing peroxisomes [Veenhuis et al., 1983; Tuttle et al., 1993]. It has been shown that alcohol oxidase crystalloids are not degraded on a shift to glucose medium unless they are inside peroxisomes, suggesting that a peroxisomal membrane protein is the recognition factor in this selective autophagy [van der Klei et al., 1991]. *S. cerevisiae* cells accumulate peroxisomes when grown with oleic acid as carbon source [Veenhuis et al., 1987]. Whether the peroxisomes undergo selective autophagy when glucose is restored to the cells has not been reported.

Other possible cases of autophagy. The rate of degradation of autophagic bodies in vacuoles appears to be quite fast. In wild-type (proteinase-competent) *S. cerevisiae* cells, autophagic bodies are rarely observed, even under stress conditions. In addition, when PMSF is washed out of cultures of wild-type cells, the autophagic bodies within the cells are completely degraded within 3–4 hours even though new PrB must be resynthesized to replace the irreversibly inactivated molecules [Takeshige et al., 1992]. These results raise the possibility that several other known degradative events may be mediated by autophagy, although autophagy has not yet been observed.

Several instances of degradation in yeast show indications of autophagy. The degradation of FBPase [Gancedo, 1971; Chiang and Schekman, 1991], the inactivation of the enzymes of reserved carbohydrate metabolism [Francois et al., 1991], and the degradation of the virus-like particles containing killer toxin ds RNA [Clare and Oliver, 1979] are all sensitive to cycloheximide, which is known to inhibit autophagic body formation as well as protein synthesis. The degradation of FBPase shows several other possible indications of autophagy. Some of the mutants isolated as defective in degradation of FBPase proved to be *pas* mutants, which are defective in peroxisomal assembly, and peroxisomes are also taken up into the vacuole upon shift to glucose in proteinase-deficient (*pep*4) cells. It has been suggested that FBPase may be first targeted to peroxisomes that are then taken up into the vacuole [Hoffman et al., 1993]. Degradation of peroxisomal thiolase has been shown to be *PEP*4 dependent in the experimental regimen employed in these studies (H.-L. Chiang, personal communication).

We speculate that yeast sporulation is likely to involve autophagy. Sporulation is associated with massive amounts of PrA- and PrB-dependent protein turnover [Zubenko and Jones, 1981; Teichert et al., 1989]. General autophagy could facilitate both of these requirements. Autophagy is likewise an attractive candidate as the mechanism for the degradation of excess mitochondrial capacity when glucose replaces acetate or ethanol as carbon source. In fact, it seems possible that autophagy is a very common mechanism for degradation in the yeast vacuole. Further work with proteinase-deficient strains may clarify exactly how many cases of degradation in the vacuole occur by autophagy.

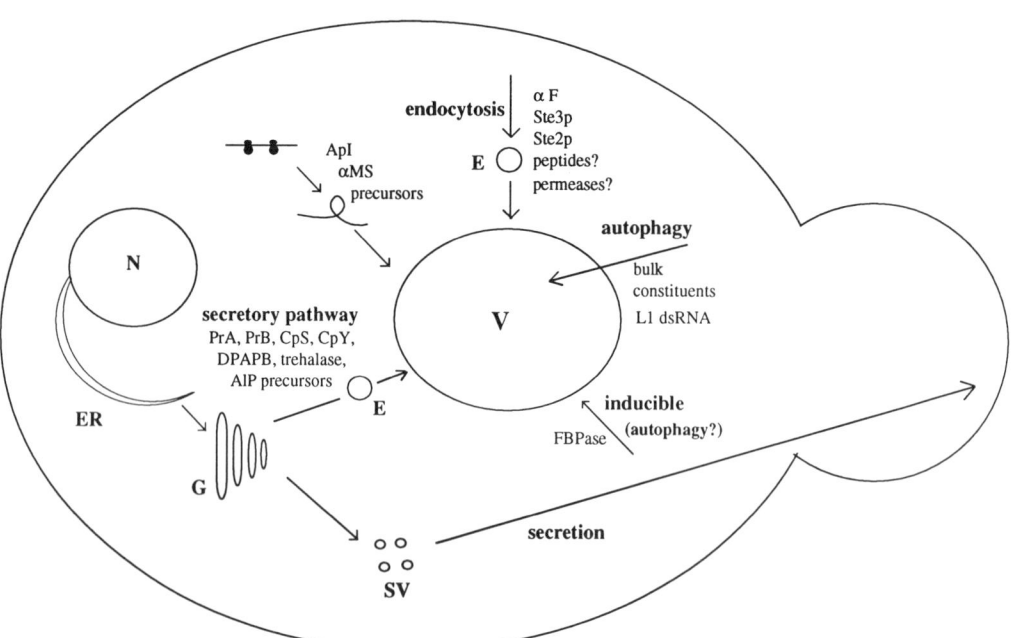

Fig. 2. *Delivery pathways for proteins destined for vacuolar proteolysis. αF, α-factor; Ste2p and Ste3p, the α- and **a**-factor receptors, respectively; N, nucleus; ER, endoplasmic reticulum; G, Golgi membranes; V, vacuole; E, endosome; SV, secretion vesicles. The relation between the two endosomes is not known. See text for details.*

DELIVERY ROUTES FOR VACUOLAR PROTEOLYSIS

In Figure 2 are summarized the several paths that proteins may take en route to the vacuole and vacuolar proteolysis. There are at least three constitutive routes: delivery of enzyme precursors and peptide by-products of processing reactions via the early stages of the secretory pathway and through an endosome-like compartment; endocytosis, be it receptor mediated or not, ligand triggered or not, but probably through an endosome-like compartment; and apparent direct uptake of hydrolase precursors from the cytosol. Inducible pathways thus far identified include the formation of autophagic bodies for digestion of bulk constituents, which is triggered by various starvations, and the uptake of FBPase for digestion in vacuoles, which is triggered by a switch from gluconeogenesis to glycolysis.

PROSPECTS

Figure 2 raises more questions than it answers. We do not know whether the *PEP*4-dependent degradation of peroxisomes and the uptake of peroxisomes into vacuoles are indicative of selective autophagy and whether there is a comparable mechanism for mitochondria. We do not know whether autophagy will prove to be much more widely used by this yeast in responding to metabolic changes like those embodied in stationary phase, sporulation, or shifts in carbon source and whether it will prove to be the underlying mechanism for uptake and degradation of FBPase, for example. We do not know what the signal(s) for induction of autophagy are or how they are transmitted. Are there receptors on vacuolar membranes for taking up selected hydrolase precursors? or FBPase? Do these two systems share functions? We have no idea where ubiquitinated proteins fit into this picture. How are they taken into vacuoles? Via autophagy? Receptor-mediated uptake? Is uptake of ubiquitinated proteins selective? In situations in which both proteasomes and vacuolar proteases are implicated in degradation, corresponding to situations in which the amount to be degraded is large—during nitrogen starvation or in the presence of amino acid analogs—what determines that both systems will be involved? Is one used preferentially and the second brought into play when the first is saturated? Is the proteasome a cytosolic, log phase system and the vacuole a stationary phase, starvation-triggered system? Is autophagy always implicated when both systems are participants? Will the vacuole prove to be the primary site for turnover of membrane proteins? Does catabolite inactivation of sugar transporters involve selective endocytosis and vacuolar degradation triggered by glucose? A host of lesser questions abound. The next decade should be fruitful.

ACKNOWLEDGMENTS

This work was supported by grant DK18090 from the National Institutes of Health.

REFERENCES

Achstetter T, Ehmann C, Osaki A, Wolf DH (1984): Proteolysis in eukaryotic cells: Proteinase yscE, a new yeast peptidase. J Biol Chem 259:13344–13348.

Achstetter T, Ehmann C, Wolf DH (1982): Aminopeptidase Co, a new yeast peptidase. Biochem Biophys Res Commun 109:341–347.

Anraku Y (1987): Unveiling the mechanism of ATP-dependent energization of yeast vacuolar membranes: Discovery of a third type of H^+-translocating ATPase. In Ozawa T, Papa S (eds): Bioenergetics: Structure and Function of Energy Transducing Systems. Tokyo: Springer-Verlag, pp 249–262.

Bachmair A, Finley D, Varshavsky A (1986): In vivo half-life of a protein is a function of its amino-terminal residue. Science 234:179–186.

Bakalkin G, Kalnov S, Zubatov A, Luzikov V (1976): Degradation of total cell protein at different stages of *Saccharomyces cerevisiae* yeast growth. FEBS Lett 63:218–221.

Balzi E, Chen W, Capieaux E, McCusker JH, Haber JE, Goffeau A (1989): The suppressor gene *scl1*[+] of *Saccharomyces cerevisiae* is essential for growth. Gene 83:271–279. Corrigendum (1990) Gene 89:151.

Becker J, Naider F (1980): Transport and utilization of peptides in yeast. In Payne J (ed): Microorganisms and Nitrogen Sources. New York: Wiley, pp 257–275.

Berkower C, Michaelis S (1993): Assembly and metabolic turnover of STE6, the a-factor transporter. In Davis T, Douglas M, Rose M (eds): Yeast Cell Biology. Cold Spring Harbor, NY: Cold Spring Harbor Laboratory, p 248.

Betz H, Weiser U (1976): Protein degradation and proteinases during yeast sporulation. Eur J Biochem 62:65–76.

Blachly-Dyson E, Stevens TH (1987): Yeast carboxypeptidase Y can be translocated and glycosylated without its amino-terminal signal sequence. J Cell Biol 104:1183–1191.

Chang A, Fink GR (1993): A novel protein is involved in stabilizing the yeast plasma membrane ATPase. In Davis T, Douglas M, Rose M (eds): Yeast Cell Biology. Cold Spring Harbor, NY: Cold Spring Harbor Laboratory, p 168.

Chapman C, Bartley W (1968): The kinetics of enzyme changes in yeast under conditions that cause the loss of mitochondria. Biochem J 107:455–465.

Chau V, Tobias JW, Bachmair A, Marriot D, Ecker DJ, Gonda DK, Varshavsky A (1989): A multiubiquitin chain is confined to a specific lysine in a targeted short-lived protein. Science 243:1576–1586.

Chen P, Johnson P, Sommer T, Jentsch S, Hochstrasser M (1993): Multiple ubiquitin-conjugating enzymes participate in the in vivo degradation of the yeast MAT α2 repressor. Cell 74:357–369.

Chiang H-L, Dice JF (1988): Peptide sequences that target proteins for enhanced degradation during serum withdrawal. J Biol Chem 263:6797–6805.

Chiang H-L, Schekman RW (1991): Regulated import and degradation of a cytosolic protein in the yeast vacuole. Nature 350:313–318.

Chiang H-L, Terlecky SR, Plant CR, Dice JF (1989): A role for a 70-kilodalton heat shock protein in lysosomal degradation of intracellular proteins. Science 246:382–385.

Chvatchko Y, Howald I, Riezman H (1986): Two yeast mutants defective in endocytosis are defective in pheromone response. Cell 46:355–364.

Clare JJ, Oliver SG (1979): The regulation of RNA synthesis in yeast IV. Synthesis of double-stranded RNA. Mol Gen Genet 171:161–166.

Davis NG, Horecka JL, Sprague Jr GF (1993): Cis and trans-acting functions required for endocytosis of the yeast pheromone receptors. J Cell Biol 122:53–65.

Dice JF, Chiang HL (1989): Peptide signals for protein degradation within lysosomes. Biochem Soc Symp 55:45–55.

Dice JF, Chiang H-L, Spencer EP, Backer JM (1986): Regulation of catabolism of microinjected ribonuclease A: Identification of residues 7-11 as the essential pentapeptide. J Biol Chem 261:6853–6859.

Düntze W, Neumann W, Gancedo JM, Atzpodien W, Holzer H (1969): Studies on the regulation and localization of the glyoxylate cycle enzymes in Saccharomyces cerevisiae. Eur J Biochem 10:83–89.

Egner R, Thumm M, Straub M, Simeon A, Schüller H-J, Wolf DH (1993): Tracing intracellular proteolytic pathways: Proteolysis of fatty acid synthase and other cytoplasmic proteins in the yeast Saccharomyces cerevisiae. J Biol Chem 268:27269–27276.

Emori Y, Tsukahara T, Kawasaki H, Ishiura S, Sugita H, Suzuki K (1991): Molecular cloning and functional analysis of three subunits of yeast proteasome. Mol Cell Biol 11:344–353.

Ferguson Jr J, Boll M, Holzer H (1967): Yeast malate dehydrogenase enzyme inactivation in catabolite repression. Eur J Biochem 1:21–25.

Finley D, Özaynak E, Varshavsky A (1987): The yeast polyubiquitin gene is essential for resistance to high temperature, starvation, and other stresses. Cell 48:1035–1046.

Francois J, Neves M-J, Hers H-G (1991): The control of trehalose biosynthesis in Saccharomyces cerevisiae: Evidence for a catabolite inactivation and repression of trehalose-6-phosphate synthase and trehalose-6-phosphate phosphatase. Yeast 7:575–587.

Frey J, Röhm K-H (1979): The glucose-induced inactivation of aminopeptidase I in Saccharomyces cerevisiae. FEBS Lett 100:261–264.

Friedman H, Goebl M, Snyder M (1992): A homolog of the proteasome-related RING10 gene is essential for yeast cell growth. Gene 122:203–206.

Fujiwara T, Tanaka K, Orino E, Yoshimura T, Kumatori A, Tamura T, Chung CH, Nakai T, Yamaguchi K, Shin S, Kakizuka A, Nakanishi S, Ichihara A (1990): Proteasomes are essential for yeast proliferation: cDNA cloning and gene disruption of two major subunits. J Biol Chem 265:16604–16613.

Funayama S, Gancedo J, Gancedo C (1980): Turnover of yeast fructosebisphosphatase in different metabolic conditions. Eur J Biochem 109:61–66.

Funaguma T, Toyoda Y, Sy J (1985): Catabolite inactivation of fructose 1,6-bisphosphatase and cytoplasmic malate dehydrogenase in yeast. Biochem Biophys Res Commun 130:467–471.

Gancedo C (1971): Inactivation of fructose 1,6-diphosphatase by glucose in yeast. J Bacteriol 107:401–405.

Gancedo C, Schwerzmann J (1976): Inactivation by glucose of phosphoenolpyruvate carboxykinase from Saccharomyces cerevisiae. Arch Microbiol 109: 221–225.

Georgatsou E, Georgakopoulos T, Thireos G (1992): Molecular cloning of an essential yeast gene encoding a proteasomal subunit. FEBS Lett 299:39–43.

Goldberg AL (1992): The mechanism and functions of ATP-dependent proteases in bacterial and animal cells. Eur J Biochem 203:9–23.

Gorts C (1969): Effect of glucose on the activity and the kinetics of the maltose uptake system and of α-glucosidase in Saccharomyces cerevisiae. Biochim Biophys Acta 184:299–305.

Gropper R, Brandt RA, Elias S, Bearer CF, Mayer A,

Schwartz AL, Ciechanover A (1991): The ubiquitin-activating enzyme, E1, is required for stress induced lysosomal degradation of cellular proteins. J Biol Chem 266:3602–3610.

Haffter P, Fox TD (1991): Nucleotide sequence of *PUP*1 encoding a putative proteasome subunit in *Saccharomyces cerevisiae*. Nucl Acid Res 19:5075.

Hägele E, Neefe J, Mecke D (1978): The malate dehydrogenase isoenzymes of *Saccharomyces cerevisiae*. Eur J Biochem 83:67–76.

Halvorson H (1958a): Intracellular protein and nucleic acid turnover in resting yeast cells. Biochim Biophys Acta 27:255–266.

Halvorson H (1958b): Studies on protein and nucleic acid turnover in growing cultures of yeast. Biochim Biophys Acta 27:267–276.

Hampton RY, Rine JD (1993): Regulated Turnover of HMG-CoA Reductase, an Integral ER Membrane Protein, in Yeast. Madison, WI: Yeast Genetics and Molecular Biology Meeting, p 20.

Hansen R, Switzer R, Hinze H, Holzer H (1977): Effects of glucose and nitrogen source on the levels of proteinases, peptidases, and proteinase inhibitors in yeast. Biochim Biophys Acta 196:103–114.

Harris SD, Cotter DA (1987): Vacuolar (lysosomal) trehalase of *Saccharomyces cerevisiae*. Curr Microbiol 15:247–249.

Harris SD, Cotter DA (1988): Transport of yeast vacuolar trehalase to the vacuole. Can J Microbiol 34:835–838.

Hasilik A, Tanner W (1978): Biosynthesis of the vacuolar yeast glycoprotein carboxypeptidase Y. Conversion of precursor into enzyme. Eur J Biochem 91:567–575.

Heinemeyer W, Gruhler A, Möhrle V, Mahé Y, Wolf DH (1993): *PRE*2, highly homologous to the human major histo-compatibility complex-linked *RING*10 gene, codes for a yeast proteasome subunit necessary for chymotryptic activity and degradation of ubiquinated proteins. J Biol Chem 268:5115–5120.

Heinemeyer W, Kleinschmidt JA, Sadowsky J, Escher C, Wolf DH (1991): Proteinase yscE, the yeast proteasome/multicatalytic-multifunctional proteinase: Mutants unravel its function in stress induced proteolysis and uncover its necessity for cell survival. EMBO J 10:555–562.

Hemmings BA, Zubenko GS, Hasilik A, Jones EW (1981): Mutant defective in processing of an enzyme located in the lysosome-like vacuole of *Saccharomyces cerevisiae*. Proc Natl Acad Sci USA 78:435–439.

Hemmings B, Zubenko GS, Jones EW (1980): Proteolytic inactivation of the NADP-dependent glutamate dehydrogenase in proteinase-deficient mutants of *Saccharomyces cerevisiae*. Arch Biochem Biophys 202:657–660.

Hilt W, Enenkel C, Gruhler A, Singer T, Wolf DH (1993): The *PRE*4 gene codes for a subunit of the yeast proteasome necessary for peptidylglutamyl-peptide-hydrolyzing activity. Mutations link the proteasome to stress and ubiquitin-dependent proteolysis. J Biol Chem 268:3479–3486.

Hirsch HH, Schiffer HH, Müller H, Wolf DH (1992a): Biogenesis of the yeast vacuole (lysosome): Mutation in the active site of the vacuolar serine proteinase yscB abolishes proteolytic maturation of its 73-kDa precursor to the 41.5 kDa pro-enzyme and a newly detected 41 kDa peptide. Eur J Biochem 203:641–653.

Hirsch HH, Schiffer HH, Wolf DH (1992b): Biogenesis of the yeast vacuole (lysosome): Proteinase yscB contributes molecularly and kinetically to vacuolar hydrolase maturation. Eur J Biochem 207:867–876.

Hochstrasser M, Varshavsky A (1990): In vivo degradation of a transcriptional regulator: The yeast α2 repressor. Cell 61:697–708.

Hoffman M, Hamamoto S, Schekman R, Chiang H-L (1993): Molecular studies of protein targeting and degradation in the yeast vacuole, abstracted. In Davis T, Douglas M, Rose M (eds): Yeast Cell Biology. Cold Spring Harbor, NY: Cold Spring Harbor Laboratory.

Holzer H (1976): Catabolite inactivation in yeast. TIBS 1:178–181.

Hopper A, Magee P, Welch S, Friedman M, Hall B (1974): Macromolecule synthesis and breakdown in relation to sporulation and meiosis in yeast. J Bacteriol 119:619–628.

Island MD, Perry JR, Naider F, Becker JM (1991): Isolation and characterization of *S. cerevisiae* mutants deficient in amino acid-inducible peptide transport. Curr Genet 20:457–463.

Jenness DD, Spatrick P (1986): Down regulation of the α-factor pheromone receptor in *Saccharomyces cerevisiae*. Cell 46:345–353.

Jenness D, Spatrick S, Kane T (1993): Genes controlling elimination of defective plasma membrane proteins. In Yeast Cell Biology. Cold Spring Harbor, NY: Cold Spring Harbor Laboratory. p 247.

Jones EW (1977): Proteinase mutants of *Saccharomyces cerevisiae*. Genetics 85:23–33.

Jones EW (1991a): Tackling the protease problem in *Saccharomyces cerevisiae*. Methods Enzymol 194:428–453.

Jones EW (1991b): Three proteolytic systems in the yeast *Saccharomyces cerevisiae*. J Biol Chem 266:7963–7966.

Jones EW, Woolford CA, Moehle CM, Noble JA, Innis MA (1989): Genes, zymogens and activation cascades of yeast vacuolar proteases. In Hugli TE (ed): Cellular Proteases and Control Mechanisms. New York: Alan R. Liss, Inc., pp 141–147.

Jones, EW, Zubenko GS, Parker RR (1982): *PEP*4 gene function is required for expression of several vacuolar hydrolases in *Saccharomyces cerevisiae*. Genetics 102:665–677.

Julius D, Blair L, Brake A, Sprague G, Thorner J (1983): Yeast α-factor is processed from a larger precursor polypeptide: The essential role of a membrane-bound dipeptidyl aminopeptidase. Cell 32:839–852.

Julius D, Schekman R, Thorner J (1984): Glycosylation and processing of *prepro-α-factor* through the yeast secretory pathway. Cell 36:309–318.

Klar A, Halvorson H (1975): Proteinase activities of *Saccharomyces cerevisiae*. J Bacteriol 124:863–869.

Kleinschmidt JA, Escher C, Wolf DH (1988): Proteinase yscE of yeast shows homology with the 20S cylinder particles of *Xenopus laevis*. FEBS Lett 239:35–42.

Klionsky DJ, Banta LM, Emr SD (1988): Intracellular sorting and processing of a yeast vacuolar hydrolase: Proteinase A propeptide contains vacuolar targeting information. Mol Cell Biol 8:2105–2116.

Klionsky DJ, Cueva R, Yaver DS (1992): Aminopeptidase I of *Saccharomyces cerevisiae* is localized to the vacuole independent of the secretory pathway. J Cell Biol 119:287–299.

Klionsky DJ, Emr SD (1989): Membrane protein sorting: Biosynthesis, transport and processing of yeast vacuolar alkaline phosphatase. EMBO J 8:2241–2250.

Lee DH, Tanaka K, Tamura T, Chung CH, Ichihara A (1992): *PRS3* encoding an essential subunit of yeast proteasomes homologous to mammalian proteasome subunit C5. Biochem Biophys Res Commun 182:452–460.

Lenk SE, Dunn Jr WA, Trausch JS, Ciechanover A, Schwartz AL (1992): Ubiquitin-activating enzyme, E1, is associated with maturation of autophagic vacuoles. J Cell Biol 118:301–308.

Lenney JF, Matile P, Wiemken A, Schellenberg M, Meyer J (1974): Activities and cellular localization of yeast proteases and their inhibitors. Biochem Biophys Res Commun 60:1378–1383.

Londesborough J, Varimo K (1984): Characterization of two trehalases in bakers' yeast. Biochem J 291:511–518.

Lopez S, Gancedo JM (1979): Effect of metabolic conditions on protein turnover in yeast. Biochem J 178:769–776.

Malhotra V, Orci L, Glick BS, Block MR, Rothman JE (1988): Role of an *N*-ethylmaleimide-sensitive transport component in promoting fusion of transport vesicles with cisternal of the Golgi stack. Cell 54:221–227.

Matern H, Holzer H (1977): Catabolite inactivation of the galactose uptake system in yeast. J Biol Chem 252:6399–6402.

Matile P (1975): Cell biology monographs. In The Lytic Compartment of Plant Cells, vol 1. New York: Springer-Verlag, p86.

Mazon M (1978): Effect of glucose starvation on the nicotinamide adenine dinucleotide phosphate-dependent glutamate dehydrogenase of yeast. J Bacteriol 133:780–785.

Mazon M, Hemmings B (1979): Regulation of *Saccharomyces cerevisiae* nicotinamide adenine dinucleotide phosphate-dependent glutamate dehydrogenase by proteolysis during carbon starvation. J Bacteriol 139:686–689.

Mechler B, Hirsch HH, Muller H, Wolf DH (1988): Biogenesis of the yeast lysosome (vacuole): Biosynthesis and maturation of proteinase yscB. EMBO J 7:1705–1710.

Mechler B, Müller H, Wolf DH (1987): Maturation of vacuolar (lysosomal) enzymes in yeast: Proteinase yscA and proteinase yscB are catalysts of the processing and activation event of carboxypeptidase yscY. EMBO J 6:2157–2163.

Mechler B, Wolf DH (1981): Analysis of proteinase A function in yeast. Eur J Biochem 121:47–52.

Metz G, Röhm K-H (1976): Yeast aminopeptidase I. Chemical composition and catalytic properties. Biochim Biophys Acta 429:933–949.

Mittenbühler K, Holzer H (1988): Purification and characterization of acid trehalase from the yeast *suc2* mutant. J Biol Chem 263:8537–8543.

Mittenbühler K, Holzer H (1991): Characterization of different forms of yeast acid trehalase in the secretory pathway. Arch Microbiol 155:217–220.

Moehle CM, Dixon C, Jones EW (1989): Processing pathway for protease B of *Saccharomyces cerevisiae*. J Cell Biol 108:309–324.

Moehle CM, Jones EW (1990): Consequences of growth media, gene copy number and regulatory mutations on the expression of the *PRB1* gene of *Saccharomyces cerevisiae*. Genetics 124:39–55.

Nebes VL, Jones EW (1991): Activation of the proteinase B precursor of the yeast *Saccharomyces cerevisiae* by autocatalysis and by an internal sequence. J Biol Chem 268:22851–22857.

Neeff J, Hägele E, Nauhaus J, Heer U, Mecke D (1978): Evidence for catabolite degradation in the glucose-dependent inactivation of yeast cytoplasmic malate dehydrogenase. Eur J Biochem 87:489–495.

Orlowski M (1990): The multicatalytic proteinase complex, a major extralysosomal proteolytic system. Biochemistry 29:10289–10297.

Perlman P, Mahler HR (1974): Derepression of mitochondria and their enzymes in yeast: Regulatory aspects. Arch Biochem Biophys 162:248–271.

Polakis ES, Bartley W (1965): Changes in the enzyme activities of *Saccharomyces cerevisiae* during aerobic growth on different carbon sources. Biochem J 97:284–297.

Polakis ES, Bartley W, Meek GA (1965): Changes in the activities of respiratory enzymes during the aerobic growth of yeast on different carbon sources. Biochem J 97:298–302.

Rechsteiner M, Hoffman L, Dubiel W (1993): The multicatalytic and 26S proteases. J Biol Chem 268:6065–6068.

Redding K, Holcomb C, Fuller RS (1991): Immunolocalization of Kex2 protease identifies a putative late

Golgi compartment in the yeast *Saccharomyces cerevisiae*. J Cell Biol 113:527–538.

Rendueles MP, Schwenke J, Garcia Alvarez N, Gascon S (1981): A new X-prolyl-dipeptidyl aminopeptidase from yeast associated with a particulate fraction. FEBS Lett 131:296–300.

Rendueles P, Wolf DH (1988): Proteinase function in yeast: Biochemical and genetic approaches to a central mechanism of post-translational control in the eukaryotic cell. FEMS Microbiol Rev 54:17–46.

Richter-Ruoff B, Heinemeyer W, Wolf H (1992): The proteasome/multicatalytic-multifunctional proteinase: In vivo function in the ubiquitin-dependent N-end rule pathway of protein degradation in eukaryotes. FEBS Lett. 302:192–196.

Roberts CJ, Pohlig G, Rothman JH, Stevens TH (1989): Structure, biosynthesis, and localization of dipeptidyl aminopeptidase B, an integral membrane glycoprotein of the yeast vacuole. J Cell Biol 108:1363–1373.

Robertson J, Halvorson H (1957): The components of maltozymase in yeast and their behavior during deadaptation. J Bacteriol 73:186–198.

Rupp S, Hirsch HH, Wolf DH (1991): Biogenesis of the yeast vacuole (lysosome). Active site mutation in the vacuolar aspartate proteinase A blocks maturation of vacuolar proteinases. FEBS Lett 283:62–66.

Schäfer W, Kalisz H, Holzer H (1987): Evidence for non-vacuolar proteolytic catabolite inactivation of yeast fructose 1,6-bisphosphatase. Biochim Biophys Acta 925:150–155.

Schüller H-J, Fortsch B, Rautenstrauss B, Wolf DH, Schweizer E (1992): Differential proteolytic sensitivity of yeast fatty acid synthetase subunits α and β contributing to a balanced ratio of both fatty acid synthetase components. Eur J Biochem 203:607–614.

Schwartz AL, Ciechanover A, Brandt RA, Geuze HJ (1988): Immunoelectron microscopic localization of ubiquitin in hepatoma cells. EMBO J 7:2961–2966.

Simeon A, van der Klei IJ, Veenhuis M, Wolf DH (1992): Ubiquitin, a central component of selective cytoplasmic proteolysis is linked to proteins residing at the locus of non-selective proteolysis, the vacuole. FEBS Lett 301:231–235.

Singer B, Riezman H (1990): Detection of an intermediate compartment involved in transport of α-factor from the plasma membrane to the vacuole in yeast. J Cell Biol 110:1911–1922.

Spormann DO, Heim J, Wolf DH (1992): Biogenesis of the yeast vacuole (lysosome): The precursor forms of the soluble hydrolase carboxypeptidase yscS are associated with the vacuolar membrane. J Biol Chem 267:8021–8029.

Stevens TH, Esmon B, Schekman RW (1982): Early stages in the yeast secretory pathway are required for transport of carboxypeptidase Y to the vacuole. Cell 30:439–448.

Takeshige K, Baba M, Tsuboi S, Noda T, Ohsumi Y (1992): Autophagy in yeast demonstrated with proteinase-deficient mutants and conditions for its induction. J Cell Biol 19:301–311.

Tan P, Davis N, Sprague G (1993): Clathrin facilitates the internalization of the seven transmembrane segment yeast mating pheromone receptors. In Yeast Cell Biology. Cold Spring Harbor, NY. Cold Spring Harbor Laboratory, p 249.

Tanaka K, Yoshimura T, Kutamori A, Ichihara A, Ikai A, Nishigai M, Kameyama K, Takogi T (1988): Proteasomes (multi-protease complexes) as 20S ring-shaped particles in a variety of eukaryotic cells. J Biol Chem 263:16209–16217.

Teichert U, Mechler B, Müller H, Wolf DH (1989): Lysosomal (vacuolar) proteinases of yeast are essential catalysts for protein degradation, differentiation, and cell survival. J Biol Chem 264:16037–16045.

Terlecky SR, Chiang H-L, Olson TS, Dice JF (1992): Protein and peptide binding and stimulation of in vitro lysosomal proteolysis by the 73 kDa heat shock cognate protein. J Biol Chem 267:9202–9209.

Trumbly R, Bradley G (1983): Isolation and characterization of aminopeptidase mutants of *Saccharomyces cerevisiae*. J Bacteriol 156:36–48.

Tsay Y-F, Thompson JR, Rotenberg MO, Larkin JC, Woolford Jr JL (1988): Ribosomal protein synthesis is not regulated at the translational level in *Saccharomyces cerevisiae*: Balanced accumulation of ribosomal proteins L16 and rp59 is mediated by turnover of excess protein. Genes Dev 2:664–676.

Tuttle DL, Lewin AS, Dunn Jr WA (1993): Selective autophagy of peroxisomes in methylotrophic yeasts. Eur. J Cell Biol 60:283–290.

van den Hazel HB, Kielland-Brandt MC, Winther JR (1992): Autoactivation of proteinase A initiates activation of yeast vacuolar zymogens. Eur J Biochem 207:277–283.

van der Klei IJ, Harder W, Veenhuis M (1991): Selective inactivation of alcohol oxidase in two peroxisome-deficient mutants of the yeast *Hansenula polymorpha*. Yeast 7:813–821.

Veenhuis M, Dijken JP van, Pilon SAF, Harder W (1978): Development of crystalline peroxisomes in methanol-grown cells of the yeast *Hansenula polymorpha* and its relation to environmental conditions. Arch Microbiol 117:153–163.

Veenhuis M, Douma A, Harder W, Osumi M (1983): Degradation and turnover of peroxisomes in the yeast *Hansenula polymorpha* induced by selective inactivation of peroxisomal enzymes. Arch Microbiol 134:193–203.

Veenhuis M, Keizer I, Harder W (1979): Characterization of peroxisomes in glucose-grown *Hansenula polymorpha* and their development after the transfer of cells into methanol-containing media. Arch Microbiol 120:165–175.

Veenhuis M, Mateblowski M, Kunau WH, Harder W

(1987): Proliferation of microbodies in *Saccharomyces cerevisiae*. Yeast 3:77–84.

Wales DS, Cartledge TG, Lloyd D (1980): Effects of glucose repression and anaerobiosis on the activities and subcellular distribution of tricarboxylic acid cycle and associated enzymes in *Saccharomyces carlsbergensis*. J Gen Microbiol 116:93–98.

Wiemken A, Dürr M (1974): Characterization of amino acid pools in the vacuolar compartment of *Saccharomyces cerevisiae*. Arch Microbiol 101:45–57.

Wiemken A, Schellenberg M, Urech K (1979): Vacuoles: The sole compartments of digestive enzymes in yeast (*Saccharomyces cerevisiae*). Arch Microbiol 123:23–35.

Wilcox CA, Fuller RS (1991): Posttranslational processing of the prohormone-cleaving Kex2 protease in the *Saccharomyces cerevisiae* secretory pathway. J Cell Biol 115:297–307.

Wilcox CA, Redding K, Wright R, Fuller RS (1992): Mutation of a tyrosine localization signal in the cytosolic tail of yeast Kex2 protease disrupts Golgi retention and results in default transport to the vacuole. Mol Biol Cell 3:1353–1371.

Wilson DW, Wilcox CA, Flynn GC, Chen E, Kuang W-J, Henzel WJ, Block MR, Ullrich A, Rothman JE (1989): A fusion protein required for vesicle-mediated transport in both mammalian cells and yeast. Nature 339:355–359.

Witt I, Kronau R, Holzer H (1966): Isoezym der malatdehydrogenase und ihre regulation in *Saccharomyces cerevisiae*. Biochim Biophys Acta 128:63–73.

Wolf DH, Ehmann C (1978): Carboxypeptidase S from yeast: Regulation of its activity during vegetative growth and sporulation. FEBS Lett 91:59–62.

Wolf DH, Ehmann C (1979): Studies on a proteinase B deficient mutant of yeast. Eur J Biochem 98:375–384.

Wolf DH, Ehmann C (1981): Carboxypeptidase S- and carboxypeptidase Y–deficient mutants of *Saccharomyces cerevisiae*. J Bacteriol 147:418–426.

Wolf DH, Weiser U (1977): Studies on a carboxypeptidase Y mutant of yeast and evidence for a second carboxypeptidase activity. Eur J Biochem 83:553–556.

Woolford CA, Noble JA, Garman JD, Tam MF, Innis MA, Jones EW (1993): Phenotypic analysis of proteinase A mutants: Implications for autoactivation and the maturation pathway of the vacuolar hydrolases of *Saccharomyces cerevisiae*. J Biol Chem 268:8990–8998.

Wright R, Basson M, D'Ari L, Rine J (1988): Increased amounts of HMG-CoA reductase induce "karmellae," a proliferation of stacked membrane pairs surrounding the yeast nucleus. J Cell Biol 107:101–114.

Yoshihisa T, Anraku Y (1990): A novel pathway of import of α-mannosidase, a marker enzyme of vacuolar membrane, in *Saccharomyces cerevisiae*. J Biol Chem 265:22418–22425.

Zubenko GS, Jones EW (1981): Protein degradation, meiosis and sporulation in proteinase-deficient mutants of *Saccharomyces cerevisiae*. Genetics 97:45–64.

Zubenko GS, Park FJ, Jones EW (1982): Genetic properties of mutations at the *PEP4* locus in *Saccharomyces cerevisiae*. Genetics 102:679–690.

Zwickl P, Grziwa A, Pühler G, Dahlmann B, Lottspeich F, Baumeister W (1992): Primary structure of the thermoplasma proteasome and its implications for the structure, function and evolution of the multicatalytic proteinase. Biochemistry 31:964–972.

ABOUT THE AUTHORS

ELIZABETH W. JONES is Professor of Biological Sciences at Carnegie Mellon University in Pittsburgh, PA, where she teaches undergraduate and graduate courses in genetics and molecular biology. She received a B.S. magna cum laude in chemistry in 1960 and a Ph.D. in genetics in 1964 from the University of Washington in Seattle. Her doctoral work was on genetic fine structural analysis in yeast and was carried out in the laboratory of Herschel Roman. Her postdoctoral work on the biochemistry of tetrahydrofolate metabolism in yeast was performed in the laboratory of Boris Magasanik at Massachusetts Institute of Technology in Boston. Since 1973 Dr. Jones's research has involved molecular genetic analysis of vacuolar biogenesis and vacuolar proteases in yeast. She received the Julius Ashkin Award for Teaching Excellence in 1984. She served as President of the Genetics Society of America and currently serves on the Council of the American Society for Cell Biology. She is co-editor of two monographs: "The Molecular Biology of the Yeast *Saccharomyces*" and "The Molecular and Cellular Biology of the Yeast *Saccharomyces*," is Associated Editor of *Genetics, Yeast*, and *Annual Review of Genetics*, and is on the editorial board of *Molecular Biology of the Cell*.

DEBORAH G. MURDOCK is a graduate student at Carnegie Mellon University. She received a B.S. in biology at the University of Georgia and is currently pursuing doctoral research on vacuolar targeting and intramolecular chaperone function of yeast proteinase B in the laboratory of Elizabeth W. Jones.

PROTEOLYSIS IN THE ENDOPLASMIC RETICULUM

Degradation of Proteins Retained in the Endoplasmic Reticulum

Juan S. Bonifacino and Richard D. Klausner

INTRODUCTION

The endoplasmic reticulum (ER) is the most extensive membrane-bound organelle of eukaryotic cells. Biochemical reactions take place within its numerous cisternae and tubules that are essential for maintaining cellular structure and function [reviewed by Vertel et al., 1992; Lippincott-Schwartz, 1994]. The inactivation of toxic substances and the maintenance of Ca^{2+} homeostasis are among the functions in which the ER and related compartments play a central role. In addition, the ER is the major site of synthesis of proteins, lipids, and carbohydrates destined for the different organelles of the secretory pathway or for export into the extracellular space. Nascent polypeptide chains containing hydrophobic signal sequences are targeted to the ER membrane and are subsequently translocated through the lipid bilayer, in a process that involves a complex molecular machinery [reviewed by Sanders and Schekman, 1992; Rapoport, 1992]. Once the nascent chains emerge in the ER lumen, they undergo a series of co- and post-translational modifications, including glycosylation, folding, formation of disulfide bonds, and oligomerization [reviewed by Rose and Doms, 1988; Hurtley and Helenius, 1989]. All of these functions are carried out by resident ER proteins, many of which are known to be localized to this organelle by virtue of specific targeting signals in their structure [Pelham, 1989; Nilsson et al., 1989]. In addition to ER resident proteins, many aberrant newly synthesized proteins become retained in the ER, including proteins that are intrinsically abnormal in structure or proteins that fail to undergo any of the co- and post-translational modifications described above [Rose and Doms, 1988; Hurtley and Helenius, 1989].

In the past few years, numerous observations have suggested the existence of an additional activity potentially associated with the ER: the degradation of a set of proteins localized to this compartment. Some of the proteins shown to undergo degradation are normal constituents of the ER (e.g., the enzyme 3-hydroxy-3-methylglutaryl-coenzyme A [HMG-CoA] reductase), whereas others are newly synthesized proteins normally destined for export from the biosynthetic compartment. Among the latter are secretory proteins that, for reasons that are not entirely clear, are inefficiently exported from the ER (e.g., apolipoprotein B [ApoB]) and various structurally abnormal proteins (e.g., some unassembled subunits of the T-cell antigen receptor [TCR]). Table I lists examples of proteins that have been reported to be degraded when retained in the ER or related compartments. While at present it is unclear whether all of these proteins are degraded in the same compartment and by the same pathway, their degradative processes share some common characteristics that are distinguishable from those of lysosomal degradation. Both biochemical and morphological approaches have been used to show that newly synthesized proteins

TABLE I. Examples of Proteins Reported To Be Degraded in the ER or Related Compartments

Proteins	Conditions	References
Acetylcholine receptor subunits	Unassembled	Blount and Merlie [1990]
Acetylcholinesterase	Normal*	Rotundo [1988], Rotundo et al. [1989]
α_1-Antitrypsin	Mutant	Le et al. [1990, 1992]
Apolipoprotein B	Normal*	Dixon et al. [1991], Furukawa et al. [1992], Borchardt and Davis [1987], Sato et al. [1990], Davis et al. [1990]
Asialoglycoprotein receptor H2a subunit	Unassembled	Amara et al. [1989], Lederkremer and Lodish [1991], Wikström and Lodish [1991, 1992, 1993]
CD4	Induced by Vpu gene product	Chen et al. [1993], Willey et al. [1992a,b], Vincent et al. [1993]
Class I MHC	Unassembled	Moore and Spiro [1993]
Class II MHC	Mutant, unassembled	Koppelman and Creswell [1990], Cotner [1992]
Cyclic nucleotide phosphodiesterase	Induced by Ca^{2+} depletion	Coukell et al. [1992]
Cystic fibrosis tranmembrane conductance regulator	Normal*, mutant	Cheng et al. [1990]
Cytochrome P_{450} (2E1)	Normal* and glucagon-induced	Eliasson et al. [1992]
Dipeptidyl peptidase-IV	Mutant	Erickson et al. [1992], Tsuji et al. [1992]
Fcγ III receptor α-subunit	Unassembled	Kurosaki et al. [1991], Lobell et al. [1993]
Fibrinogen chains	Unassembled	Danishefsky et al. [1990], Roy et al. [1992]
β-Hexosaminidase	Mutant	Lau and Neufeld [1989]
HMG-CoA reductase	Normal*	Edwards and Gould [1972], Chin et al. [1985], Faust et al. [1982]
Immunoglobulin chains	Normal*, unassembled, mutant	Sidman [1981], Dulis et al. [1982], Sitia et al. [1987], Bachhawat and Pillai [1991]
Influenza hemagglutinin	Mutant	Doyle et al. [1986]
β-lactamase—α-globin chimera	Mislocalized	Stoller and Shields [1989]
LDL receptors	Mutant	Esser and Russell [1988]
Prepro-α-factor (yeast)	Expressed in mammalian cells	Su et al. [1993]
Lysozyme	Mutant	Omura et al. [1992a,b]
Retinol binding protein	Vitamin A deficiency	Tosetti et al. [1992]
Ribophorin I	Mutant	Tsao et al. [1992]
TCR chains	Unassembled	Hannum et al. [1987], Bonifacino et al. [1989], Chen et al. [1988], Lippincott-Schwartz et al. [1988], Wileman et al. [1990a,c, 1991]
Transferrin receptor	Mutant	Hoe and Hunt [1992], Yang et al. [1993]

*Normal refers to no obvious abnormality in primary structure or oligomeric assembly.

targeted for degradation are localized to the ER or other pre-Golgi compartments prior to degradation. In addition, it has been extensively documented that this type of degradation is not sensitive to inhibitors of lysosomal proteolysis. These observations and others discussed below have suggested the existence of a novel pathway for the degradation of proteins retained within the ER, which has been referred to as *pre-Golgi degradation* or *ER degradation*. In this chapter, we describe some of the general characteristics of this process and discuss its role in the degradation of several specific proteins retained in the ER. We also review recent approaches aimed at identifying the components of the degradative machinery and its localization within cells.

TURNOVER OF ER RESIDENT PROTEINS

Most ER resident proteins have relatively long half lives, in the range of 2–6 days [Omura et al., 1967; Arias et al., 1969]. Despite the rather stable nature of most ER resident proteins, however, early studies of the turnover of organellar proteins suggested that the ER is not a static environment [Omura et al., 1967; Arias et al., 1969; Schimke, 1975]. Certain specific constituents of the ER, including several enzymatic activities, were found to be constantly changing, with their levels being the result of a dynamic balance of synthesis and degradation. The enzyme HMG-CoA reductase is a classic example of a protein for which the turnover rate is faster than that of most other ER resident proteins and for which stability is modulated by the metabolic status of the cells [Edwards and Gould, 1972; Faust et al., 1982]. For instance, when sterols and isoprenoid compounds are abundant, the half-life of HMG-CoA reductase can be as short as 1 hour [Chin et al.,1985]. HMG-CoA reductase may actually be the prototype of a group of ER resident proteins, perhaps also including fatty acid desturases [Oshino and Sato, 1972; Bossie and Martin, 1989], for which turnover rates fluctuate in response to changes in the intracellular or extracellular milieus. From these observations, it has become increasingly clear that under normal conditions the ER does not turn over en bloc, but rather displays heterogeneity of degradation rates among its different constituent proteins. This "mosaic" model for the normal turnover of ER resident proteins [Omura et al., 1967; Arias et al., 1969] has to be contrasted with a generalized destruction of ER resident proteins that may occur in very specific situations, such as during certain cell differentiation processes or autophagy.

DESTRUCTION OF NEWLY SYNTHESIZED SECRETORY PROTEINS

Early studies on the biosynthesis of secretory proteins resulted in the unexpected observation that many normal secretory proteins, such as parathyroid hormone and immunoglobulins, are to some extent destroyed intracellularly shortly after synthesis [reviewed by Bienkowski, 1983]. Although the processes involved in the degradation of most of these proteins remain poorly understood, some of the original studies on this subject revealed that this type of degradation had a nonlysosomal component [Morrissey and Cohn, 1979; Sidman, 1981; Dulis et al., 1982]. The studies of Sidman [1981] uncovered a connection between the fate of newly synthesized secretory proteins and cell differentiation by showing that resting B cells synthesize but do not secrete soluble forms of IgM, whereas more differentiated B-cell lines secrete IgM very efficiently. Dulis et al. [1982] extended these studies to non-secreting B lymphoma cells in which they found that both secretory and transmembrane forms of IgM were rapidly catabolized intracellularly. The intracellular IgM molecules did not acquire carbohydrate modifications characteristic of the Golgi system, indicating that degradation took place within an early compartment of the secretory pathway such as the ER or the cis-Golgi.

For other secretory proteins, retention and degradation in the ER appear to be controlled by the metabolic status of the cells, as is the case for ApoB [Borchardt and Davis, 1987; Davis et al., 1990; Sato et al., 1990]. Why a fraction of newly made secretory proteins would be degraded in the ER is not known. A probable explanation is that, under certain physiological or developmental conditions, some of the newly synthesized molecules fail to mature structurally, despite their having a normal amino acid sequence. The immature proteins would have properties that make them incompetent for transport out of the ER, and that increase their susceptibility to degradation.

DEGRADATION OF ABNORMAL PROTEINS RETAINED IN THE ER

By far the most numerous group of proteins shown to be degraded in the ER or related compartments corresponds to newly synthesized proteins with various types of structural abnormalities (see Table I). Sidman [1981] and colleagues were the first investigators to show that structurally abnormal IgM molecules produced in mutant B-cell hybridoma lines were rapidly degraded intracellularly. Furthermore, inhibition of N-glycan addition by treatment with tunicamycin caused complete degradation of unglycosylated secretory IgM, even in normal B-cell hybridomas [Sidman, 1981]. In hybridomas that produced both normal and abnormal IgM, only the normal molecules were secreted, whereas the abnormal ones were rapidly degraded [Sidman, 1981], suggesting that cells had means of distinguishing proteins on the basis of their structure and selectively degrading proteins recognized as abnormal in structure. The characteristics of the degradation process defined by these original studies were similar to those that would later be described for many other structurally abnormal proteins.

CHARACTERISTICS OF THE DEGRADATIVE PROCESSES

The exact nature of the pathways responsible for the degradation of ER resident proteins, newly synthesized secretory products, and abnormal polypeptides retained in the ER has not been elucidated in most cases. However, most of the available information points to a nonlysosomal process associated with the ER or an ER related compartment. The following sections summarize some of the characteristics of the degradation of various proteins retained in the ER.

Subcellular Localization of the Substrates

Analyses of glycoproteins that are subjected to degradation have demonstrated that the proteins have post-translational modifications characteristic of the ER, such as the presence of high mannose chains [see, for example, Dulis et al., 1982; Lippincott-Schwartz et al., 1988; Amara et al., 1989]. In addition, immunofluorescence and immunoelectron microscopy studies have clearly shown the presence of such proteins in the cisternae of the ER, including the nuclear envelope [Sitia et al., 1987; Lippincott-Schwartz et al., 1988; Bonifacino et al., 1989]. Finally, degradation has been shown to occur in permeabilized cells in the absence of transport between the ER and the Golgi system [Stafford and Bonifacino, 1991; Wikström and Lodish, 1992; Meigs and Simoni, 1992; Young et al., 1993]. Taken together, these observations argue that the ER itself may be a site of degradation for a selected group of proteins. Since the ER is composed of several interconnected subcompartments [Vertel et al., 1992], it still remains possible that proteins have to migrate to a specialized area of the ER prior to degradation.

A few studies have suggested that degradation could also take place in a post-ER but pre-Golgi nonlysosomal compartment [Amitay et al., 1992; Shachar et al., 1992; Tsao et al., 1992] or to occur by nonautophagic delivery of lysosomal proteins into elements of the ER [Noda and Farquhar, 1992]. This latter pathway was described in a study of the fate of intra-ER granules in cells overproducing thyrotrophic hormones. The granules were found to accumulate in a region of the ER that progressively acquires lysosomal membrane and content proteins. The granule proteins are thus exposed to proteolytic enzymes without transiting to lysosomes via the Golgi system and without being captured into autophagic structures.

Substrate Selectivity

One of the most remarkable characteristics of degradative processes occurring from the ER is their exquisite selectivity. Most ER resident proteins, as well as many abnormal proteins retained in the ER, are in fact quite stable, and only a limited number of proteins undergo what could be considered rapid degradation in this compartment (e.g., those listed in Table I). Furthermore, even different components of the same multisubunit complex or structurally homologous proteins are sometimes degraded at vastly different rates. The fate of different subunits of the TCR retained in the ER serves as a good example of the selectivity of the degradation process. For instance, whereas the unassembled TCR-α, TCR-β, and CD3-δ subunits of the receptor are degraded, the unassembled CD3-ε and TCR-ζ subunits are stable [Lippincott-Schwartz et al., 1988; Bonifacino et al., 1989; Wileman et al., 1990a,c]. Another example of selectivity is observed in analyses of the fate of two variants of the asialoglycoprotein receptor subunit H2, produced from alternatively spliced forms of the message. The two variants, known as H2a and H2b, differ only in the presence of an additional five amino sequence near the boundary of the lumenal and transmembrane domains of H2a [Lederkremer and Lodish, 1991]. When H2a is expressed by transfection into fibroblasts, the protein is totally retained and degraded in the ER [Amara et al., 1989]. In contrast, a fraction of H2b can exit the ER and is delivered to the plasma-membrane [Lederkremer and Lodish, 1991; Wikström and Lodish, 1993]. The underlying cause for the selectivity of degradation is not known, but is likely to be related to structural features of the proteins, such as the presence of certain sequence or structural motifs that predispose proteins to degradation, or to a global abnormality of protein structure (e.g., folding, aggregation state, association with the ER membrane, and so forth).

Effect of Weak Bases and Other Inhibitors of Lysosomal Proteolysis

Weak bases (e.g., ammonium chloride, chloroquine) and certain protease inhibitors (e.g., leupeptin, E64) have been shown to block effectively the lysosomal degradation of proteins delivered from both the endocytic and biosynthetic pathways in intact cells [Libby and Goldberg, 1978; Seglen, 1983; Ohkuma et al., 1986; Mehdi, 1991]. In contrast, studies of the degradation of proteins retained in the ER have demonstrated that none of these agents affects the degradative process in vivo [for example, see Lippincott-Schwartz et al., 1988; Chen et al., 1988; Amara et al., 1989]. The failure of weak bases to inhibit degradation in the ER is likely due to the fact that the ER cisternae, as well as other early compartments of the secretory pathway, have a neutral pH [Anderson et al., 1984] in contrast to the lysosomal lumen, which is acidic [Ohkuma and Poole, 1978]. The lack of an effect of various protease inhibitors in intact cells, on the other hand, could be due to an inability of the compounds to reach the compartments where degradation takes place. While the evidence provided by these tests is essentially negative, the failure of weak bases and inhibitors of lysosomal proteases to block degradation in intact cells has been helpful to distinguish lysosomal from nonlysosomal processes.

Effects of Other Protease Inhibitors

More detailed analyses of the effect of a wide variety of protease inhibitors on the degradation of proteins retained in the ER has produced a list of compounds that to a greater or lesser degree inhibit ER degradation in intact cells. Most numerous among these compounds are inhibitors of cysteine proteases. The thiol-oxidizing agent diamide, for example, has been shown to inhibit degradation of TCR chains and chimeric proteins derived from them [Stafford and Bonifacino, 1991; Inoue et al., 1991] and of HMG CoA reductase [Inoue and Simoni, 1992]. Other nonspecific thiol reagents, such as N-ethylmaleimide, iodoacetamide and 1,10-phenanthroline complexed with copper, likewise inhibit degradation of some TCR chains [Wileman et al., 1991] and of a truncated form of ribophorin I [Tsao et al., 1992]. More specific peptidyl inhibitors of cysteine proteases such as the calpain inhibitors N-acetyl-leucyl-leucyl-methioninal (ALLM) and N-acetyl-leucyl-leucyl-norleucinal (ALLN) and E64-D, a hydrophobic analog of E64, have also been shown to slow the degradation of TCR chains and HMG CoA reductase [Wileman et al., 1991;

Inoue et al., 1991; Inoue and Simoni, 1992]. Finally, other inhibitors of cysteine and serine proteases such as the chloromethyl ketones *N*-tosyl-L-phenylalanine chloromethyl ketone (TPCK), *N*-carbobenzoxy-L-phenylalanine chloromethyl ketone (ZPCK), and *N*-tosyl-L-lysine chloromethyl ketone (TLCK) have also been shown to inhibit the degradation of some ER retained proteins [Wikström and Lodish, 1992; Wileman et al., 1991; Inoue and Simoni, 1992]. All of these observations are consistent with a role for cysteine proteases, and perhaps also serine proteases, in the degradation of proteins retained in the ER. However, at present it cannot be ruled out that these agents act not by inhibiting specific proteases, but by modifying free suflhydryls or other active groups in the substrates or in other protein factors involved in the degradative process. This latter possibility is supported by the recent finding that the stability of proteins retained in the ER is strongly dependent on redox conditions [Young et al., 1993].

Effects of Brefeldin A

The fungal metabolite brefeldin A causes redistribution of Golgi-resident proteins into the ER and eventually leads to the virtual disappearance of the Golgi stacks [reviewed by Klausner et al., 1992]. As a consequence, newly synthesized proteins are unable to exit the resulting ER–Golgi hybrid compartment and fail to be transported into more distal organelles of the secretory pathway. Because of its unique effects on organelle structure and function, brefeldin A has been extensively used in studies of intracellular protein trafficking. The application of brefeldin A to the study of protein degradation in the secretory pathway has produced different results for different proteins, which makes the data difficult to fit into a single model. For most proteins, treatment with brefeldin A has either no effect or a marginally inhibitory effect on degradation [Bonifacino et al., 1990a,b; Sato et al., 1990; Wileman et al., 1990c; Chun et al., 1990; Inoue and Simoni, 1992; Le et al., 1992; Tosetti et al., 1992; Furukawa et al., 1992; Su et al., 1993; Omura et al., 1992b; Tsuji et al., 1992]. In contrast, studies of the degradation of secretory IgM in a B-cell line demonstrated a strong inhibitory effect of brefeldin A on degradation [Amitay et al., 1992; Shachar et al., 1992]. This effect was interpreted as meaning that degradation of secretory IgM occurs in a post-ER, but still nonlysosomal, compartment. The complexity of brefeldin A effects on degradation of some proteins is perhaps best illustrated by the study of Tsao et al. [1992], who showed that the degradation of truncated forms of ribophorin I occurs with biphasic kinetics. Addition of brefeldin A converts the degradation into a monophasic process, with a rate constant intermediate between those of the two original phases. On the bases of these and other observations, Tsao et al. [1992] proposed the existence of two compartments where degradation takes place: one being the ER cisternae and the other a post-ER, but pre-Golgi compartment. How can these different results be interpreted? In the majority of cases in which brefeldin A has only a minor effect, it is safe to conclude that the degradative process is probably contained within the hybrid organelle generated by brefeldin A–induced merging of the ER and the Golgi complex, thus emphasizing the nonlysosomal nature of the process. In the few cases in which brefeldin A inhibits degradation, on the other hand, it could be exerting its inhibitory effects either by preventing movement into a post-Golgi degradative compartment or by altering the environment of the degradative compartment by the merging of the ER and the Golgi system.

REGULATED TURNOVER OF HMG-CoA REDUCTASE

The enzyme HMG-CoA reductase is an ER resident protein that catalyzes the formation of mevalonate, an intermediate in the synthesis of sterol and isoprenoid compounds [Brown and Goldstein, 1980]. The enzyme is a 97-kDa transmembrane glycoprotein consisting of a membrane-bound domain in which the polypeptide chain spans the membrane seven or eight times, and a large cytoplasmic domain that harbors the catalytic activity of the enzyme [Liscum et al., 1985; Sengstag et al., 1990; Roitelman et al., 1992] (Fig. 1). The only sequences exposed on the lumenal side of the ER membrane are hydrophilic loops that connect the membrane-spanning segments [Liscum et al., 1985; Sengstag et al., 1990; Roitelman et al., 1992]. The expression of HMG-CoA reductase activity is controlled at several levels, including the transcription of the gene, the translation of its messenger RNA, the stability of the protein, and the modulation of its enzymatic activity [reviewed by Luskey, 1988]. This complex regulatory mechanism is set to provide a swift response to changes in the concentration of mevalonate, cholesterol, and probably other products of mevalonate metabolism, all of which contribute to suppressing the levels of HMG-CoA reductase activity when present in high amounts [Brown and Goldstein, 1980; Luskey, 1988].

HMG-CoA reductase is notable for its relatively short half-life in various cultured cells and organs (1–4 hours) [Faust et al., 1982; Chin et al., 1985; Jingami et al., 1987] compared with most other ER resident proteins that have half-lives in excess of 2 days [Omura et al., 1967; Arias et al., 1969; Schimke, 1975]. Another distinctive feature of HMG-CoA reductase is that its half-life varies with the availability of mevalonate metabolites. The turnover of HMG-CoA reductase is accelerated by the presence of high concentrations of sterol and isoprenoid compounds and slowed by depletion of sterols from the medium or by the addition of HMG-CoA reductase inhibitors such as lovastatin or compactin [Chang et al., 1981; Chin et al., 1985; Faust et al., 1982; Edwards et al., 1983a,b; Orci et al., 1984; Nakanishi et al., 1988]. The regulation of HMG-CoA reductase turnover appears to be mediated by the noncatalytic, membrane-bound domain of the enzyme [Gil et al., 1985; Jingami et al., 1987]. Gil et al. [1985] demonstrated that deletion of the membrane-bound domain resulted in a soluble protein with full enzymatic activity. However, the half-life of the protein was increased by about fivefold, and its degradation was no longer modulated by sterols. Further evidence for a role of the membrane-bound domain in the regulated degradation of HMG-CoA reductase was obtained by Jingami et al. [1987], who showed that deletion of two of the membrane-spanning sequences produced a protein that was still membrane bound but was rapidly degraded independently of the levels of sterol compounds. In addition, a hybrid protein composed of the membrane-bound domain of HMG-CoA reductase fused to *Escherichia coli* β-galactosidase was found to be degraded in a sterol-dependent fashion [Skalnik et al., 1988; Chun et al., 1990]. These results are consistent with a model in which the membrane-bound domain senses, either directly or indirectly, the availability of products of mevalonate metabolism. High concentrations of sterols or other compounds derived from mevalonate could cause a conformational change or other modifications of the membrane-bound domain, which in turn would trigger the degradation of the enzyme.

The processes responsible for the regulated degradation of HMG-CoA reductase are not well understood. Tanaka et al. [1986] showed that incubation of cells with ammonium chloride decreased the basal rate of HMG-CoA reductase degradation by about twofold. However, the accelerated degradation induced by sterols was unaffected by treatment with ammonium chloride, suggesting the involvement of a nonlysosomal pathway in the regulated component of this process [Tanaka et al., 1986].

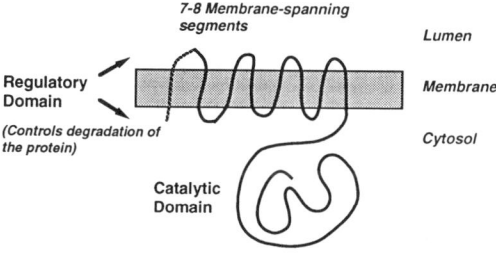

Fig. 1. *Schematic representation of the structure of HMG-CoA reductase.*

This suggestion was further supported by the studies of Simoni and colleagues [Inoue et al., 1991], who demonstrated that various inhibitors of lysosomal degradation failed to inhibit the mevalonate-induced degradation of the enzyme. In addition, treatment of cells with brefeldin A had no effect on the mevalonate-dependent component of the degradation [Chun et al., 1990]. All of these observations suggested that the accelerated degradation of HMG-CoA reductase does not require transport to lysosomes, but rather occurs directly in the ER. This interpretation is consistent with the finding that HMG-CoA reductase is degraded in digitonin-permeabilized cells in the absence of added cytosol and ATP, i.e., under conditions that do not allow transport out of the ER [Meigs and Simoni, 1992]. Thus, these studies seem to imply that the regulated degradation of HMG-CoA reductase is effected by a yet unidentified proteolytic system associated with the ER.

Some of the characteristics of the degradative system were defined by examining the effect of pharmacological agents. Treatment of cells with cycloheximide prevents the induction of degradation caused by sterols, suggesting a role for a short-lived protein in this process [Chen et al., 1982; Chang et al., 1981; Chun et al., 1990]. Various cysteine protease inhibitors, such as ALLN, ALLM, and diamide, block both the regulated and the basal degradation of HMG-CoA reductase, raising the possibility that cysteine proteases are involved in both processes [Inoue et al., 1991; Inoue and Simoni, 1992]. Another manipulation shown to inhibit the regulated degradation of HMG-CoA reductase is the perturbation of intracellular Ca^{2+} levels with ionomycin, thapsigargin, or Co^{2+} [Roitelman et al., 1991; Inoue and Simoni, 1992].

A recent development that promises to shed light on the mechanisms involved in the regulated turnover of HMG-CoA reductase has been the demonstration of this phenomenon in yeast cells [Hampton and Rine, 1994]. The yeast *Saccharomyces cerevisiae* has two genes encoding HMG-CoA reductase isozymes that are known as Hmg1p and Hmg2p [Basson et al., 1988]. Analyses of the turnover of the two isozymes have shown that whereas Hmg1p is stable, Hmg2p is rapidly degraded [Hampton and Rine, 1994]. Biochemical and genetic characterizations of Hmg2p degradation have revealed striking similarities with the degradation of HMG-CoA reductase in mammalian cells. For instance, degradation of Hmg2p does not require transport through the secretory pathway, is independent of vacuolar proteases, and is abrogated by lowering intracellular levels of mevalonate. As in mammalian cells, the instability of Hmg2p and the response to mevalonate are mediated by the membrane-bound N-terminal domain [Hampton and Rine, 1994]. Disruption of a gene encoding a subunit of farnesyl transferase has been found to result in stabilization of Hmg2p, which suggests a role of farnesylation in signaling Hmg2p degradation [Hampton and Rine, 1994]. This observation may explain the sensitivity of Hmg2p degradation to depletion of mevalonate, which is a precursor of farnesyl groups. These experiments demonstrate that the characteristics of the regulated turnover of HMG-CoA reductase have been remarkably conserved throughout evolution and open the way to the use of yeast genetic approaches to study degradation of proteins in the ER (see Nonlysosomal Degradation in the Yeast Secretory Pathway, below).

REGULATED RETENTION AND DEGRADATION OF APOB IN THE ER

ApoB is the major protein component of very-low-density lipoproteins (VLDL) synthesized in liver and intestinal cells [reviewed by Olofsson et al., 1987]. ApoB exists as two major forms, known as ApoB-100 and ApoB-48, that are derived by post-transcriptional editing of a single mRNA [Powell et al., 1987; Chen et al., 1987]. As with other secretory proteins, the nascent ApoB polypeptides are translocated into the ER lumen, and their signal peptides are cleaved. VLDL particles are then assembled by interaction of the newly synthesized ApoB with lipids (Fig. 2). This assembly is presumed to occur in the ER from a pool of membrane-

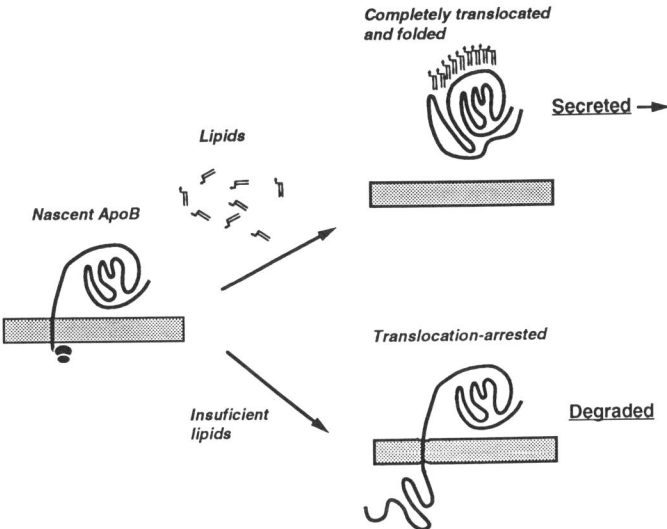

Fig. 2. Model depicting the different fates of ApoB pools in the ER. Interaction of nascent ApoB with lipids promotes folding and complete translocation of ApoB polypeptides in the ER. The lipoprotein complexes are transported out of the ER and eventually released into the extracellular space. In the absence of sufficient lipids, the partially translocated ApoB polypeptides are retained and degraded in the ER.

associated ApoB [reviewed by Olofsson et al., 1987]. The assembled VLDL particles are eventually secreted into the extracellular medium. An aspect of ApoB biogenesis that distinguishes it from most other secretory proteins is that a fraction of ApoB is retained in the ER and undergoes degradation by a nonlysosomal pathway [Borchardt and Davis, 1987; Davis et al., 1990; Sato et al., 1990]. The degradation of this fraction of ApoB has all the pharmacological hallmarks of the degradation of other ER retained proteins [Sato et al., 1990; Furukawa et al., 1992]. Several proteolytic fragments derived from ApoB can be identified in ER but not in Golgi fractions, further supporting the idea that degradation occurs in the ER [Davis et al., 1990]. Degradation of both forms of ApoB (ApoB-100 and ApoB-48) is observed in primary cultures of hepatocytes [Borchardt and Davis, 1987; Sparks and Sparks, 1990] and in hepatoma cell lines [Sato et al., 1990; Borën et al., 1990; Dixon et al., 1991; Yao et al., 1991]. Other shorter, C-terminal truncated forms of ApoB are also partially retained and degraded in the ER [Yao et al., 1991; White et al., 1992].

What determines that a fraction of newly synthesized ApoB is retained and degraded in the ER, while the rest is secreted into the medium? One possibility is that there are two different pools of ApoB with different physicochemical characteristics. This explanation was first postulated by Davis and colleagues [Davis et al., 1990; Thrift et al., 1992], who observed that about half of the ApoB in ER fractions from cultured hepatocytes was membrane bound and had sequences exposed on both the cytoplasmic and lumenal sides of the ER membrane. This pool was suggested to account for the ApoB that is degraded intracellularly. The remaining half was completely translocated into the ER lumen and probably corresponds to the protein that becomes assembled into VLDL particles and that is eventually secreted. The association of a fraction of ApoB molecules with the ER membrane may be due to an arrest in the translocation of the nascent ApoB chains, as suggested by the observations of Chuck et al. [1990]. These authors observed that the translocation of ApoB occurs by stepwise conversion of transmembrane intermediates. This explanation is consistent with a model in which the availability of lipids controls the translocation and folding of the nascent ApoB molecules. In the absence of sufficient lipids with which to assemble VLDL particles, ApoB would remain partially translocated and incompletely folded, which would render it susceptible to degradation (Fig. 2).

In this scenario, it is easy to imagine that the secretion or ER retention of ApoB would be highly sensitive to nutritional factors, such as the availability of certain lipids. Many studies have proven that this is indeed the case. Addition of oleate to cultured cells, for instance, has been shown to stimulate secretion of ApoB-containing particles without an increase in mRNA levels [Pullinger et al., 1989; Moberly et al., 1990]. Dixon et al. [1991] and White et al. [1992] demonstrated that the stimulation of ApoB secretion by oleate is clearly a post-translational event due to protection of newly synthesized ApoB from degradation in the ER. In contrast to oleate, addition of insulin to cells reduces secretion of ApoB-containing particles. Although this effect is due in part to decreased synthesis of ApoB, the ApoB protein that is made is also more extensively degraded in insulin-treated cells [Sparks and Sparks, 1990]. Similarly, incubation of cells with low-density lipoproteins (LDL) increases the degradation of ApoB [Sato et al., 1990].

RETENTION IN THE ER AND DEGRADATION OF UNASSEMBLED SUBUNITS OF THE TCR

The TCR is one of the best-characterized examples of a multiprotein complex whose subunits can undergo retention and degradation in the ER if they are unable to assemble into a complete complex [reviewed by Klausner et al., 1990; Exley et al., 1991]. Although recent studies have demonstrated a considerable diversity in the structure of the TCR, the most common form of the receptor expressed on the surface of mature T cells consists of the products of six genetic loci. Two of the chains encoded by these loci (TCR-α and -β) are variable, whereas the other four chains (CD3-γ, -δ, and -ϵ, and TCR-ζ) are invariant. For simplicity, in this chapter these chains are referred to as α, β, γ, δ, ϵ, and ζ. The exact stoichiometry of the subunits in the complex has not been definitively established, although the ζ-chain is known to occur as a disulfide-linked homodimer [Samelson et al., 1985] and the ϵ chain as a partially disulfide-linked homodimer [Jin et al., 1990; de la Hera et al., 1991; Blumberg et al., 1990b]. In T cells that fail to synthesize some of the receptor chains, and in fibroblasts transfected with genes that encode single chains, expression of TCR chains at the cell surface is comparatively lower than in normal T cells. While some partial complexes have been shown to be expressed at the cell surface [Kappes and Tonegawa, 1991], the majority of newly synthesized chains seem to be retained intracellularly unless assembly is complete [reviewed by Klausner et al., 1990; Exley et al., 1991]. Analyses of the fate of free chains or incompletely assembled complexes have demonstrated that the newly synthesized proteins are either delivered to lysosomes or retained within the ER (Fig. 3). Delivery to lysosomes has been shown to occur for complexes composed of α, β, γ, δ, and ϵ in the absence of the ζ-chain [Sussman et al., 1988; Minami et al., 1987], as well as for a fraction of δ-chains expressed in fibroblasts [Wileman et al., 1990c]. Most other incompletely assembled species were found to be retained in the ER [Bonifacino et al., 1988, 1989; Chen et al., 1988; Lippincott-Schwartz et al., 1988; Wileman et al., 1990a,c].

Among the ER retained chains, the α-, β-, and δ-chains of both mouse and human origin were found to have relatively short half-lives [Bonifacino et al., 1989; Chen et al., 1988; Lippincott-Schwartz et al., 1988; Wileman et al., 1990a,c]. In contrast, the ϵ-chain was much more stable [Bonifacino et al., 1989; Wileman et al., 1990a,c]. The mouse γ-chain is unstable but becomes much more stable when it assembles with ϵ [Bonifacino et al., 1989]. Similarly, assembly of α and β with γ and ϵ protected these chains from degradation [Wileman et al., 1990a]. On the bases of these observations, it

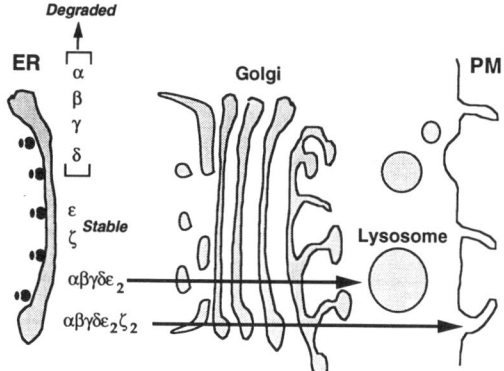

Fig. 3. *Fates of the mouse TCR chains. The variable chains α and β and the invariant chains γ and δ are retained and degraded in the ER when expressed individually in transfected fibroblasts. The unassembled ε-chain is also retained in the ER but has a relatively long half-life. The localization of unassembled ζ-chains has not been established. However, when newly synthesized ζ-chains are retained in the ER–Golgi system by treatment with brefeldin A, they form ζ_2 homodimers that are relatively stable. $\alpha\beta\gamma\delta\epsilon_2$ complexes are able to leave the ER but are mostly degraded in lysosomes. Complete $\alpha\beta\gamma\delta\epsilon_2\zeta_2$ complexes are transported to the plasma membrane (PM).*

was proposed that the tendency of the TCR chains to be degraded in the ER depends on an intrinsic susceptibility of each chain to degradation and on their state of assembly.

The possibility that degradation was specified by discrete regions of the TCR chains was examined by molecular dissection of the α- and β-chains [Bonifacino et al., 1990a,b; Wileman et al., 1990b, 1993; Shin et al., 1993]. These studies produced the surprising finding that the putative membrane anchors of α and β were responsible for their degradation phenotype. Potentially charged amino acid residues within the hydrophobic region of these sequences were essential for degradation [Bonifacino et al., 1990a; Wileman et al., 1993]. The ability of charged residues to cause retention and degradation in the ER was dependent on the nature of the charged amino acid residue, its position within the putative membrane-spanning domain, and the length of the hydrophobic sequence [Bonifacino et al., 1991; Lankford et al., 1993]. A similar role of the membrane anchors in retention and degradation in the ER was demonstrated for the α-subunit of immunoglobulin Fc receptors [Kurosaki et al., 1991] and membrane-bound forms of IgM [Williams et al., 1990; Bachhawat and Pillai, 1991]. As possible explanations for the role of the membrane anchors in degradation, it was hypothesized that the membrane anchors could either mediate interactions with some component of the degradation machinery or otherwise cause changes in the global physicochemical properties of the proteins that render them sensitive to degradation [Bonifacino et al., 1990a, 1991]. The latter possibility is supported by the observation that the presence of charged residues within putative membrane-spanning domains reduces the stability of association with the membrane [Bonifacino et al., 1991]. In addition, CD4–TCR-α chimeric proteins with substitutions in their cytoplasmic domains have been found to be released into the ER lumen [Shin et al., 1993]. Finally, mutations that would be predicted to decrease the stability of association with membranes increase the probability of ER degradation [Lankford et al., 1993]. These observations raise the possibility that release into the ER lumen could be the first step in a chain of events that leads to the degradation of some TCR chains and of chimeric proteins derived from them.

While the mechanism by which certain polypeptide membrane anchors predispose proteins to degradation still remains to be explained, it has become clear that the membrane anchors will mainly manifest this activity in the context of an unassembled chain or of some partial complexes. This is probably due to the fact that the same sequences that predispose proteins for degradation are also involved in subunit interactions. Indeed, the putative transmembrane domains of the TCR-α and -β chains mediate interactions with other subunits of the complex in a process that is also dependent on the potentially charged transmembrane residues [Cosson et al., 1991; Manolios et al., 1990; Wileman et al., 1993; Tan et al., 1991]. Assembly of the TCR-α or -β chains, or of chimeric proteins having the TCR-α or -β transmembrane domains, with some of the CD3 chains results in stabilization of the partial complexes [Bonifacino et al., 1990a; Wileman et al., 1993]. Interactions involving potentially charged transmembrane residues have not only been demonstrated to be sufficient for assembly of pairs of TCR chains or chimeric proteins, but also have been shown to be essential for the assembly of complete complexes that are competent for transport to the plasma membrane [Morley et al., 1988; John et al., 1989; Blumberg et al., 1990a; Alcover et al., 1990]. Recently, Wileman et al. [1993] showed that the ectodomain of the TCR-β chain is also involved in interactions with CD3 chains and that these interactions protect the chains from degradation in the ER. Taken together, the above observations have been incorporated into a model that suggests that certain sites of subunit assembly have evolved also to encompass determinants of protein instability. The colocalization of sites of assembly and instability determinants ensures that the activity of the determinants is only manifested in the context of an unassembled subunit and not of a properly assembled oligomer. This model is likely to apply to other multiprotein complexes and perhaps even to some single chain molecules for which interactions between domains of the same protein can likewise hide instability determinants. For a very complex structure like the TCR, we envision the existence of multiple determinants of protein instability or of intracellular retention that are fully neutralized only upon assembly of a complete complex.

DEGRADATION OF CD4 INDUCED BY THE HIV-1 VPU GENE PRODUCT

The CD4 molecule, a transmembrane glycoprotein expressed at the surface of a subset of T cells and macrophages, functions as a coreceptor during T-cell activation and also acts as the receptor for the human immunodeficiency virus-1 (HIV-1) envelope glycoprotein subunit gp120 [Dalgleish et al., 1984; Klatzmann et al., 1984]. In infected cells, gp120 is generated by proteolytic cleavage of a larger membrane-bound precursor protein known as gp160 [Allan et al., 1985; Veronese et al., 1985]. Both CD4 and gp160 are synthesized in the ER, where they rapidly acquire the ability to interact with each other. Formation of CD4-gp160 complexes in the ER prevents transport of CD4 to the cell surface [Crise et al., 1990; Jabbar and Nayak, 1990]. As a consequence, overproduction of gp160 during HIV-1 infection ultimately leads to a drastic reduction of surface CD4 levels. Recent studies with HeLa cells expressing CD4 have shown that transfection of HIV-1 clones results in a dramatic decrease in total CD4 without a detectable change in its synthesis, suggesting that expression of viral genes leads to destabilization of CD4 molecules [Willey et al., 1992a,b]. This destabilization requires synthesis of gp160 and is therefore probably related to the ability of gp160 to cause retention of CD4 in the ER. However, the reduction of CD4 levels is also dependent on a second viral gene product, known as Vpu [Strebel et al., 1988; Willey et al., 1992a,b].

Vpu is a nonstructural protein of HIV-1 believed to regulate virus assembly or release [Strebel et al., 1989]. The Vpu open reading frame overlaps with the 5´ end of the gp160 gene and encodes a 16-kDa cytoplasmically oriented, membrane-bound phosphoprotein [Strebel et al., 1988; Maldarelli et al., 1993]. Interestingly, Willey and colleagues [1992a,b] have recently reported that Vpu induces rapid degradation of endo H–sensitive CD4 complexed with gp160 in the ER. The effects of Vpu on CD4 degradation require that CD4 is retained in the ER, either by interaction with gp160 or by introduction of an ER retention signal such as the cytoplasmic C-terminal sequence KKTC [Nilsson et al., 1989; Shin et al., 1991]. Similarly, blocking CD4 transport out of the ER–Golgi system by using the drug brefeldin A causes degradation of CD4 in the presence of Vpu but not in its absence [Willey et al., 1992b]. The surface expression and stability of other proteins, such as gp160 and CD8 in cells treated with brefeldin A, are not affected by Vpu, suggesting that its effects are specific for CD4.

How Vpu affects the stability of CD4 in the ER is not known. Its sequence reveals no homology to known proteases. Furthermore, the bulk of the Vpu molecule is on the cytoplasmic side of the membrane, separated from the lumenal domain of CD4 by the lipid bilayer [Maldarelli et al., 1993]. The Vpu sequence shows some homology to the M2 protein of influenza virus, which probably functions as an ion channel [Strebel et al., 1988]. This raises the possibility that Vpu might alter the ionic environment of the ER lumen, thus enhancing the susceptibility of CD4 to degradation in this compartment. Another possibility is that targeting of CD4 for degradation is mediated by a direct interaction with Vpu. This latter possibility is supported by the recent studies of Chen et al. [1993] and Vincent et al. [1993]. Chen et al. [1993] have shown that, when an mRNA encoding CD4 is translated in a rabbit reticulocyte system in the presence of dog pancreas microsomes, the newly synthesized CD4 is very stable. In contrast, cotranslation with the Vpu mRNA results in degradation of CD4. This degradation only occurs in the presence of the microsomal membranes and requires the insertion of both CD4 and Vpu into the same membranes [Chen et al., 1993]. The effects of Vpu on CD4 stability are also specific in this system, as yeast pre-pro-α-factor cotranslated with CD4 is stable in both the presence and the absence of Vpu. Deletion of 32 residues from the cytoplasmic tail of CD4 abolishes degradation induced by Vpu, suggesting that this domain of the CD4 molecule contains a recognition element for Vpu [Chen et al., 1993]. On the other hand, both an internal deletion and a C-terminal truncation in the hydrophilic cytoplasmic domain of Vpu suppress the ability of Vpu to cause enhanced degradation of CD4, suggesting that this domain is also involved in destabilizing CD4 [Chen et al., 1993]. Vincent et al. [1993] constructed chimeric proteins having the ectodomain of the gp160 envelope precursor, the transmembrane domain of CD4, and various portions of the CD4 cytoplasmic tail. Chimeras having 24–38 residues from the CD4 tail were found to be degraded in the ER in a Vpu-dependent fashion. Shortening the CD4 tail to the 18 membrane-proximal residues, however, resulted in a protein that was stable in both the presence and the absence of Vpu [Vincent et al., 1993]. These studies also support the idea that the cytoplasmic tail of CD4 contains sequences that can cause retention and degradation in the ER upon expression of Vpu. Further analyses of this system promise to provide additional insights into the mechanisms that control susceptibility to degradation in ER compartments.

PERMEABILIZED CELLS AND CELL-FREE ASSAYS FOR DEGRADATION

Most of the observations reported to date on the degradation of proteins retained in the ER were made by following the fate of metabolically labeled proteins in intact cells. To gain a better understanding of this process, it will be necessary to develop more isolated systems that retain the ability to degrade proteins with the same characteristics as in the intact cells. The use of systems such as permeabilized cells or isolated subcellular fractions has several advantages over intact cells. First, using these systems it is possible to examine whether degradation depends on transport from the ER into some other intracellular compartment. The use of drugs or temperature manipulations to block transport out of the ER in intact cells often produces ambiguous results, due to the fact that most of these manipulations are not specific for ER to Golgi transport but also affect other processes. Permeabilization or disruption of cells, on the other hand, causes complete cessation of interorganellar trafficking under more controlled conditions. The biochemical requirements for vesicular transport between the ER and the cis-Golgi compartment have been extensively characterized using such cell-free systems. These studies have shown that transport from the donor ER fractions requires ATP, Ca^{2+}, and cytosolic factors and is inhibited by GTPγS [Balch, 1989]. A comparison of these requirements with those of degradation of ER retained proteins should test whether degradation is dependent on transport. The second advantage of the in vitro systems is that, by breaking the permeability barrier imposed by the plasma membrane, the cytoplasmic aspect of the intracellular organelles becomes accessible to a large number of agents that would otherwise be unable to penetrate cells. Perturbation of degradative processes by some of these agents could reveal additional characteristics of the degradative process. Finally, and perhaps most importantly, the ability to observe degradation in in vitro systems is a necessary step in the identification of components of the degradation machinery.

Several groups have exploited the ability of pore-forming compounds such as streptolysin O [Stafford and Bonifacino, 1991], and digitonin [Leonard and Chen, 1987; Meigs and Simoni, 1992] to examine the degradation of proteins retained in the ER. Streptolysin O is a protein produced by β-hemolytic streptococci that has the property of creating large pores in the plasma membrane of eukaryotic cells. The pores are large enough to allow exchange of most low-molecular-weight substances, as well as proteins up to molecular weights of about 140,000 [Ahnert-Hilger et al., 1989]. Permeabilization can be limited to the plasma membrane by binding the toxin to cells at 4°C and, following removal of the unbound toxin, shifting the temperature to 37°C [Ahnert-Hilger et al., 1989]. Digitonin is a detergent-like steroid compound that selectively permeabilizes cholesterol-rich membranes, such as the plasma membrane [Fiskum et al., 1980]. Its effects on membrane permeability are very similar to streptolysin O. Using streptolysin O, Stafford and Bonifacino [1991] demonstrated that ER retained chimeric proteins were degraded in permeabilized cells in a process that did not require exogenously added ATP and cytosol and that was not blocked by GTPγS or EDTA. The process maintained the specificity previously reported for intact cells, as a protein that is normally transported to the cell surface was degraded at a much slower rate. An additional finding of this study was that reducing agents, such as DTT, accelerate degradation of the chimeric proteins. Meigs and Simoni [1992] found that HMG CoA reductase was degraded in a mevalonate-dependent fashion in digitonin-permeabilized cells, by a process that closely mimicked the characteristics of the in vivo phenomenon. Also in this case, ATP and cytosol were not required to observe degradation. Similar findings were made by Wikström and Lodish [1992] in the study of the degradation of the asialoglycoprotein receptor H2a subunit in mechanically permeabilized cells. Degradation in this system also replicated the characteristics of the process in intact cells, including the appearance of a degradation intermediate and the inhibition of the initial cleavage by TLCK and TPCK [Wikström and Lodish, 1992]. Taken together, these studies suggested that transport out of the ER is not an absolute requirement for degradation.

While approaches based on the use of permeabilized cells have proven useful to examine the dependence on intracellular transport and other characteristics of the process, permeabilized cells are not readily amenable to biochemical dissection of the degradation machinery. This could be best accomplished by using in vitro assays with isolated subcellular fractions. Selective degradation of a Tac–TCR-α chimeric protein was shown to occur in a microsomal fraction isolated from metabolically labeled cells [Stafford and Bonifacino, 1991], but the process was not further characterized. Furukawa et al. [1992] also showed that ApoB was degraded in purified ER vesicles, but not in Golgi fractions. Eliasson et al. [1992] examined the degradation of an ethanol-inducible cytochrome P_{450} (2E1) in isolated liver microsomes. Analysis by SDS-PAGE and immunoblotting of microsomes incubated at 37°C showed that the P_{450} (2E1) molecule was rapidly degraded in the presence of CCl_4, an agent that was previously shown to induce loss of the P_{450} (2E1) molecule in vivo. This effect is probably due to free-radical damage to the protein induced by CCl_4. The protein was protected from degradation by the addition of the substrate imidazole. The degradation did not require cytosol but, unlike the examples discussed above, was accelerated by the inclusion of MgATP in the assay buffer [Eliasson et al., 1992]. ATP was probably used to phosphorylate the P_{450} (2E1) molecule, thus rendering it susceptible to degradation. Several lower molecular weight degradation fragments were detected in the gels. The proteolytic activity cofractionated with both rough and smooth ER membranes and could not be extracted by treatment of the microsomal fraction with low concentrations of sodium cholate, suggesting that it is tightly bound to the membranes.

All of the experiments described above rely on the ability of antibody reagents to detect the degradation substrates by either immunoprecipitation or immunoblotting. In some situations it would be desirable to examine the degradation of the proteins directly, without a step involving antibody recognition. This has recently been achieved by Chen et al. [1993] in the study of the degradation of CD4 induced by the HIV-1 Vpu gene product (see above). These authors utilized an in vitro translation system consisting of a rabbit reticulocyte system and dog pancreas microsomes as a source of ER membranes. CD4 and Vpu mRNAs were cotranslated in the presence of ^{35}S-methionine, and the fate of the labeled proteins was directly analyzed by SDS-PAGE. Under these conditions, CD4 was found to disappear with a halftime of 40 minutes at 37°C [Chen et al., 1993], thus confirming that the ER itself is capable of degrading newly synthesized proteins.

PROTEOLYTIC ACTIVITIES IN THE ER

All of the above observations imply the existence of proteolytic activities associated with the ER. To date, the only well-characterized proteolytic enzyme in the ER is signal peptidase [Shelness et al., 1988], an endopeptidase that cleaves signal peptides upon translocation of nascent polypeptide chains into the ER lumen. However, a generalized role for signal peptidase in ER degradation can be discounted because of the rather restricted specificity of cleavage exhibited by this enzyme. Over the years, other proteolytic activities have been found in association with ER fractions, but their possible role in the degradation of ER-retained proteins remains to be established. Mumford et al. [1980] described an endopeptidase activity in solubilized dog pancreas microsomes that hydrolyzes different fluorogenic peptide substrates at Ala-Phe, Ala-Ala, and Gly-Ala peptide bonds. The pH optimum of the enzyme is between 7.0 and 7.5, consistent with a role in a neutral environment like the ER. The enzymatic activity is inhibited by 1,10-phenanthroline and phosphoramidon, suggesting that the enzyme is a Zn^{2+} metallo-endopeptidase. In contrast, serine protease inhibitors had no effect on the activity of the enzyme [Mumford et al., 1980].

Urade and colleagues [Urade and Kito, 1992; Urade et al., 1992] have recently identified a proteolytic activity in purified fractions of an ER protein with homology to thiol:protein disulfide oxidoreductases. This protein contains a KDEL sequence and is known to be localized to the ER cisternae [Urade et al., 1992], but shares no homology with members of the known families of proteases. Purified preparations of this protein contain eight immunologically related species that, owing to their molecular weights of around 60,000, are referred to as ER60A–ER60H. All of the isolated species of ER60 can catalyze their own cleavage and can also cleave other ER proteins, such as protein disulfide isomerase and calreticulin. Autocatalytic cleavage of ER60 is optimal at pH 6–7 and is enhanced by urea and β-mercaptoethanol [Urade et al., 1992]. Degradation by preparations of ER60 was inhibited by pCMB, ALLN, ALLM, leupeptin, and E64, but not by DFP, PMSF, or EGTA, suggesting that it functions as a cysteine protease [Urade et al., 1992; Urade and Kito, 1992]. A curious property of this enzyme is that it is inhibited by acidic phospholipids such as phosphatidylinositol, phosphatidylinositol 4,5-biphosphate, and phosphatidylserine [Urade and Kito, 1992]. The preparations of ER60 shown to have proteolytic activity seem to be of high purity, as judged by silver staining of SDS-PAGE analyses. However, the lack of homology with other proteases will make it necessary to use more stringent tests in order to establish whether the proteolytic activity is intrinsic to ER60 or corresponds to another protein in the preparations.

Another candidate for a protein that could potentially play a role in the degradation of proteins retained in the ER is a sulfhydryl endopeptidase (SH-EP) recently described in seeds from certain leguminosae plants [Mitsuhashi et al., 1988]. A major site of action of this enzyme is probably the storage granules, which are organelles derived directly from the ER. Molecular cloning of the gene encoding SH-EP revealed the presence of the sequence KDEL at the C terminus of the protein [Akasofu et al., 1989]. As in mammalian cells, the KDEL motif has been shown to be active for ER retention in plants [Napier et al., 1992]. Although there is presently no evidence that this protein is involved in protein degradation in the ER, it is of interest as the first protease known to have an ER retention signal. This finding should stimulate a search for similar proteases in other organisms.

A possibility that also has to be considered is that the proteases responsible for degrading proteins in the ER are not specific for this compartment but are found in other locations of the secretory pathway. It is unlikely that the major lysosomal cathepsins are involved, since they are synthesized as inactive precursors that only become activated after cleavage in a post-Golgi compartment. There is evidence, however, that cathepsin E is not localized to lysosomes, but is instead prominent in microvilli, ER cisternae, and nuclear envelope [Yamamoto, 1992].

NONLYSOSOMAL DEGRADATION IN THE YEAST SECRETORY PATHWAY

The generation of yeast mutants deficient in different steps of the secretory pathway has proven an invaluable approach to identify gene products involved in intracellular protein transport and sorting [Pryer et al., 1992]. If yeast cells were shown to be capable of degrading proteins retained in early compartments of the Golgi system, then it would be possible to devise screening assays for mutants of the degradative process. In conjunction with biochemical methods, the yeast genetic approach could lead to the identification of proteases and other factors involved in degradation. Hampton and Rine (1994) have recently shown that an isozyme of yeast HMG-CoA reductase (Hmg2p) is rapidly degraded in yeast cells by a pathway with the characteristics of that previously described in mammalian cells (see Regulated Turnover of HMG-CoA Reductase, above). In addition, two groups have independently demonstrated that abnormal newly synthesized polypeptides can also be degraded in yeast cells [McCracken and Karpichev, 1992; McCracken and Kruse, 1993; Finger et al., 1993].

McCracken and colleagues studied the degradation of normal and mutant forms of human α_1-protease inhibitor (also called α_1-antitrypsin or A1Pi) expressed in the yeast *Saccharomyces cerevisiae* [McCracken and Kruse, 1993; McCracken and Karpichev, 1992]. A1Pi is a serine protease inhibitor produced and secreted by liver cells. A naturally occurring point mutation in human A1Pi generates the variant known as A1PiZ, which misfolds and is retained in the ER. The ER retained protein aggregates and is largely degraded by a nonlysosomal pathway in mammalian cells [Graham et al., 1990; Le et al., 1990, 1992].

When the normal A1Pi or the mutant A1PiZ molecules were expressed in *S. cerevisiae*, the proteins were only poorly secreted into the culture medium and for the most part became retained in the ER [McCracken and Kruse, 1993]. Interestingly, both forms of A1Pi were found to be degraded, albeit at different rates ($t_{1/2}$ ~3 hours vs. 30 minutes for A1Pi and A1PiZ, respectively). Degradation was unchanged in a yeast mutant deficient in the vacuolar enzymes proteinase A and carboxypeptidase Y (see chapter by Jones, this volume), suggesting that the process was independent of vacuolar proteases [McCracken and Kruse, 1993]. Inhibition of vesicular transport between the ER and the Golgi system, which occurs in temperature-sensitive mutants of the sec17 and sec18 genes at the nonpermissive temperature, partially inhibited degradation of A1PiZ. These observations open the possibility that degradation takes place not only in the ER but also in some post-ER, but pre-trans-Golgi compartment [McCracken and Kruse, 1993]. This idea was previously raised in studies of degradation of abnormal proteins in mammalian cells [Le et al., 1990; Amitay et al., 1992; Shachar et al., 1992; Tsao et al., 1992].

Finger et al. [1993] reported similar findings for mutant forms of two vacuolar enzymes, proteinase A and carboxypeptidase Y. Mutants of both proteins were found to be degraded with half-times of 15–20 minutes after a lag period of approximately 10 minutes. The mutated proteins were localized to the ER by both subcellular fractionation and immunofluorescence microscopy and had carbohydrate modifications typical of the ER [Finger et al., 1993]. Degradation was not inhibited in sec18 mutants, which suggests that the process takes place in the ER or some other pre-Golgi location. The degradation process was also shown to be independent of the peptidylglutamylpeptide splitting activity of the proteasome [Finger et al., 1993].

McCracken and Karpichev [1992] have taken advantage of the ability to observe degradation of proteins retained in the ER of yeast cells to develop a colony blotting screening for mutants deficient in this process. Clones that accumulate high levels of A1PiZ were selected and further tested for their ability to degrade the protein on pulse-chase experiments. Several clones were isolated in which A1PiZ became stable after synthesis, suggesting that they carry mutations in genes that are important for degradation of the abnormal protein [McCracken and Karpichev, 1992]. Molecular cloning of these genes will hopefully result in the identification of components of the degradation machinery.

ASSOCIATION OF THE YEAST UBIQUITIN-CONJUGATING ENZYME UBC6 WITH THE ER MEMBRANE

The *S. cerevisiae* gene *UBC6* encodes a cytoplasmically oriented ubiquitin-conjugating enzyme having a C terminal hydrophobic extension that functions as a membrane anchor [Sommer and Jentsch, 1993]. Subcellular fractionation and immunofluorescence microscopy experiments have shown that the enzyme is associated with the ER membrane [Sommer and Jentsch, 1993]. Interestingly, disruption of the *UBC6* gene suppresses a mutation of the SEC61 gene that blocks protein translocation across the ER membrane. SEC61 is a component of a multiprotein complex involved in the translocation of nascent chains in the ER. On the bases of these observations, Sommer and Jentsch [1993] have proposed that mutation of the sec61 gene product may result in abnormal assembly of the translocation apparatus, which would render some of the subunits susceptible to degradation by the ubiquitin pathway. When UBC6 is absent from the cells, the abnormally assembled but now stable translocation apparatus would be capable of mediating translocation of nascent chains. While still indirect, this finding strongly suggests that the ubiquitin pathway might be involved in degrading proteins bound to the ER membrane and is consistent with the observation of ubiquitinated forms of rapidly degraded cytochrome P_{450} 2E1 molecules [Tierney et al., 1992].

CONCLUSION

The ability to degrade organellar proteins is an essential component of mechanisms that function to maintain cellular homeostasis and to regulate the levels of critical proteins in metabolic and differentiation pathways. While much of the degradation of cellular proteins is carried out by cytosolic and lysosomal proteolytic systems, certain organelles, such as mitochondria, may have resident degradative systems that play specific roles in the organellar environment. The experimental evidence reviewed in this chapter indicates that the ER or a related compartment may similarly be the site of a specific degradative pathway, involved in the normal turnover of ER resident proteins, the disposal of newly synthesized proteins with various types of structural abnormalities, and the post-translational regulation of the expression of certain gene products. Some of the distinctive characteristics of this pathway have been inferred from studies on the degradation of a large number of proteins retained in the ER. However, many important questions remain to be answered regarding the nature of the degradative process, its molecular components, and the exact intracellular localization of the degradative machinery. It is to be hoped that ongoing research using cell-free systems and yeast genetic approaches will soon elucidate the underlying mechanism of this commonly observed process.

ACKNOWLEDGMENTS

We thank Ardythe McCracken, Randolph Hampton, Klaus Strebel, Thomas Wileman, and Dieter Wolf for sending us preprints of some of their publications. We also thank Martha Delahunty and Michael Marks for comments on the manuscript and especially all of our colleagues at the Cell Biology and Metabolism Branch who contributed to the work reviewed in this chapter.

REFERENCES

Ahnert-Hilger G, Mach W, Föhr KJ, Gratzl M (1989): Poration by alpha-toxin and streptolysin O: An approach to analyze intracellular processes. Methods Cell Biol 31:63–90.

Akasofu H, Yamauchi D, Mitsuhashi W, Minamikawa T (1989): Nucleotide sequence of cDNA for sulfhydryl-endopeptidase (SH-EP) from cotyledons of germinating Vigna mungo seeds. Nucleic Acids Res 17:6733.

Alcover A, Mariuzza RA, Ermonval M, Acuto O (1990): Lysine 271 in the transmembrane domain of the T-cell antigen receptor beta chain is necessary for its assembly with the CD3 complex but not for alpha/beta dimerization. J Biol Chem 265:4131–4135.

Allan JS, Colligan JE, Barin F, McLane MF, Sodroski JG, Rosen CA, Haseltine WA, Essex M (1985): Major glycoprotein antigens that induce antibodies in AIDS patients are encoded by HTLV-III. Science 228:1091–1094.

Amara JF, Lederkremer G, Lodish HF (1989): Intracellular degradation of unassembled asialoglycoprotein receptor subunits: A pre-Golgi, nonlysosomal endoproteolytic cleavage. J Cell Biol 109:3315–3324.

Amitay R, Shachar I, Rabinovich E, Haimovich J, Bar-Nun S (1992): Degradation of secretory immunoglobulin M in B lymphocytes occurs in a postendoplasmic reticulum compartment and is mediated by a cysteine protease. J Biol Chem 267:20694–20700.

Anderson RG, Falck JR, Goldstein JL, Brown MS (1984): Visualization of acidic organelles in intact cells by electron microscopy. Proc Natl Acad Sci USA 81:4838–4842.

Arias IM, Doyle D, Schimke RT (1969): Studies on the synthesis and degradation of proteins of the endoplasmic reticulum of rat liver. J Biol Chem 244:3303–3315.

Bachawat AK, Pillai S (1991): Distinct intracellular fates of membrane and secretory immunoglobulin heavy chains in a pre-B cell line. J Cell Biol 115:619–624.

Balch WE (1989): Biochemistry of interorganelle transport. A new frontier in enzymology emerges from versatile in vitro model systems. J Biol Chem 264:16965–16968.

Basson ME, Thorsness M, Finer MJ, Stroud RM, Rine J (1988): Structural and functional conservation between yeast and human 3-hydroxy-3-methylglutaryl coenzyme A reductases, the rate-limiting enzymes of sterol biosynthesis. Mol Cell Biol 8:3797–3808.

Bienkowski RS (1983): Intracellular degradation of newly synthesized secretory proteins. Biochem J 214:1–10.

Blount P, Merlie JP (1990): Mutational analysis of muscle nicotinic acetylcholine receptor subunit assembly. J Cell Biol 111:2613–2622.

Blumberg RS, Alarcon D, Sancho J, McDermott FV, Lopez P, Breitmeyer J, Terhorst C (1990a): Assembly and function of the T cell antigen receptor. Requirement of either the lysine or arginine residues in the transmembrane region of the alpha chain. J Biol Chem 265:14036–14043.

Blumberg RS, Ley S, Sancho J, Lonberg N, Lacy E, McDermott F, Schad V, Greenstein JL, Terhorst C (1990a): Structure of the T-cell antigen receptor: Evidence for two CD3 epsilon subunits in the T-cell receptor–CD3 complex. Proc Natl Acad Sci USA 87:7220–7224.

Bonifacino JS, Cosson P, Klausner RD (1990a): Colocalized transmembrane determinants for ER degradation and subunit assembly explain the intracellular fate of TCR chains. Cell 63:503–513.

Bonifacino JS, Suzuki CK, Klausner RD (1990b): A peptide sequence confers retention and rapid degradation in the endoplasmic reticulum. Science 247:79–82.

Bonifacino JS, Cosson P, Shah N, Klausner RD (1991): Role of potentially charged transmembrane residues in targeting proteins for retention and degradation within the endoplasmic reticulum. EMBO J 10:2783–2793.

Bonifacino JS, Lippincott-Schwartz J, Chen C, Antusch D, Samelson LE, Klausner RD (1988): Association and dissociation of the murine T cell receptor associated protein (TRAP). Early events in the biosynthesis of a multisubunit receptor. J Biol Chem 263:8965–8971.

Bonifacino JS, Suzuki CK, Lippincott-Schwartz J, Weissman AM, Klausner RD (1989): Pre-Golgi degradation of newly synthesized T-cell antigen receptor chains: Intrinsic sensitivity and the role of subunit assembly. J Cell Biol 109:73–83.

Borchardt RA, Davis RA (1987): Intrahepatic assembly of very low density lipoproteins. Rate of transport out of the endoplasmic reticulum determines rate of secretion. J Biol Chem 262:16394–16402.

Borën J, Wettesten M, Sjöberg A, Thorlin T, Bondjers G, Wiklund O, Olofsson SO (1990): The assembly and secretion of apoB 100 containing lipoproteins in Hep G2 cells. Evidence for different sites for protein synthesis and lipoprotein assembly. J Biol Chem 265:10556–10564.

Bossie MA, Martin CE (1989): Nutritional regulation of yeast delta-9 fatty acid desaturase activity. J Bacteriol 171:6409–6413.

Brown MS, Goldstein JL (1980): Multivalent feedback regulation of HMG CoA reductase, a control mechanism coordinating isoprenoid synthesis and cell growth. J Lipid Res 21:505–517.

Chang TY, Limanek JS, Chang CC (1981): Evidence indicating that inactivation of 3-hydroxy-3-methylglutaryl coenzyme A reductase by low density lipoprotein or by 25-hydroxycholesterol requires mediator protein(s) with rapid turnover rate. J Biol Chem 256:6174–6180.

Chen C, Bonifacino JS, Yuan LC, Klausner RD (1988): Selective degradation of T cell antigen receptor chains retained in a pre-Golgi compartment. J Cell Biol 107:2149–2161.

Chen MY, Maldarelli F, Karczewski MK, Willey RL, Strebel K (1993): Human immunodeficiency virus type 1 vpu protein induces degradation of CD4 in vitro: The cytoplasmic domain of CD4 contributes to vpu sensitivity. J Virol 67:3877–3884.

Chen SH, Habib G, Yang CY, Gu ZW, Lee BR, Weng SA, Silberman SR, Cai SJ, Deslypere JP, Rosseneu M, et al. (1987): Apolipoprotein B-48 is the product of a messenger RNA with an organ-specific in-frame stop codon. Science 238:363–366.

Chen HW, Richards BA, Kandutsch AA (1982): Inhibition of protein synthesis blocks the response to 25-hydroxycholesterol by inhibiting degradation of hydroxymethylglutaryl-CoA reductase. Biochim Biophys Acta 712:484–489.

Cheng SH, Gregory RJ, Marshall J, Paul S, Souza DW, White GA, O'Riordan CR, Smith AE (1990): Defective intracellular transport and processing of CFTR is the molecular basis of most cystic fibrosis. Cell 63:827–834.

Chin DJ, Gil G, Faust JR, Goldstein JL, Brown MS, Luskey KL (1985): Sterols accelerate degradation of hamster 3-hydroxy-3-methylglutaryl coenzyme A reductase encoded by a constitutively expressed cDNA. Mol Cell Biol 5:634–641.

Chuck SL, Yao Z, Blackhart BD, McCarthy BJ, Lingappa VR (1990): New variation on the translocation of proteins during early biogenesis of apolipoprotein B. Nature 346:382–385.

Chun KT, Bar-Nun S, Simoni RD (1990): The regulated degradation of 3-hydroxy-3-methylglutaryl-CoA reductase requires a short-lived protein and occurs in the endoplasmic reticulum. J Biol Chem 265:22004–22010.

Cosson P, Lankford SP, Bonifacino JS, Klausner RD (1991): Membrane protein association by potential intramembrane charge pairs. Nature 351:414–416.

Cotner T (1992): Unassembled HLA-DR beta monomers are degraded rapidly by a nonlysosomal mechanism. J Immunol 148:2163–2168.

Coukell MB, Cameron AM, Adames NR (1992): Involvement of intracellular calcium in protein secretion in *Dictyostelium discoideum*. J Cell Sci 103:371–380.

Crise B, Buonocore L, Rose JK (1990): CD4 is retained in the endoplasmic reticulum by the human immunodeficiency virus type 1 glycoprotein precursor. J Virol 64:5585–5593.

Dalgleish AG, Beverley PC, Clapham PR, Crawford DH, Greaves MF, Weiss RA (1984): The CD4 (T4) antigen is an essential component of the receptor for the AIDS retrovirus. Nature 312:763–767.

Danishefsky K, Hartwig R, Banerjee D, Redman C (1990): Intracellular fate of fibrinogen B beta chain expressed in COS cells. Biochim Biophys Acta 1048:202–208.

Davis RA, Thrift RN, Wu CC, Howell KE (1990): Apolipoprotein B is both integrated into and translocated across the endoplasmic reticulum membrane. Evidence for two functionally distinct pools. J Biol Chem 265:10005–10011.

de la Hera A, Müller U, Olsson C, Isaaz S, Tunnacliffe A (1991): Structure of the T cell antigen receptor (TCR): Two CD3 epsilon subunits in a functional TCR/CD3 complex. J Exp Med 173:7–17.

Dixon JL, Furukawa S, Ginsberg HN (1991): Oleate stimulates secretion of apolipoprotein B-containing lipoproteins from Hep G2 cells by inhibiting early intracellular degradation of apolipoprotein B. J Biol Chem 266:5080–5086.

Doyle C, Sambrook J, Gething MJ (1986): Analysis of progressive deletions of the transmembrane and cytoplasmic domains of influenza hemagglutinin. J Cell Biol 103:1193–1204.

Dulis BH, Kloppel TM, Grey HM, Kubo RT (1982): Regulation of catabolism of IgM heavy chains in a B lymphoma cell line. J Biol Chem 257:4369–4374.

Edwards PA, Gould RG (1972): Turnover rate of hepatic 3-hydroxy-3-methylglutaryl coenzyme A reductase as determined by use of cycloheximide. J Biol Chem 247:1520–1524.

Edwards PA, Lan SF, Fogelman AM (1983a): Alterations

in the rates of synthesis and degradation of rat liver 3-hydroxy-3-methylglutaryl coenzyme A reductase produced by cholestyramine and mevinolin. J Biol Chem 258:10219–10222.

Edwards PA, Lan SF, Tanaka RD, Fogelman AM (1983b): Mevalonolactone inhibits the rate of synthesis and enhances the rate of degradation of 3-hydroxy-3-methylglutaryl coenzyme A reductase in rat hepatocytes. J Biol Chem 258:7272–7275.

Eliasson E, Mkrtchian S, Ingelman SM (1992): Hormone- and substrate-regulated intracellular degradation of cytochrome P450 (2E1) involving MgATP-activated rapid proteolysis in the endoplasmic reticulum membranes. J Biol Chem 267:15765–15769.

Erickson RH, Suzuki Y, Sedlmayer A, Kim YS (1992): Biosynthesis and degradation of altered immature forms of intestinal dipeptidyl peptidase IV in a rat strain lacking the enzyme. J Biol Chem 267:21623–21629.

Esser V, Russell DW (1988): Transport-deficient mutations in the low density lipoprotein receptor. Alterations in the cysteine-rich and cysteine-poor regions of the protein block intracellular transport. J Biol Chem 263:13276–13281.

Exley M, Terhorst C, Wileman T (1991): Structure, assembly and intracellular transport of the T cell receptor for antigen. Semin Immunol 3:283–297.

Faust JR, Luskey KL, Chin DJ, Goldstein JL, Brown MS (1982): Regulation of synthesis and degradation of 3-hydroxy-3-methylglutaryl-coenzyme A reductase by low density lipoprotein and 25-hydroxycholesterol in UT-1 cells. Proc Natl Acad Sci USA 79:5205–5209.

Finger A, Knop M, Wolf DH (1993): Analysis of two mutated vacuolar proteins reveals a degradation pathway in the ER or an ER-related compartment of yeast. Eur J Biochem 218:565–574.

Fiskum G, Craig SW, Decker GL, Lehninger AL (1980): The cytoskeleton of digitonin-treated rat hepatocytes. Proc Natl Acad Sci USA 77:3430–3434.

Furukawa S, Sakata N, Ginsberg HN, Dixon JL (1992): Studies of the sites of intracellular degradation of apolipoprotein B in Hep G2 cells. J Biol Chem 267:22630–22638.

Gil G, Faust JR, Chin DJ, Goldstein JL, Brown MS (1985): Membrane-bound domain of HMG CoA reductase is required for sterol-enhanced degradation of the enzyme. Cell 41:249–258.

Graham KS, Le A, Sifers RN (1990): Accumulation of the insoluble PiZ variant of human alpha 1-antitrypsin within the endoplasmic reticulum does not elevate the steady-state level of grp 78/BiP. J Biol Chem 265:20463–20468.

Hannum C, Marrack P, Kubo R, Kappler J (1987): Thymocytes with the predicted properties of pre-T cells. J Exp Med 166:874–889.

Hoe MH, Hunt RC (1992): Loss of one asparagine-linked oligosaccharide from human transferrin receptors results in specific cleavage and association with the endoplasmic reticulum. J Biol Chem 267:4916–4923.

Hurtley SM, Helenius A (1989): Protein oligomerization in the endoplasmic reticulum. Annu Rev Cell Biol 5:277–307.

Inoue S, Bar-Nun S, Roitelman J, Simoni RD (1991): Inhibition of degradation of 3-hydroxy-3-methylglutaryl-coenzyme A reductase in vivo by cysteine protease inhibitors. J Biol Chem 266:13311–13317.

Inoue S, Simoni RD (1992): 3-Hydroxy-3-methylglutaryl-coenzyme A reductase and T cell receptor alpha subunit are differentially degraded in the endoplasmic reticulum. J Biol Chem 267:9080–9086.

Jabbar MA, Nayak DP (1990): Intracellular interaction of human immunodeficiency virus type 1 (ARV-2) envelope glycoprotein gp160 with CD4 blocks the movement and maturation of CD4 to the plasma membrane. J Virol 64:6297–6304.

Jin YJ, Koyasu S, Moingeon P, Steinbrich R, Tarr GE, Reinherz EL (1990: A fraction of CD3ε subunits exists as disulfide-linked dimers in both human and murine T lymphocytes. J Biol Chem 265:15850–15853.

Jingami H, Brown MS, Goldstein JL, Anderson RG, Luskey KL (1987): Partial deletion of membrane-bound domain of 3-hydroxy-3-methylglutaryl coenzyme A reductase eliminates sterol-enhanced degradation and prevents formation of crystalloid endoplasmic reticulum. J Cell Biol 104:1693–1704.

John S, Banting GS, Goodfellow PN, Owen MJ (1989): Surface expression of the T cell receptor complex requires charged residues within the alpha chain transmembrane region. Eur J Immunol 19:335–339.

Kappes DJ, Tonegawa S (1991): Surface expression of alternative forms of the TCR/CD3 complex. Proc Natl Acad Sci USA 88:10619–10623.

Klatzmann D, Champagne E, Chamaret S, Gruest J, Guetard D, Hercend T, Gluckman JC, Montagnier L (1984): T-lymphocyte T4 molecule behaves as the receptor for human retrovirus LAV. Nature 312:767–768.

Klausner RD, Donaldson JG, Lippincott-Schwartz J (1992): Brefeldin A: Insights into the control of membrane traffic and organelle structure. J Cell Biol 116:1071–1080.

Klausner RD, Lippincott-Schwartz J, Bonifacino JS (1990): The T cell antigen receptor: Insights into organelle biology. Annu Rev Cell Biol 6:403–431.

Koppelman B, Cresswell P (1990): Rapid nonlysosomal degradation of assembled class II glycoproteins incorporating a mutant DR alpha chain. J Immunol 145:2730–2736.

Kurosaki T, Gander I, Ravetch JV (1991): A subunit common to an IgG Fc receptor and the T-cell receptor mediates assembly through different interactions. Proc Natl Acad Sci USA 88:3837–3841.

Lankford SP, Cosson P, Bonifacino JS, Klausner RD (1993): Transmembrane domain length affects charge-mediated retention and degradation of proteins within the endoplasmic reticulum. J Biol Chem 268:4814–4820.

Lau MM, Neufeld EF (1989): A frameshift mutation in a

patient with Tay-Sachs disease causes premature termination and defective intracellular transport of the alpha-subunit of beta-hexosaminidase. J Biol Chem 264:21376–21380.

Le A, Ferrell GA, Dishon DS, Le QQ, Sifers RN (1992): Soluble aggregates of the human PiZ alpha 1-antitrypsin variant are degraded within the endoplasmic reticulum by a mechanism sensitive to inhibitors of protein synthesis. J Biol Chem 267:1072–1080.

Le A, Graham KS, Sifers RN (1990): Intracellular degradation of the transport-impaired human PiZ alpha 1-antitrypsin variant. Biochemical mapping of the degradative event among compartments of the secretory pathway. J Biol Chem 265:14001–14007.

Lederkremer GZ, Lodish HF (1991): An alternatively spliced miniexon alters the subcellular fate of the human asialoglycoprotein receptor H2 subunit. Endoplasmic reticulum retention and degradation or cell surface expression. J Biol Chem 266:1237–1244.

Leonard DA, Chen HW (1987): ATP-dependent degradation of 3-hydroxy-3-methylglutaryl coenzyme A reductase in permeabilized cells. J Biol Chem 262:7914–7919.

Libby P, Goldberg AL (1978): Leupeptin, a protease inhibitor, decreases protein degradation in normal and diseased muscles. Science 199:534–536.

Lippincott-Schwartz J (1994): The endoplasmic reticulum-Golgi membrane system. In The Liver: Biology and Pathology. New York: Raven Press (in press).

Lippincott-Schwartz J, Bonifacino JS, Yuan LC, Klausner RD (1988): Degradation from the endoplasmic reticulum: disposing of newly synthesized proteins. Cell 54:209–220.

Liscum L, Finer MJ, Stroud RM, Luskey KL, Brown MS, Goldstein JL (1985): Domain structure of 3-hydroxy-3-methylglutaryl coenzyme A reductase, a glycoprotein of the endoplasmic reticulum. J Biol Chem 260:522–530.

Lobell RB, Arm JP, Raizman MB, Austen KF, Katz HR (1993): Intracellular degradation of Fc gamma RIII in mouse bone marrow culture-derived progenitor mast cells prevents its surface expression and associated function. J Biol Chem 268:1207–1212.

Luskey KL (1988): Regulation of cholesterol synthesis: Mechanism for control of HMG CoA reductase. Recent Prog Horm Res 44:35–51.

Maldarelli F, Chen MY, Willey RL, Strebel K (1993): Human immunodeficiency virus type 1 Vpu protein is an oligomeric type I integral membrane protein. J Virol 67:5056–5061.

Manolios N, Bonifacino JS, Klausner RD (1990): Transmembrane helical interactions and the assembly of the T cell receptor complex. Science 249:274–277.

McCracken AA, Karpichev IV (1992): Isolation of yeast mutants deficient in the "ER" degradation pathway. Mol Biol Cell 3:34a.

McCracken AA, Kruse KB (1993): Selective protein degradation in the yeast exocytic pathway. Mol Biol Cell 4:729–736.

Mehdi S (1991): Cell-penetrating inhibitors of calpain. Trends Biochem Sci 16:150–153.

Meigs TE, Simoni RD (1992): Regulated degradation of 3-hydroxy-3-methylglutaryl-coenzyme A reductase in permeabilized cells. J Biol Chem 267:13547–13552.

Minami Y, Weissman AM, Samelson LE, Klausner RD (1987): Building a multichain receptor: Synthesis, degradation, and assembly of the T-cell antigen receptor. Proc Natl Acad Sci USA 84:2688–2692.

Mitsuhashi W, Minamikawa T (1988): Synthesis and post-translational activation of sulfhydryl-endopeptidase in cotyledons of germinating vigna mungo seeds. Plant Physiol 89:274–279.

Moberly JB, Cole TG, Alpers DH, Schonfeld G (1990): Oleic acid stimulation of apolipoprotein B secretion from HepG2 and Caco-2 cells occurs post-transcriptionally. Biochim Biophys Acta 1042:70–80.

Moore SEH, Spiro RG (1993): Inhibition of glucose trimming by castanospermine results in rapid degradation of unassembled major histocompatibility complex class I molecules. J Biol Chem 268:3809–3812.

Morley BJ, Chin KN, Newton ME, Weiss A (1988): The lysine residue in the membrane-spanning domain of the beta chain is necessary for cell surface expression of the T cell antigen receptor. J Exp Med 168:1971–1978.

Morrissey JJ, Cohn DV (1979): Secretion and degradation of parathormone as a function of intracellular maturation of hormone pools. Modulation by calcium and dibutyryl cyclic AMP. J Cell Biol 83:521–528.

Mumford RA, Strauss AW, Powers JC, Pierzchala PA, Nishino N, Zimmerman M (1980): A zinc metalloendopeptidase associated with dog pancreatic membranes. J Biol Chem 255:2227–2230.

Nakanishi M, Goldstein JL, Brown MS (1988): Multivalent control of 3-hydroxy-3-methylglutaryl coenzyme A reductase. Mevalonate-derived product inhibits translation of mRNA and accelerates degradation of enzyme. J Biol Chem 263:8929–8937.

Napier RM, Fowke LC, Hawes C, Lewis M, Pelham HR (1992): Immunological evidence that plants use both HDEL and KDEL for targeting proteins to the endoplasmic reticulum. J Cell Sci 102:261–271.

Nilsson T, Jackson MR, Peterson PA (1989): Short cytoplasmic sequences serve as retention signals for transmembrane proteins in the endoplasmic reticulum. Cell 58:707–718.

Noda T, Farquhar MG (1992): A non-autophagic pathway for diversion of ER secretory proteins to lysosomes. J Cell Biol 119:85–97.

Ohkuma S, Chudzik J, Poole B (1986): The effects of basic substances and acidic ionophores on the digestion of exogenous and endogenous proteins in mouse peritoneal macrophages. J Cell Biol 102:959–9566.

Ohkuma S, Poole B (1978): Fluorescence probe measurement of the intralysosomal pH in living cells and the perturbation of pH by various agents. Proc Natl Acad Sci USA 75:3327–3331.

Olofsson SO, Bjursell G, Boström K, Carlsson P, Elovson J, Protter AA, Reuben MA, Bondjers G (1987): Apolipoprotein B: Structure, biosynthesis and role in the lipoprotein assembly process. Atherosclerosis 68:1–17.

Omura F, Otsu M, Kikuchi M (1992a): Accelerated secretion of human lysozyme with a disulfide bond mutation. Eur J Biochem 205:551–559.

Omura F, Otsu M, Yoshimori T, Tashiro Y, Kikuchi M (1992b): Non-lyosomal degradation of misfolded human lysozymes with and without an asparagine-linked glycosylation site. Eur J Biochem 210:591–599.

Omura T, Siekevitz P, Palade GE (1967): Turnover of constituents of the endoplasmic reticulum membranes of rat hepatocytes. J Biol Chem 242:2389–2396.

Orci L, Brown MS, Goldstein JL, Garcia SL, Anderson RG (1984): Increase in membrane cholesterol: A possible trigger for degradation of HMG CoA reductase and crystalloid endoplasmic reticulum in UT-1 cells. Cell 36:835–8345.

Oshino N, Sato R (1972): The dietary control of the microsomal stearyl CoA desaturation enzyme system in rat liver. Arch Biochem Biophys 149:369–377.

Pelham HR (1989): The selectivity of secretion: Protein sorting in the endoplasmic reticulum. Biochem Soc Trans 17:795–802.

Powell LM, Wallis SC, Pease RJ, Edwards YH, Knott TJ, Scott J (1987): A novel form of tissue-specific RNA processing produces apolipoprotein-B48 in intestine. Cell 50:831–840.

Pryer NK, Wuestehube LJ, Schekman R (1992): Vesicle-mediated protein sorting. Annu Rev Biochem 61:471–516.

Pullinger CR, North JD, Teng BB, Rifici VA, Ronhild de Brito AE, Scott J (1989): The apolipoprotein B gene is constitutively expressed in HepG2 cells: Regulation of secretion by oleic acid, albumin, and insulin, and measurement of the mRNA half-life. J Lipid Res 30:1065–1077.

Rapoport T (1992): Transport of proteins across the endoplasmic reticulum membrane. Science 258:931–936.

Roitelman J, Bar-Nun NS, Inoue S, Simoni RD (1991): Involvement of calcium in the mevalonate-accelerated degradation of 3-hydroxy-3-methylglutaryl-CoA reductase. J Biol Chem 266:16085–16091.

Roitelman J, Olender EH, Bar-Nun NS, Dunn WJ, Simoni RD (1992): Immunological evidence for eight spans in the membrane domain of 3-hydroxy-3-methylglutaryl coenzyme A reductase: Implications for enzyme degradation in the endoplasmic reticulum. J Cell Biol 117:959–973.

Rose JK, Doms RW (1988): Regulation of protein export from the endoplasmic reticulum. Annu Rev Cell Biol 4:257–288.

Rotundo RL (1988): Biogenesis of acetylcholinesterase molecular forms in muscle. Evidence for a rapidly turning over, catalytically inactive precursor pool. J Biol Chem 263:19398–19406.

Rotundo RL, Thomas K, Porter JK, Benson RJ, Fernandez VC, Fine RE (1989): Intracellular transport, sorting, and turnover of acetylcholinesterase. Evidence for an endoglycosidase H-sensitive form in Golgi apparatus, sarcoplasmic reticulum, and clathrin-coated vesicles and its rapid degradation by a non-lysosomal mechanism. J Biol Chem 264:3146–3152.

Roy S, Yu S, Banerjee D, Overton O, Mukhopadhyay G, Oddoux C, Grieninger G, Redman C (1992): Assembly and secretion of fibrinogen. Degradation of individual chains. J Biol Chem 267:23151–23158.

Samelson LE, Harford JB, Klausner RD (1985): Identification of the components of the murine T cell antigen receptor complex. Cell 43:223–231.

Sanders SL, Schekman R (1992): Polypeptide translocation across the endoplasmic reticulum membrane. J Biol Chem 267:13791–13794.

Sato R, Imanaka T, Takatsuki A, Takano T (1990): Degradation of newly synthesized apolipoprotein B-100 in a pre-Golgi compartment. J Biol Chem 265:11880–11884.

Schimke RT (1975): Turnover of membrane proteins in animal cells. Methods Membrane Biol 3:201–236.

Seglen PO (1983): Inhibitors of lysosomal function. Methods Enzymol 96:737–764.

Sengstag C, Stirling C, Schekman R, Rine J (1990): Genetic and biochemical evaluation of eucaryotic membrane protein topology: Multiple transmembrane domains of Saccharomyces cerevisiae 3-hydroxy-3-methylglutaryl coenzyme A reductase. Mol Cell Biol 10:672–680.

Shachar I, Amitay R, Rabinovich E, Haimovich J, Bar-Nun NS (1992): Polymerization of secretory IgM in B lymphocytes is prevented by a prior targeting to a degradation pathway. J Biol Chem 267:24241–24247.

Shelness GS, Kanwar YS, Blobel G (1988): cDNA-derived primary structure of the glycoprotein component of canine microsomal signal peptidase complex. J Biol Chem 263:17063–17070.

Shin J, Dunbrack RL, Strominger JL (1991): Signals for retention of transmembrane proteins in the endoplasmic reticulum studied with CD4 truncation mutants. Proc Natl Acad Sci USA 88:1918–1922.

Shin J, Lee S, Strominger JL (1993): Translocation of TCRα chains into the lumen of the endoplasmic reticulum and their degradation. Science 259:1901–1904.

Sidman C (1981): B lymphocyte differentiation and the control of IgM mu chain expression. Cell 23:379–389.

Sitia R, Neuberger MS, Milstein C (1987): Regulation of membrane IgM expression in secretory B cells: Translational and post-translational events. EMBO J 6:3969–3977.

Skalnik DG, Narita H, Kent C, Simoni RD (1988): The membrane domain of 3-hydroxy-3-methylglutaryl-coenzyme A reductase confers endoplasmic reticulum localization and sterol-regulated degradation onto beta-galactosidase. J Biol Chem 263:6836–6841.

Sommer T, Jentsch S (1993): A protein translocation defect linked to ubiquitin conjugation at the endoplasmic reticulum. Nature 365:176–179.

Sparks JD, Sparks CE (1990): Insulin modulation of hepatic synthesis and secretion of apolipoprotein B by rat hepatocytes. J Biol Chem 265:8854–8862.

Stafford FJ, Bonifacino JS (1991): A permeabilized cell system identifies the endoplasmic reticulum as a site of protein degradation. J Cell Biol 115:1225–1236.

Stoller TJ, Shields D (1989): The propeptide of preprosomatostatin mediates intracellular transport and secretion of α-globin from mammalian cells. J Cell Biol 108:1647–1655.

Strebel K, Klimkait T, Maldarelli F, Martin MA (1989): Molecular and biochemical analyses of human immunodeficiency virus type I vpu protein. J Virol 63:3784–3791.

Strebel K, Klimkait T, Martin MA (1988): A novel gene of HIV-1, vpu, and its 16-kilodalton product. Science 241:1221–1223.

Su K, Stoller T, Rocco J, Zemsky J, Green R (1993): Pre-Golgi degradation of yeast prepro-alpha-factor expressed in a mammalian cell. J Biol Chem 268:14301–14309.

Sussman JJ, Bonifacino JS, Lippincott-Schwartz J, Weissman AM, Saito T, Klausner RD, Ashwell JD (1988): Failure to synthesize the T cell CD3-zeta chain: Structure and function of a partial T cell receptor complex. Cell 52:85–95.

Tan L, Turner J, Weiss A (1991): Regions of the T cell receptor alpha and beta chains that are responsible for interactions with CD3. J Exp Med 173:1247–1256.

Tanaka RD, Li AC, Fogelman AM, Edwards PA (1986): Inhibition of lysosomal protein degradation inhibits the basal degradation of 3-hydroxy-3-methylglutaryl coenzyme A reductase. J Lipid Res 27:261–273.

Thrift RN, Drisko J, Dueland S, Trawick JD, Davis RA (1992): Translocation of apolipoprotein B across the endoplasmic reticulum is blocked in a nonhepatic cell line. Proc Natl Acad Sci USA 89:9161–9165.

Tierney DJ, Haas AL, Koop DR (1992): Degradation of cytochrome P450 2E1: Selective loss after labilization of the enzyme. Arch Biochem Biophys 293:9–16.

Tosetti F, Ferrari N, Pfeffer U, Brigati C, Vidali G (1992): Regulation of plasma retinol binding protein secretion in human HepG2 cells. Exp Cell Res 200:467–472.

Tsao YS, Ivessa NE, Adesnik M, Sabatini DD, Kreibich G (1992): Carboxy terminally truncated forms of ribophorin I are degraded in pre-Golgi compartments by a calcium-dependent process. J Cell Biol 116:57–67.

Tsuji E, Misumi Y, Fujiwara T, Takami N, Ogata S, Ikehara Y (1992): An active-site mutation (Gly633→Arg) of dipeptidyl peptidase IV causes its retention and rapid degradation in the endoplasmic reticulum. Biochemistry 31:11921–11927.

Urade R, Kito M (1992): Inhibition by acidic phospholipids of protein degradation by ER-60 protease, a novel cysteine protease, of endoplasmic reticulum. FEBS Lett 312:83–86.

Urade R, Nasu M, Moriyama T, Wada K, Kito M (1992): Protein degradation by the phosphoinositide-specific phospholipase C-alpha family from rat liver endoplasmic reticulum. J Biol Chem 267:15152–15159.

Veronese FD, DeVico AL, Copeland TD, Oroszlan S, Gallo RC, Sarngadharan MG (1985): Characterization of gp41 as the transmembrane protein coded by the HTLV-III/LAV envelope gene. Science 229:1402–1405.

Vertel BM, Walters LM, Mills D (1992): Subcompartments of the endoplasmic reticulum. Semin Cell Biol 3:325–341.

Vincent MJ, Raja NU, Jabbar MA (1993): Human immunodeficiency virus type 1 protein induces degradation of chimeric envelope glycoproteins bearing the cytoplasmic and anchor domains of CD4: Role of the cytoplasmic domain in Vpu-induced degradation in the endoplasmic reticulum. J Virol 67:5538–5549.

White AL, Graham DL, LeGros J, Pease RJ, Scott J (1992): Oleate-mediated stimulation of apolipoprotein B secretion from rat hepatoma cells. A function of the ability of apolipoprotein B to direct lipoprotein assembly and escape presecretory degradation. J Biol Chem 267:15657–15664.

Wikström L, Lodish HF (1991): Nonlysosomal, pre-Golgi degradation of unassembled asialoglycoprotein receptor subunits: A TLCK- and TPCK-sensitive cleavage within the ER. J Cell Biol 113:997–1007.

Wikström L, Lodish HF (1992): Endoplasmic reticulum degradation of a subunit of the asialoglycoprotein receptor in vitro. Vesicular transport from endoplasmic reticulum is unnecessary. J Biol Chem 267:5–8.

Wikström L, Lodish HF (1993): Unfolded H2b asialoglycoprotein receptor subunit polypeptides are selectively degraded within the endoplasmic reticulum. J Biol Chem 268:14412–14416.

Wileman T, Carson GR, Concino M, Ahmed A, Terhorst C (1990a): The gamma and epsilon subunits of the CD3 complex inhibit pre-Golgi degradation of newly synthesized T cell antigen receptors. J Cell Biol 110:973–986.

Wileman T, Carson GR, Shih FF, Concino MF, Terhorst C (1990b): The transmembrane anchor of the T-cell antigen receptor beta chain contains a structural determinant of pre-Golgi proteolysis. Cell Regul 1:907–919.

Wileman T, Kane LP, Terhorst C (1991): Degradation of T-cell receptor chains in the endoplasmic reticulum is inhibited by inhibitors of cysteine proteases. Cell Regul 2:753–765.

Wileman T, Kane LP, Young J, Carson GR, Terhorst C (1993): Associations between subunit ectodomains promote T cell antigen receptor assembly and protect against degradation in the ER. J Cell Biol 122:67–78.

Wileman T, Pettey C, Terhorst C (1990c): Recognition for degradation in the endoplasmic reticulum and lysosomes prevents the transport of single TCR beta and CD3 delta subunits of the T-cell antigen receptor to the surface of cells. Int Immunol 2:743–754.

Willey RL, Maldarelli F, Martin MA, Strebel K (1992a): Human immunodeficiency virus type 1 Vpu protein induces rapid degradation of CD4. J Virol 66:7193–7200.

Willey RL, Maldarelli F, Martin MA, Strebel K (1992b): Human immunodeficiency virus type 1 Vpu protein regulates the formation of intracellular gp160–CD4 complexes. J Virol 66:226–234.

Williams GT, Venkitaraman AR, Gilmore DJ, Neuberger MS (1990): The sequence of the mu transmembrane segment determines the tissue specificity of the transport of immunoglobulin M to the cell surface. J Exp Med 171:947–952.

Yamamoto K (1992): Synthesis and localization of cathepsin E. ICOP Newslett (January issue), pp 1–2.

Yang B, Hoe MH, Black P, Hunt RC (1993): Role of oligosaccharides in the processing and function of human transferrin receptors. Effect of the loss of the three N-glycosyl oligosaccharides individually or together. J Biol Chem 268:7435–7441.

Yao ZM, Blackhart BD, Linton MF, Taylor SM, Young SG, McCarthy BJ (1991): Expression of carboxyl-terminally truncated forms of human apolipoprotein B in rat hepatoma cells. Evidence that the length of apolipoprotein B has a major effect on the buoyant density of the secreted lipoproteins. J Biol Chem 266:3300–3308.

Young J, Kane LP, Exley M, Wileman T (1993): Regulation of selective protein degradation in the endoplasmic reticulum by redox potential. J Biol Chem 268:19810–19818.

ABOUT THE AUTHORS

RICHARD D. KLAUSNER is the Chief of the Cell Biology and Metabolism Branch, National Institute of Child Health and Human Development, National Institutes of Health, where he investigates various aspects of intracellular protein trafficking and metal metabolism. Dr. Klausner received a Bachelor of Science degree *summa cum laude* from Yale University with honors in molecular biophysics and biochemistry. He received a medical degree from Duke University and trained in internal medicine at Massachusetts General Hospital. A member of numerous panels and boards, Dr. Klausner has been executive editor of *Analytical Biochemistry* and a member of the editorial boards of *Cell, Journal of Cell Biology, Chemistry and Biology, The New Biologist,* and *The Annual Review of Cell Biology*. In 1993, he was elected to the National Academy of Sciences and currently chairs the National Committee on Science Education Standards and Assessment. He is currently the president of the American Society of Clinical Investigation.

JUAN S. BONIFACINO is an investigator at the Cell Biology and Metabolism Branch, National Institute of Child Health and Human Development, National Institutes of Health, where he conducts research on intracellular protein trafficking in the secretory pathway. He did his undergraduate and graduate studies at the University of Buenos Aires, Argentina, where he pursued doctoral research in the laboratory of Alejandro Paladini. In 1981, he earned a doctoral degree in biochemistry for work on the characteristics of prolactin and growth hormone receptors in mammalian tissues. In 1982, he moved to the NIH, where he did postdoctoral research in the laboratories of Maria Dufau, Ignacio Sandoval, and Richard Klausner on various aspects of the structure, intracellular transport and fate of integral membrane proteins. His current research focuses on the mechanisms involved in protein localization and sorting in late compartments of the Golgi system.

PROTEOLYSIS RELATED TO ANTIGEN PRESENTATION

Protein Catabolism and Antigen Processing

Clifford V. Harding

INTRODUCTION

Antigen processing forms the basis for recognition of antigenic epitopes by T lymphocytes, and a central feature of this mechanism is the proteolysis of antigens to produce the immunogenic peptides that are recognized by T cells. Whereas B lymphocytes express an antigen receptor (immunoglobulin) that recognizes antigens in their native conformation, T cells do not recognize conformational aspects of antigen structure. Instead, T cells recognize antigens as short, linear peptides associated with major histocompatibility complex (MHC) proteins; these peptide–MHC complexes are "presented" by antigen-presenting cells (APC). Antigen processing may be defined as the conversion of native protein antigens to peptide–MHC complexes, which are displayed on the plasma membrane, available for T-cell recognition.

Peptide–MHC complexes on APC are recognized by the clonotypic T-cell antigen receptor (TCR), which is specific for the combination of a particular antigenic peptide bound to a particular MHC allele. T cells are remarkably sensitive and can recognize as few as 100–300 specific peptide–MHC complexes [Harding and Unanue, 1990b; Demotz et al., 1990]. This molecular recognition event, accompanied by appropriate interactions between adhesion molecules and "costimulator" molecules on the APC and T cell, leads to T-cell activation events including proliferation and cytokine secretion. The resulting cytokine effects can provide "help" for B-cell responses (immunoglobulin production) and activate or recruit other leukocytes, such as macrophages, to mediate important functions of cellular immunity. In addition, cytotoxic T cells can attack and kill cells expressing foreign peptide–MHC complexes, which may be infected by viruses or intracellular bacteria. Thus, T-cell responses are central to specific immunity, both cellular and humoral, to protein antigens (which are termed *T-dependent antigens*), and antigen processing lies at the heart of this recognition pathway. Carbohydrate antigens, which do not bind to MHC molecules [Harding et al., 1991c, Ishioka et al., 1992] are T independent and are not processed by these mechanisms, although glycopeptides can bind to MHC molecules for presentation to T cells [Ishioka et al., 1992; Harding et al., 1994].

Two types of MHC molecules exist, class I MHC (MHC-I) and class II MHC (MHC-II). These classes differ in both structure and function, although they also share certain important features. The mechanisms of antigen processing also differ significantly for the two classes, and the distinct mechanisms of the class I pathway and class II antigen processing pathways are discussed below. Both pathways appear to utilize mechanisms of protein degradation that have general utility in cell physiology (which have been described in the preceding chapters), as opposed to a completely separate system of proteolysis specifically for antigen processing. However, there are also specialized molecular mechanisms involved in the intracellular transport of antigens and MHC molecules that adapt and harness these basic degradative mechanisms to achieve antigen presentation. MHC-II molecules are transported to late endocytic compartments, where they bind peptides catabolically derived from exogenous (extracellular) antigens internalized by endocytosis or phagocytosis. The resulting peptide–MHC-II complexes are then transported to the plasma membrane, where they are available for recognition by T cells. In contrast, peptides presented by MHC-I molecules are proteolytically generated in the cytosol and are then transported by a recently characterized peptide transporter into the ER, where they bind the MHC-I molecules. Thus, MHC-II molecules present antigenic peptides derived from proteins that target to endocytic compartments, whereas MHC-I molecules generally present proteins that are synthesized in the cytosol or penetrate into the cytosol [Morrison et al., 1986; Bevan, 1987; Germain, 1986].

While this general rule describes the major routes of processing, some exceptions imply interesting alternative processing pathways. For example, MHC-I molecules can present exogenous antigens that have no known mechanism for escaping from endocytic organelles into the cytosol [Pfeifer et al., 1993; Collins et al., 1992; Carbone and Bevan, 1990; Debrick et al., 1991; Reddy et al., 1991; Gooding, 1980; Bevan, 1976; Grant and Rock, 1992; Rock et al., 1990, 1993; Aggarwal et al., 1990; Flynn et al., 1990], at least in some cells [Rock et al., 1993]. MHC-II molecules may also present cytosolic proteins [Malnati et al., 1992; Long, 1992; Nuchtern et al., 1990; Jaraquemada et al., 1990; Hackett et al., 1991; Jin et al., 1988; Jacobson et al., 1989]. While the mechanisms of this are unclear and may involve more than one pathway [Malnati et al., 1992], hypothetical possibilities are that peptides are transported from the cytosol into lysosomal class II processing compartments, in a manner analogous to that described by Dice and colleagues for peptides containing the KFERQ motif [Chiang et al., 1989; Isenman and Dice, 1989; Dice, 1990; Dice et al., 1990] or that autophagic mechanisms serve to introduce cytosolic antigens into vesicular proteolytic compartments [Dunn, 1990; Marzella and Glaumann, 1987]. MHC-II molecules may also present proteins or peptides that are present in the endoplasmic reticulum (ER) [Weiss and Bogen, 1991; Brooks et al., 1991], as well as endogenous proteins that target to endocytic compartments [Weiss and Bogen, 1989; Yurin et al., 1989; Bikoff and Eckhardt, 1989] (also indicated by sequencing of peptides derived from MHC-II molecules; see below).

In the subsequent sections we discuss processing pathways for both MHC-I and MHC-II antigen presentation, but we place somewhat greater emphasis on endoycytic processing mechanisms. The basic principles of cytosolic and vacuolar proteolysis have been discussed in previous chapters of this volume, and these principles form the basis for the proteolytic processing of antigens for both the class I and class II pathways.

MHC-I AND MHC-II MOLECULES

The genetics of the major histocompatibility complex is a complicated topic that is not central to this discussion, and thus only an abbreviated outline is provided. In fact, the nomenclature that defines these loci and alleles is somewhat daunting to those not familiar with the field, and after this section we largely revert to more general terms. The MHC is a gene complex that is specifically referred to as *HLA* in humans and *H-2* in mouse. It encodes three class I loci in both humans (HLA-A, HLA-B and HLA-C) and mouse (H-2K, D, and L). There are also additional nonclassic class I loci that encode molecules that are less understood [Hedrick, 1992; Porcelli et al., 1992; Kurlander et al., 1992; Pamer et al., 1992]. There are three class II loci in humans (HLA-DR, HLA-DP, and HLA-DQ) and two in the mouse (I-A and I-E, sometimes collectively termed *Ia*). The MHC molecules are extremely polymorphic, i.e., a large number of different alleles exist within the gene pool. These alleles are designated by a suffix in humans (e.g. HLA-B27) and a superscript in mouse (e.g., $I-A^k$).

MHC-I molecules are composed of a single transmembrane heavy chain (45 kDa) that includes a small cytoplasmic domain and three extracellular domains (α_1, α_2, and α_3). β_2-Microglobulin (12 kDa) is a soluble protein that is noncovalently associated with the extracellular domains. In contrast, MHC-II molecules are composed of two transmembrane polypeptide chains, α (34 kDa) and β (28 kDa), that are noncovalently associated. The structure of MHC-I has been defined by x-ray crystallography [Bjorkman et la., 1987a,b; Fremont et al., 1992; Matsumura et al., 1992; Zhang et al., 1992; Madden et al., 1992; Guo et al., 1992; Silver et al., 1992], and, although the crystallographic structure of MHC-II has not yet been published, its structure has been modeled on that of MHC-I [Brown et al., 1988; Gorga et al., 1989], despite the structural differences described above. Both molecules appear to form a peptide-binding groove (formed by α_1 and α_2 domains of MHC-I and the α_1 and β_1 domains of MHC-II). This groove is formed between two α-helices, with an underlying floor of β-pleated sheet. The highly polymorphic residues that produce the extreme allelic polymorphism of the MHC molecules are clustered around the peptide-binding groove; in this position they can influence both the spectrum of peptides that an MHC molecule can bind and the composite structure of peptide plus MHC molecule that is recognized by the TCR. Between the peptide-binding domain and the membrane lie the immunoglobulin-like domains of MHC-II (α_2 and β_2) or class I (α_3 plus β_2-microglobulin).

THE NATURE OF THE PROCESSED PEPTIDES THAT BIND TO MHC MOLECULES

Since the initial demonstration of peptide binding to MHC-II molecules [Babbit et al., 1985, 1986; Buus et al., 1986a,b] it has become clear that each MHC-I or MHC-II allele binds a particular spectrum of peptides that can fit within the groove and its specific pockets (which accommodate amino acid side chains). The selection of peptides by MHC molecules does not discriminate between self and nonself.

Thus, peptide binding to MHC molecules is specific in the chemical sense but nonspecific in the immunological sense. Binding to MHC is required but not sufficient for immunogenicity, and immunologically specific recognition and self/nonself discrimination must be mediated by T cells based on the TCR repertoire of the individual.

The sequencing of peptides eluted from purified MHC-I molecules has revealed sequence motifs that allow peptide binding to a particular MHC-I allele (each allele has a different motif) [Van Bleek and Nathenson, 1990; Rotzschke et al., 1990; Falk et al., 1991; Jardetzky et al., 1991; Wei and Cresswell, 1992; Hunt et al., 1992a; Henderson et al., 1992]. This approach has also confirmed that peptides bound to MHC-I derive from cytosolic proteins. MHC-I molecules preferentially bind nonamer peptides (nine amino acids in length), which allows the N and C termini of the peptide to fit into fixed sites (apparently constant between different alleles) at either end of the MHC-I peptide-binding groove. For a given immunogenic peptide epitope, the MHC-I–associated peptides show precise terminal truncations (i.e., the terminal residues are constant, and there is no variation in the length of the peptide). Among different epitopes, some variation in peptide size is permissible with more or less extended conformations of the peptides (e.g., "bowing" in the middle) that still allow the terminal residues to fit within the fixed sites. Crystallographic analysis of peptide–MHC-I complexes has contributed to this understanding and has revealed allele-specific pockets, determined by the polymorphic residues in the peptide-binding groove, that accommodate specific amino acid side chains at certain positions along the peptide [Fremont et al., 1992; Matsumura et al., 1992; Zhang et al., 1992; Madden et al., 1992; Guo et al., 1992; Silver et al., 1992].

Peptides have also been eluted from MHC-II molecules and sequenced. In the case of MHC-II, the sequence motifs that permit peptide binding to a given allele are still less clear, and the structure of MHC-II has not yet been solved crystallographically. MHC-II–associated peptides are longer and more variable, generally 13–25 amino acids in length [Rudensky et al., 1991, 1992; Hunt et al., 1992b; Nelson et al., 1992; Chicz et al., 1992; Newcomb and Cresswell, 1993]. Their sequence exhibits heterogeneous N and C termini, such that a given epitope is often represented by a nested set of multiple peptides with varying N and C extensions beyond the core epitope. As expected, the peptides bound to MHC-II molecules generally derive from cellular proteins that target to endocytic compartments or extracellular proteins that can be internalized and processed within these compartments. Interestingly, peptides derived from heat shock proteins have also been reported as associated with MHC-II molecules [Newcomb and Cresswell, 1993], but the mechanisms and significance of this are still unclear.

ANTIGEN CATABOLISM AND THE CLASS II ANTIGEN PROCESSING PATHWAY

During processing for presentation by MHC-II molecules, antigens must first be internalized by endocytosis (nonspecific or receptor-enhanced) [Gosselin et al., 1992; Chu and Pizzo, 1993; Lorenz et al., 1990; Lanzavecchia, 1987, 1990] or phagocytosis [Pfeifer et al., 1993], and they must then be proteolytically processed within intracellular compartments. Processing is inhibited by weak base amines [Ziegler and Unanue, 1982], which disrupt pH gradients and many functions of acidified endocytic compartments. Antigen processing is also deficient in mutant cells with abnormal endosomal acidification [McCoy et al., 1989].

APC that are lightly fixed with paraformaldehyde or glutaraldehyde generally cannot process and present native antigens, although some determinants can bind MHC-II molecules on fixed cells and be recognized by T cells without any processing [Lee et al., 1988] or after only reduction and denaturation without proteolysis [Streicher et al., 1984; Allen and Unanue, 1984; Sette et al., 1989], presumably due to loose conformational restraints on the epitope in these cases. However, cells that are fixed after incubation with antigen for a sufficient period to generate peptide–MHC-II expression on the plasma membrane are capable of presenting this antigen to T cells. In addition, in vitro digestion of antigens with a wide range of proteases or CNBr produces peptide fragments that, like synthetic peptides, can be presented by pre-fixed cells [Shimonkevitz et al., 1984a,b; Kovac and Schwartz, 1985]. Different epitopes vary with respect to the proteases that are most efficient in this regard. Failure of an enzyme to generate a given epitope may result from specificity of the enzyme for a cleavage site within the T-cell epitope, destroying the determinant, or lack of appropriate cleavage sites adjacent to the epitope. For example, both aspartyl proteases (e.g., cathepsin D) and thiol proteases (e.g., cathepsin B) are effective in some cases for in vitro generation of peptides for presentation, but ineffective for other epitopes [Collins et al., 1991; Diment, 1990; Rodriguez and Diment, 1992; Van der Drift et al., 1990]. However, the efficacy of an enzyme in the in vitro ("cell-free") generation of peptides from intact antigen does not establish a true role for that enzyme in cellular antigen processing (microbial proteases, e.g., pronase, are also effective in this regard).

The enzymes involved in antigen processing have been probed using various protease inhibitors, which will inhibit antigen processing when added with antigen to cells [Puri and Factorovich, 1988; Takahashi et al., 1989; Berzofsky et al., 1988; Werdelin et al., 1988; Vidard et al., 1992; Yoshikawa et al., 1987; Chain et al., 1988]. Effective compounds include inhibitors of both aspartyl and thiol proteases. While these and other results suggest that the processing of some epitopes may be dependent on a specific protease (e.g., cathepsin D, B, or E) [Diment, 1990; Rodriquez and Diment, 1992; Van der Drift et al., 1990; Van Noort and Van der Drift, 1989; Van der Noort et al., 1991; Bennet et al., 1992] the enzyme implicated varies with different epitopes

[Vidard et al., 1992; Berzofsky et al., 1988; Takahashi et al., 1989], i.e., the sensitivity to a particular inhibitor varies with the antigen studied. Overall, antigen processing appears to result from the contributions of many different enzymes that function in general lysosomal protein catabolism, and there does not seem to be one enzyme or set of enzymes that is specific to antigen processing. Differences in the antigen processing capabilities of different cell lines [Michalek et al., 1989] may reflect differing levels of certain enzymes within endocytic organelles. Another point is that results with protease inhibitors must be interpreted with caution, since they may also affect antigen processing by mechanisms other than directly interfering with the proteolysis of an antigenic substrate. For example, leupeptin inhibits invariant chain cleavage [Neefjes and Pleogh, 1992; Blum and Cresswell, 1988], thereby interfering with antigen presentation [Neefjes and Pleogh, 1992].

Antigen processing must create a balance in proteolytic processing sufficient to generate immunogenic peptides from native antigens but avoiding more extensive proteolysis, which could destroy the same immunogenic peptides. One answer to this problem may be that immunogenic peptides, once generated, are rescued from further catabolism by binding to MHC molecules. Donermeyer and Allen [1989] and Mouritsen et al. [1992b] have demonstrated that the binding of a peptide to MHC-II protects it from further degradation, except for some trimming of terminal extensions, e.g., by aminopeptidases. The pattern of proteolysis in antigen processing may also determine which epitopes are efficiently generated for presentation. For example, proteolytic cleavage in the middle of a potential epitope would destroy that epitope, or the lack of appropriate adjacent cleavage sites could make generation of an immunogenic peptide inefficient. Thus, a number of explanations exist for the existence of "cryptic" epitopes, which do not generate immune responses when introduced as native antigen, but are immunogenic as peptides [Gammon et al., 1987].

Another aspect of antigen catabolism in this pathway is the reduction of disulfide bonds in antigens. Disulfide reduction appears necessary to generate certain antigenic peptides that contain reduced cysteine residues, where these residues formed disulfide bonds in the native antigen. For example, an antigenic peptide from insulin requires cysteine reduction prior to recognition by T cells [Hampl et al., 1992; Jensen, 1991a]. Disulfide reduction also allows the unfolding of antigens, increased access of proteases to susceptible sites, and vastly enhanced rates of proteolysis [Collins et al., 1991; Mego, 1984]. The mechanisms of disulfide reduction remain somewhat unclear. Pisoni et al. [1990] have proposed that a specific transporter (distinct from the transporter for oxidized cystine, which is defective in cystinosis) [Gahl et al., 1982] may deliver reduced cysteine from the cytosol into lysosomes, where internalized antigens may be processed (below). Hypothetically, cysteine could then serve as an intralysosomal reductant, ultimately generating oxidized cystine and reduced substrate molecules, as depicted in the model shown in Figure 1 [Pisoni et al., 1990; Lloyd, 1986; but see discussion on transport stoichiometry in Lloyd, 1992].

SUBCELLULAR COMPARTMENTS INVOLVED IN ANTIGEN CATABOLISM IN THE CLASS II PATHWAY

Processing of exogenous antigens occurs in endocytic/phagocytic vesicular compartments, and recent research has addressed more specifically the nature of these compartments. In the endocytic pathway the potential compartments for processing are endosomes and lysosomes, each of which is a heterogeneous class of vesicles with differences in maturity and composition. In some cases, there may also be some confusion in actually defining the demarcation between a late endosome and a lysosome. Lysosomes are defined here as kinetically late endocytic structures with high density upon subcellular fractionation, high levels of lysosomal proteases and marker proteins such as Lamp 1, and an absence of mannose-6-phos-

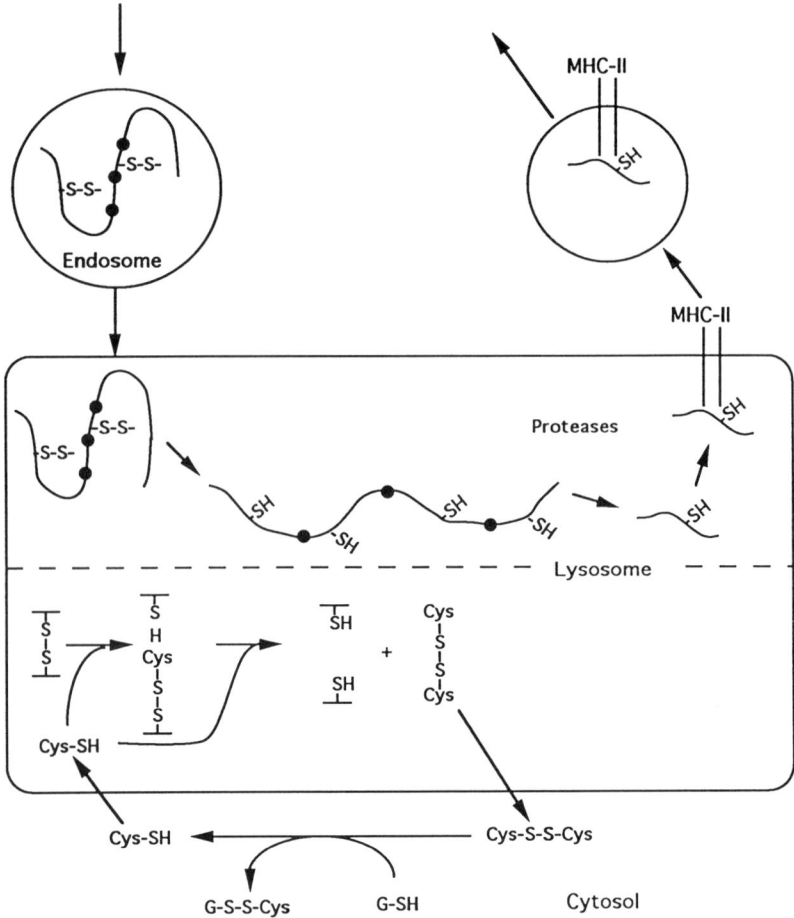

Fig. 1. *Model for the reduction of disulfide bonds in lysosomes in the course of antigen processing (see text). [Adapted from Collins et al., 1991, with permission of the publisher.]*

phate receptor. Endosomes also contain some proteases, e.g., cathepsin D, albeit at lower levels [Diment and Stahl, 1985; Diment et al., 1988; Guagliardi et al., 1989]. Endosomal proteolysis of some substrates has also been demonstrated [Diment et al., 1989]. Thus, some immunogenic peptides could theoretically be produced by endosomal antigen proteolysis, and these peptides could then bind to MHC-II molecules in endosomes. This model is appealing, since mechanisms for recycling from endosomes to the plasma membrane are well demonstrated. However, lysosomal compartments appear to play an important role in antigen processing, as revealed by recent experiments.

Antigen processing is blocked at 18°C [Harding and Unanue, 1990a], a condition that preferentially blocks certain lysosomal functions, suggesting a lysosomal contribution. However, certain deficits may also occur in endosomal functions under these conditions (particularly late endosomal functions), making this observation inconclusive. However, disulfide reduction appears to be mediated in high-density lysosome-like compartments in macrophages [Collins et al., 1991], suggesting a necessary role for lysosomes in the processing of at least those antigenic epitopes that require disulfide reduction for their processing (different results have been obtained in other cells where the intracellular compartment for disulfide reduction is less clear) [Feener et al., 1990; Chen et al., 1985]. In addition, liposome-encapsulated antigens can be sequestered by liposomal membranes of certain compositions until delivery to dense lysosomes, preventing any potential endosomal processing events, and this antigen is then even more efficiently processed than antigens encapsulated in other acid-sensitive liposomes, which release their contents into endosomes [Harding et al., 1991a,b]. This demonstrates the capacity of lysosomes to process antigens for subsequent presentation at the plasma membrane. Processing of phagocytic antigens also appears to involve the action of phagolysosomes, and processing of bacteria leads to the presentation of antigens from either the bacterial cytosol or surface [Pfeifer et al., 1992; Wick et al., 1994], consistent with the extensive degradative capacity of phagolysosomes. Figure 2 shows a model for class II antigen processing that includes important contributions by lysosomal compartments.

The previous data suggest that antigens can be processed in lysosomal compartments, defined by density. However, it is also important to define the ability of MHC-II molecules to target to these compartments and potentially to bind peptides therein. MHC-II molecules have been demonstrated in lysosomal compartments in both B cells and macrophages by immunoelectron microscopy [Peters et al., 1991; Harding and Geuze, 1992]. Using this approach, compartments were defined as lysosomal based on a lack of labeling for mannose-6-phosphate receptor, late kinetic position on the endocytic pathway using endocytic tracers, and high level labeling for lysosomal markers such as cathepsin D and Lamp 1 (also present in endosomes at lower levels). It is clear that further study is necessary to clarify the nature of this compartment and its relationship to other late endocytic compartments that have been defined in other studies, such as tubular lysosomes, prelysosomes, and late endosomes [Rabinowitz et al., 1992; Swanson et al., 1987]. Lysosomes are a heterogeneous set of organelles, and the lysosomal compartments containing high levels of MHC-II appear to represent an early lysosomal compartment, distinct from terminal, late lysosomes [Harding and Geuze, 1992]. The high levels of MHC-II present within this lysosomal compartment suggest the possibility that peptide–MHC-II complexes are formed therein, but MHC-II molecules have also been localized to endosomes [Guagliardi et al., 1989;

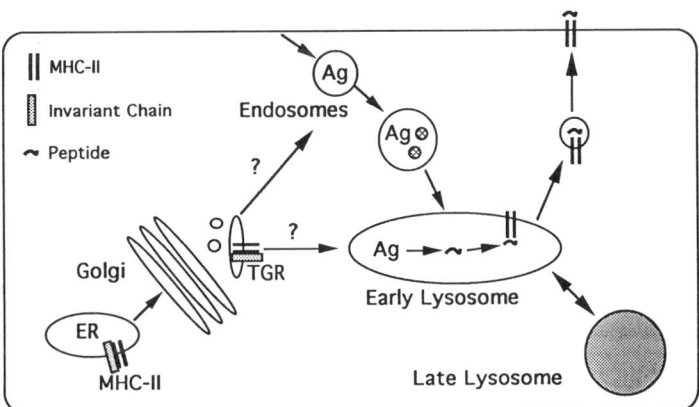

Fig. 2. *Model for antigen processing for MHC-II presentation (see text). [Adapted from Harding, 1993, with permission of the publisher.]*

Harding et al., 1990; Pieters et al., 1991]. Endocytosis of MHC-II molecules also delivers MHC-II to endosomes, albeit at relatively low levels [Harding and Unanue, 1989; Reid and Watts, 1990; Machy et al., 1990; Salamero et al., 1990; Weber et al., 1991]. Thus, while other data indicate that peptide–MHC-II complexes are irreversibly formed within an intracellular compartment using nascent MHC-II [Davidson et al., 1991; Lanzavecchia et al., 1992], the exact intracellular site where peptides bind to MHC-II molecules remains an important issue. A number of studies have revealed enhanced binding of peptides to MHC-II at acid pH [Jensen, 1990, 1991b; Harding et al., 1991c; Wettstein et al., 1991; Reay et al., 1992; Sette et al., 1992; Nag et al., 1992; Mouritsen et al., 1992a; Stern and Wiley, 1992], but this is consistent with either endosomal or lysosomal complexing, due to the broad pH range within which effective binding occurs.

MHC-II transport represents an important paradigm in intracellular protein transport and sorting. MHC-II molecules must target to endocytic compartments, and perhaps even to lysosomes, prior to their functional assembly with peptide and later expression on the plasma membrane. MHC-II and MHC-I are sorted and differentially targeted, with their postbiosynthetic transport diverging in the TGR [Peters et al., 1991]. Of the two, only MHC-II is transported to intersect efficiently with the endocytic pathway [Neefjes et al., 1990]. The targeting or retention of MHC-II to endocytic compartments is dependent on invariant chain, which forms complexes with nascent MHC-II in the ER [Jones et al., 1979; Kvist et al., 1982; Roche et al., 1991] and contains signals that direct MHC-II transport or localization [Bakke and Dobberstein, 1990; Lamb and Cresswell, 1992; Lamb et al., 1991; Lotteau et al., 1990]. Invariant chain also blocks peptide binding to MHC-II molecules [Roche and Cresswell, 1990; Teyton et al., 1990; Nguyen and Humphreys, 1989] prior to delivery to an endocytic compartment where invariant chain is proteolytically cleaved and removed from MHC-II [Roche and Cresswell, 1991; Blum and Cresswell, 1988; Nguyen and Humphreys, 1989]. Invariant chain–derived peptides have also been demonstrated bound to MHC-II molecules [Riberdy et al., 1992]. Invariant chain cleavage can be inhibited by leupeptin [Blum and Cresswell, 1988; Neefjes and Ploegh, 1992], and this partly explains the inhibition of antigen processing that is induced by leupeptin. In some systems and with some antigens, invariant chain is necessary for efficient antigen processing, although in other cases MHC-II molecules expressed in the absence of invariant chain can be functional [Peterson and Miller, 1992; Nadimi et al., 1991; Stockinger et al., 1989].

Studies of mutant APC have provided insight into the class II processing pathway. Several groups have studied mutant cell lines that are deficient in the processing of protein antigens for MHC-II presentation, although they still express MHC-II molecules and can present exogenous peptides [Ceman et al., 1992; Riberdy and Cresswell, 1992; Mellins et al., 1990]. Genetically, the defect in these cells maps to the MHC. MHC-II molecules expressed by these cells have an abnormal conformation (as defined by conformation sensitive monoclonal antibodies), and they are not resistant to treatment with SDS at room temperature. Since SDS resistance is associated with peptide binding to MHC-II molecules [Germain and Hendrix, 1991; Sadegh-Nasseri and Germain, 1991], this suggests that the mutant cells fail to load MHC-II molecules with peptides within intracellular compartments. This may be due to a defect in appropriate targeting of MHC-II molecules or a lack of an undefined function necessary for formation of peptide–MHC-II complexes. A mutation in a processing enzyme is a less likely explanation, since the loss of a single enzyme would be unlikely to produce such a global defect. Future studies should define the gene involved in these mutant cells and provide an exciting insight into class II processing mechanisms.

Another point is that MHC-II molecules expressed in some cells can exhibit a functional deficiency that is not corrected by coexpression of invariant chain [Bikoff, 1992], and this may also be explained by the existence of other proteins whose expression is regulated and necessary to antigen processing. In another system, Pierce et al. [1991] have suggested that heat shock family proteins may contribute to antigen processing, although this hypothesis requires further evaluation. Thus, it is likely that other proteins may play important but still undefined roles in antigen processing.

GENERAL ASPECTS OF THE CLASS I ANTIGEN PROCESSING PATHWAY

As discussed above, MHC-I molecules primarily present peptides derived from cytosolic antigens, such as viral antigens synthesized in infected cells. The presentation of influenza nucleoprotein, a cytosolic protein, was studied by Townsend and colleagues [Townsend et al., 1985, 1986], and these studies introduced the puzzle of antigen transport in this pathway. How was a cytosolic antigen subsequently presented at the cell surface? Later studies showed that MHC-I molecules presented cytosolic antigens whether they were synthesized by the presenting cell or they penetrated into the cytosol from the extracellular space, carried by invasive microbes such as *Listeria* [Brunt et al., 1990; Pamer et al., 1991] or via experimental techniques such as osmotic shock [Moore et al., 1988] or electroporation [Harding, 1992]. The studies of Braciale and colleagues [Morrison et al., 1986; Braciale et al., 1987] demonstrated the distinct mechanisms for class I vs. class II processing. Exogenous antigens (e.g., nonviable virus preparations) were processed for MHC-II presentation by chloroquine-sensitive mechanisms that were relatively resistant to inhibition of protein synthesis, whereas infection with live viruses produced MHC-I antigen presentation that was dependent on protein synthesis and was chloroquine resistant. Together, these and other studies suggested a model in which MHC-I molecules bound peptides derived from cytosolic antigens, whereas MHC-II molecules bound peptides derived from endocytic sources (above). Figure 3 shows a model of antigen processing that contrasts the class I pathway with the class II pathway discussed in the earlier sections.

While it is clear that this model correctly describes the major pathway of class I antigen processing, in some cases class I processing and presentation occur with particulate exogenous antigens with no known cytosolic penetration step [Pfeifer et al., 1993; Collins et al., 1992; Carbone and Bevan, 1990; Debrick et al., 1991; Reddy et al., 1991; Gooding, 1980; Bevan, 1976; Aggarwal et al., 1990; Flynn et al., 1990] or even with soluble exogenous antigen [Grant and Rock, 1992; Rock et al., 1990, 1993], but

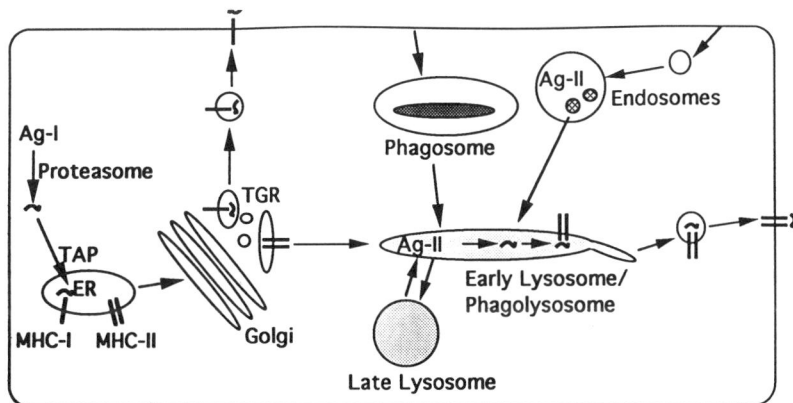

Fig. 3. *Comparison of the class I and class II antigen processing pathways (see text).*

the mechanisms involved are still unclear. In fact, this alternative class I processing pathway may involve endocytic antigen proteolysis, similar to the class II pathway, although the mechanisms whereby peptides meet and bind to MHC-I molecules in this case probably differ from the class II pathway.

CYTOSOLIC ANTIGEN PROTEOLYSIS IN THE CLASS I ANTIGEN PROCESSING PATHWAY

Given a model for cytosolic antigen processing, a number of studies have addressed mechanisms of cytosolic antigen proteolysis that may contribute to this. LMP2 and LMP7 are two proteins that have been proposed to function in cytosolic antigen processing [Martinez and Monaco, 1991; Brown et al., 1991; Glynne et al., 1991; Kelly et al., 1991; Fruh et al., 1992]. They appear to be regulated components of the proteasome, a multisubunit cytosolic proteolytic complex described in previous chapters [Goldberg and Rock, 1992]. LMP2 and LMP7 are encoded by genes located within the class II region of the MHC, and their expression is regulated along with antigen processing function by interferon-γ [Yang et al., 1992b]. However, two recent reports [Momburg et al., 1992; Arnold et al., 1992] demonstrate that cells lacking the LMP2 and LMP7 genes can still process antigens for MHC-I presentation, indicating that the subunits encoded by these genes are not absolutely necessary for this processing pathway. However, proteasomes containing alternate subunits may be able to substitute for proteasomes containing the LMP2 and LMP7 subunits, albeit less efficiently, so that a contribution of LMP2, LMP7, and proteasomes to class I processing remains a good possibility that requires further experimental testing. In fact, earlier studies by Townsend et al. [1988] showed that expression of unstable forms of influenza nucleoprotein, including a ubiquitin–nucleoprotein fusion protein, resulted in enhanced MHC-I presentation of nucleoprotein peptides. This result suggests that ubiquitin-dependent proteolysis and proteasomes, which degrade ubiquitin-conjugated proteins, may in fact contribute to class I processing mechanisms.

Proteolytic processing of cytosolic proteins for class I presentation may be influenced by a number of factors. Decreased stability and higher turnover rate appear to favor presentation [Townsend et al., 1988]. In addition, in some systems the processing of an epitope is influenced by residues that flank the actual core epitope that binds the MHC-I molecule [Del Val et al., 1991; Eisenlohr et al., 1992b], perhaps because these residues affect the efficiency of proteolysis. However, this effect has not been observed in all studies [Hahn et al., 1992].

TRANSPORT OF ANTIGENIC PEPTIDES AND MHC-I MOLECULES IN THE CLASS I PATHWAY

The next step in the class I processing pathway is the transport of peptides from the cytosol into the ER (and perhaps cis-Golgi). This transport appears to be mediated by the heterodimeric TAP1–TAP2 transporter, which is encoded by two genes in the class II region of the MHC. Cells that are lacking either of these genes are largely deficient in both class I antigen processing and surface expression of MHC-I molecules [Spies et al., 1990, 1992; Powis et al., 1991, 1992; Kelly et al., 1992; Yang et al., 1992a; Bahram et al., 1991; Cerundolo et al., 1990; Attaya et al., 1992; Deverson et al., 1990; Spies and DeMars, 1991; Trowsdale et al., 1990; Van Kaer et al., 1992; Colonna et al., 1992], although MHC-I molecules that do reach the surface in these cells can bind and present peptides [Schumacher et al., 1990; Ljunggren et al., 1990]. Immunoelectron microscopy has demonstrated that the TAP transporter is localized to the ER/cis-Golgi [Kleijmeer et al., 1992], which provides support for earlier proposals that the binding of peptides to MHC-I molecules occurs within this compartment. Peptide transport into ER-derived vesicles has been demonstrated in vitro [Levy et al., 1991; Koppelman et al., 1992; Neefjes et al., 1993; Shepherd et al., 1993].

Following transport into the ER, a trimolecular complex is assembled, comprised of peptide, β_2-microglobulin, and MHC-I heavy chains. Formation of this peptide–MHC-I complex releases MHC-I heavy chains from ER retention mediated by chaperonin molecules that bind and retain unassembled MHC-1 heavy chains in the ER [Jackson et al., 1994]. Next, the peptide–MHC-I complexes are then transported, apparently via the constitutive biosynthetic/secretory pathway, to the plasma membrane. Sorting in the trans-Golgi reticulum (TGR) separates the pathways of MHC-I and MHC-II transport [Peters et al., 1991]. In cells infected with certain viruses, MHC-I molecules can be retained within compartments along the ER/Golgi pathway by viral proteins [Del Val et al., 1992; Cox et al., 1991], a mechanism that may reduce the presentation of viral antigens.

SUMMARY

Antigen processing lies at the heart of specific immunity to protein antigens. It involves partial antigen proteolysis that produces the pool of antigenic peptides that can bind to MHC molecules for subsequent presentation to T cells. The MHC molecules bind restricted sets of peptides, based on their sequence, but this binding does not distinguish between immunogenic foreign peptides and nonimmunogenic self-peptides; this distinction is made at the level of the T cell. The MHC molecules serve a unique duel purpose. First, they serve as a peptide receptor that binds peptides within a particular subcellular compartment and then transports the peptides to the cell surface. Second, in combination with the presented peptide, they form the determinant that is recognized by the TCR.

Within this broad framework, MHC-I and MHC-II molecules differ in both structure and function. MHC-II molecules serve to bind, transport, and present peptides primarily from exogenous antigens that are internalized from the extracellular space and catabolized within endocytic/phagocytic vesicular compartments, whereas MHC-I molecules present peptides derived from endogenous, primarily cytosolic antigens. Together, the two systems provide a peptidic display of all antigens both inside and outside the presenting cell, and this forms the substrate for T-cell recognition.

Antigen proteolysis is central to both the class I and class II processing pathways. While more information is needed, the proteolytic mechanisms are largely the same as used for general protein degradation. For class II processing, a number of lysosomal proteases seem to be important; the mechanisms for cytosolic antigen proteolysis involved in class I processing remain less clear. In addition to these general mechanisms, however, additional mechanisms are necessary to transport peptides and MHC molecules to meet and bind to each other within specific intracellular compartments. After binding to MHC molecules, peptides may be protected from further proteolysis, rescuing the determinant for presentation to T cells.

REFERENCES

Aggarwal A, Kumar S, Jaffe R, Hone D, Gross M, Sadoff J (1990): Oral *Salmonella:* Malaria circumsporozoite recombinants induce specific $CD8^+$ cytotoxic T cells. J Exp Med 172:1083.

Allen PM, Unanue ER (1984): Differential requirements for antigen processing by macrophages for lysozyme-specific T cell hybridomas. J Immunol 132:1077.

Arnold D, Driscoll J, Androlewicz Hughes E, Cresswell P, Spies T (1992): Proteasome subunits encoded in the MHC are not generally required for the processing of peptides bound by MHC class I molecules. Nature 360:171–174.

Attaya M, Jameson S, Martinex CK, Hermel E, Aldrich C, Forman J, Lindahl KF, Bevan MJ, Monaco JJ (1992): Ham-2 corrects the class I antigen-processing defect in RMA-S cells. Nature 355:647–649.

Babbit BP, Allen PM, Matsueda G, Haber E, Unanue ER (1985): Binding of immunogenic peptides to Ia histocompatibility molecules. Nature 317:359–361.

Babbit BP, Matsueda G, Haber E, Unanue ER, Allen PM (1986): Antigenic competition at the level of peptide–Ia binding. Proc Natl Acad Sci USA 83:4509–4513.

Bahram S, Arnold D, Bresnahan M, Strominger JL, Spies T (1991): Two putative subunits of a peptide pump encoded in the human major histocompatibility complex class II region. Proc Natl Acad Sci USA 88:10094–10098.

Bakke O, Dobberstein B (1990): MHC class II–associated invariant chain contains a sorting signal for endosomal compartments. Cell 63:707–716.

Bennet K, Levine T, Ellis JS, Peanasky RJ, Samloff IM, Kay J, Chain BM (1992): Antigen processing for presentation by class II major histocompatibility complex requires cleavage by cathepsin E. Eur J Immunol 22:1519–1524.

Berzofsky JA, Brett SJ, Streicher HZ, Takahashi H (1988): Antigen processing for presentation to T lymphocytes: Functions, mechanisms and implications for the T cell repertoire. Immunol Rev 106:5–32.

Bevan MJ (1976): Cross-priming for a secondary cytotoxic response to minor Ha antigens with H-2 congenic cells which do not cross-react in the cytotoxicity assay. J Exp Med 143:1283.

Bevan MJ (1987): Class discrimination in the world of immunology. Nature 325:192.

Bikoff EK (1992): Formation of complexes between self-peptides and MHC class II molecules in cells defective for presentation of exogenous protein antigens. J Immunol 149:1–8.

Bikoff EK, Eckhardt LA (1989): Presentation of IgG2a antigens to class II restricted T cells by stably transfected B lymphoma cells. Eur J Immunol 19:1903–1909.

Bjorkman PJ, Saper MA, Samraoui B, Bennett WS, Strominger JL, Wiley DC (1987a): Structure of the human class I histocompatibility antigen, HLA-A2. Nature 329:506–512.

Bjorkman PJ, Saper MA, Samraoui B, Bennett WS, Strominger JL, Wiley DC (1987b): The foreign antigen binding site and T cell recognition regions of class I histocompatibility antigens. Nature 329:512–518.

Blum JS, Cresswell P (1988): Role for intracellular proteases in the processing and transport of class II HLA antigens. Proc Natl Acad Sci USA 85:3975–3979.

Braciale TJ, Morrison LA, Sweetser MT, Sambrook T, Gething MJ, Braciale VL (1987): Antigen presentation pathways to class I and class II MHC-restricted T lymphocytes. Immunol Rev 98:95.

Brooks A, Hartley S, Kjer-Nielsen I, Perera J, Goodnow CC, Basten A, McCluskey J (1991): Class II–restricted presentation of an endogenously derived immunodominant T cell determinant of hen egg lysozyme. Proc Natl Acad Sci USA 88:3290–3294.

Brown JH, Jardetzky T, Saper MA, Samraoui B, Bjorkman PJ, Wiley DC (1988): A hypothetical model of the foreign antigen binding site of class II histocompatibility molecules. Nature 332:845–850.

Brown MG, Driscoll J, Monaco JJ (1991): Structural and serological similarity of MHC-linked LMP and proteasome (multicatalytic proteinase) complexes. Nature 353:355–357.

Brunt LM, Portnoy DA, Unanue ER (1990): Presentation of Listeria monocytogenes to CD8 T cells requires secretion of hemolysin and intracellular bacterial growth. J Immunol 145:3540.

Buus S, Colon S, Smith C, Freed JH, Miles C, Grey HM (1986a): Interaction between a "processed" ovalbumin peptide and Ia molecules. Proc Natl Acad Sci USA 83:3968–3971.

Buus S, Sette A, Colon SM, Jenis DM, Grey HM (1986b): Isolation and characterization of antigen–Ia complexes involved in T cell recognition. Cell 47:1071–1077.

Carbone RR, Bevan MJ (1990): Class I-restricted processing and presentation of exogenous cell-associated antigen in vivo. J Exp Med 171:377.

Ceman S, Rudersdorf R, Long EO, Demars R (1992): MHC class II deletion mutant expresses normal levels of transgene encoded class II molecules that have abnormal conformation and impaired antigen presentation ability. J Immunol 149:754–761.

Cerundolo V, Alexander J, Anderson K, Lamb C, Cresswell P, McMichael A, Gotch F, Townsend A (1990): Presentation of viral antigen controlled by a gene in the major histocompatibility complex. Nature 345:449–452.

Chain BM, Kaye PM, Shaw M-A (1988): The biochemistry and cell biology of antigen processing. Immunol Rev 106:33–58.

Chiang H-L, Terlecky SR, Pland CP, Dice JF (1989): A role for a 70-kilodalton heat shock protein in lysosomal degradation of intracellular proteins. Science 246:382–385.

Chicz RM, Urban RG, Lane WS, Gorga JC, Stern LJ, Vignali DA, Strominger JL (1992): Predominant naturally processed peptides bound to HLA-DR1 are derived from MHC-related molecules and are heterogeneous in size. Nature 358:764–768.

Chen W-D, Reyser HJ-P, La Manna L (1985): Disulfide spacer between methotrexate and poly(D-lysine): A probe for exploring the reductive process in endocytosis. J Biol Chem 260:10905.

Chu CT, Pizzo SV (1993): Receptor-mediated antigen delivery into macrophages: Complexing antigen to alpha-2-macroglobulin enhances presentation to T cells. J Immunol 150:48–58.

Collins DS, Findlay K, Harding CV (1992): Processing of exogenous liposome-encapsulated antigens in vivo generates class I MHC-restricted T cell responses. J Immunol 148:3336–3341.

Collins D, Unanue ER, Harding CV (1991): Reduction of disulfide bonds within lysosomes is a key step in antigen processing. J Immunol 147:4054–4059.

Colonna M, Bresnahan M, Bahram S, Strominger JL, Spies T (1992): Allelic variants of the human putative peptide transporter involved in antigen processing. Proc Natl Acad Sci USA 89:3932–3926.

Cox JH, Bennink JR, Yewdell JW (1991): Retention of adenovirus E19 glycoprotein in the endoplasmic reticulum is essential to its ability to block antigen presentation. J Exp Med 174:1629–1637.

Davidson HW, Reid PA, Lanzavecchia A, Watts C (1991): Processed antigen binds to newly synthesized MHC class II molecules in antigen-specific B lymphocytes. Cell 67:105–116.

Debrick JE, Campbell PA, Staerz UD (1991): Macrophages as accessory cells for class I MHC-restricted immune responses. J Immunol 147:2846.

Del Val M, Hengel H, Hacker H, Hartlaub U, Ruppert T, Lucin P, Koszinowski UH (1992): Cytomegalovirus prevents antigen presentation by blocking the transport of peptide-loaded major histocompatibility complex class I molecules into the medial-Golgi compartment. J Exp Med 176:729–738.

Del Val M, Schlicht H-J, Ruppert T, Reddehase MJ, Koszinowski UH (1991): Efficient processing of an antigenic sequence for presentation by MHC class I molecules depends on its neighboring residues in the protein. Cell 66:1145.

Demotz S, Grey HM, Sette A (1990): The minimal number of class II MHC-antigen complexes needed for T cell activation. Science 249:1028–1030.

Deverson EV, Gow IR, Coadwell WJ, Monaco JJ, Butcher GW, Howard JC (1990): MHC class II region encoding proteins related to the multidrug resistance family of transmembrane transporters. Nature 348:738–741.

Dice JF (1990): Peptide sequences that target cytosolic proteins for lysosomal proteolysis. Trends Biochem Sci 15:305–309.

Dice JF, Terlecky SR, Chiang HL, Olson TS, Isenman LD, Short-Russell SR, Freundlieb S, Terlecky LJ (1990): A selective pathway for degradation of cytosolic proteins by lysosomes. Semin Cell Biol 1:449–455.

Diment S (1990): Different roles for thiol and aspartyl proteases in antigen presentation of ovalbumin. J Immunol 145:417.

Diment S, Leech MS, Stahl PD (1988): Cathepsin D is membrane associated in macrophage endosomes. J Biol Chem 263:6901

Diment S, Martin K, Stahl P (1989): Cleavage of parathyroid hormone in macrophage endosomes illustrates a novel pathway for intracellular processing of proteins. J Biol Chem 264:13403.

Diment S, Stahl P (1985): Macrophage endosomes contain proteases which degrade endocytosed protein ligands. J Biol Chem 260:15311–15317.

Donermeyer DL, Allen PM (1989): Binding to Ia protects an immunogenic peptide from proteolytic degradation. J Immunol 142:1063–1068.

Dunn WA Jr (1990): Studies on the mechanisms of autophagy: Formation of the autophagic vacuole. J Cell Biol 110:1923–1933.

Eisenlohr LC, Bacik I, Bennink JR, Bernstein K, Yewdell JW (1992a): Expression of a membrane protease enhances presentation of endogenous antigens to MHC class I–restricted T lymphocytes. Cell 71:963–972.

Eisenlohr LC, Yewdell JW, Bennink JR (1992b): Flanking sequences influence the presentation of an endogenously synthesized peptide to cytotoxic T lymphocytes. J Exp Med 175:481.

Falk K, Rotzschke O, Stevanovic S, Jung G, Rammensee H-G (1991): Allele-specific motifs revealed by sequencing of self-peptides eluted from MHC molecules. Nature 351:290–296.

Feener EP, Shen W-C, Reyser JJ-P (1990): Cleavage of disulfide bonds in endocytosed macromolecules: A processing not associated with lysosomes or endosomes. J Biol Chem 265:18780.

Flynn JL, Weiss WR, Norris KA, Siefert HS, Kumar S, So M (1990): Generation of a cytotoxic T-lymphocyte response using a *Salmonella* antigen-delivery system. Mol Microbiol 4:2111.

Fremont DH, Matsumura M, Stura EA, Peterson PA, Wilson IA (1992): Crystal structures of two viral peptides in complex with murine MHC class I H-2Kb. Science 257:919–927.

Früh K, Yang Y, Arnold D, Chambers J, Wu L, Waters JB, Spies T, Peterson PA (1992): Alternative exon usage and processing of the major histocompatibility complex-encoded proteasome subunits. J Biol Chem 267:22131–22140.

Gahl WA, Bashan N, Teitze F, Bernardini I, Schulman JD (1982): Cystine transport is defective in isolated leukocyte lysosomes from patients with cystinosis. Science 217:1263.

Gammon G, Shastri N, Cogswell J, Wilbur S, Sadegh-Nasseri S, Krzych U, Miller A, Sercarz E (1987): The choice of T-cell epitopes utilized on a protein antigen depends on multiple factors distant from, as well as at, the determinant site. Immunol Rev 98:53–73.

Germain RN (1986): The ins and outs of antigen processing and presentation. Nature 322:687.

Germain RN, Hendrix LR (1991): MHC class II structure, occupancy and surface expression determined by post-endoplasmic reticulum antigen binding. Nature 353:134–139.

Glynne R, Powis SH, Beck S, Kelly A, Kerr LA, Trowsdale J (1991): A proteasome-related gene between the two ABC transporter loci in the class II region of the human MHC. Nature 353:357–360.

Goldberg AL, Rock KL (1992): Proteolysis, proteasomes and antigen presentation. Nature 357:375–379.

Gooding LR (1980): H-2 antigen requirements in the in vitro induction of SV40-specific cytotoxic T lymphocytes. J Immunol 124:1258.

Gorga JC, Dong A, Manning MC, Woody RW, Caughey WS, Strominger JL (1989): Comparison of the secondary structures of human class I and class II major histocompatibility complex antigens by Fourier transform infrared and circular dichroism spectroscopy. Proc Natl Acad Sci USA 86:2321–2325.

Gosselin EJ, Wardwell K, Gosselin DR, Alter N, Fisher JL, Guyre PM (1992): Enhanced antigen presentation using human Fc-gamma receptor (monocyte/macrophage)-specific immunogens. J Immunol 149:3477–3481.

Grant EP, Rock KL (1992): MHC class I–restricted presentation of exogenous antigen by thymic antigen-presenting cells in vitro and in vivo. J Immunol 148:13–18.

Guagliardi LE, Koppelman B, Blum JS, Marks MS, Cresswell P, Brodsky FM (1989): Co-localization of molecules involved in antigen processing and presentation in an early endocytic compartment. Nature 343:133–139.

Guo H-C, Jardetzky TS, Garrett TPJ, Lane WS, Strominger JL, Wiley DC (1992): Different length peptides bind to HLA-Aw68 similarly at their ends but bulge out in the middle. Nature 360:364–366.

Hackett CJ, Yewdell JW, Bennink JR, Sysock M (1991): Class II-MHC restricted T cell determinants processed from either endosomes or the cytosol with similar requirements for host protein transport but different kinetics of production. J Immunol 146:2944–2951.

Hahn YS, Hahn CS, Braciale VL, Braciale TL, Rice CM (1992): $CD8^+$ T-cell recognition of an endogenously processed epitope is regulated primarily by residues within the epitope. J Exp Med 176:1335.

Hampl J, Gradehandt B, Kalbacher H, Rude E (1992): In vitro processing of insulin for recognition by murine T cells results in the generation of A chains with free CysSH. J Immunol 148:2664–2671.

Harding CV (1992): Electroporation of exogenous antigen into the cytosol for antigen processing and class I major histocompatibility complex (MHC) presentation: Weak base amines and hypothermia (18 degrees C) inhibit the class I MHC processing pathway. Eur J Immunol 22:1865–1869.

Harding CV (1993): Cellular and molecular aspects of antigen processing and the function of class II MHC molecules. Am J Respir Cell Mol Biol 8:461–467.

Harding CV, Collins DS, Kanagawa O, Unanue ER (1991a): Liposome-encapsulated antigens engender lysosomal processing for class II MHC presentation and cytosolic processing for class I presentation. J Immunol 147:2860–2863.

Harding CV, Collins DS, Slot JW, Geuze JJ, Unanue ER (1991b): Liposome-encapsulated antigens are processed in lysosomes, recycled, and presented to T cells. Cell 64:393–401.

Harding CV, Geuze HJ (1992): Class II MHC molecules are present in macrophage lysosomes and phagolysosomes that function in the phagocytic processing of *Listeria monocytogenes* for presentation to T cells. J Cell Biol 119:531–542.

Harding CV, Kihlberg J, Elofsson M, Magnusson G, Unanue ER (1994): Glycopeptides bind MHC molecules and elicit specific T cell responses. J Immunol 151:2419–2425.

Harding CV, Roof RW, Allen PM, Unanue ER (1991c): Effects of pH and polysaccharides on peptide binding to class II major histocompatibility complex molecules. Proc Natl Acad Sci USA 88:2740–2744.

Harding CV, Unanue ER (1989): Antigen processing and intracellular Ia. Possible roles of endocytosis and protein synthesis in Ia function. J Immunol 142:12–19.

Harding CV, Unanue ER (1990a): Low-temperature inhibition of antigen processing and iron uptake from transferrin: Deficits in endosome functions at 18°C. Eur J Immunol 20:323–329.

Harding CV, Unanue ER (1990b): Quantitation of peptide-class II MHC complexes generated in antigen presenting cells and necessary for T cell stimulation. Nature 346:574–576.

Harding CV, Unanue ER, Slot JW, Schwartz AL, Geuze JH (1990): Functional and ultrastructural evidence for intracellular formation and recycling of Ia–peptide complexes during antigen processing. Proc Natl Acad Sci USA 87:5553–5557.

Hedrick SM (1992): Dawn of the hunt for nonclassical MHC function. Cell 70:177–180.

Henderson RA, Michel H, Sakaguchi K, Shabanowitz J, Appella E, Hunt DF, Engelhard VH (1992): HLA-2.1–associated peptides from a mutant cell line: A second pathway of antigen presentation. Science 255:1264–1266.

Hunt DF, Henderson RA, Shabanowitz J, Sakaguchi K, Nichel H, Sevilir N, Cox AL, Appella E, Engelhard VH (1992a): Characterization of peptides bound to the class I MHC molecule HLA-A2.1 by mass spectrometry. Science 255:1261–1263.

Hunt DF, Michel H, Dickinson TA, Shabanowitz J, Cox AL, Sakaguchi K, Appella E, Grey HM, Sette A (1992b): Peptides presented to the immune system by the murine class II major histocompatibility complex molecule I-Ad. Science 256:1817–1820.

Isenman LD, Dice JF (1989): Secretion of intact proteins and peptide fragments by lysosomal pathways of protein degradation. J Biol Chem 264:21591–29596.

Ishioka GY, Lamont AG, Thomson D, Bulbow N, Gaeta FCA, Sette A, Grey HM (1992): MHC interaction and T cell recognition of carbohydrates and glycopeptides. J Immunol 148:2446–2451.

Jackson MR, Cohen-Doyle MF, Peterson PA, Williams DB (1994): Regulation of MHC class I transport by the molecular chaperone, calnexin (p88, IP-90). Sciences 263:384–390.

Jacobson S, Sekaly RP, Jacobson CL, McFarland HR, Loy ED (1989): HLA class II–restricted presentation of cytoplasmic measles virus antigen to cytotoxic T cells. J Virol 63:1756–1762.

Jaraquemada D, Marti M, Long EO (1990): An endogenous processing pathway in vaccinia virus-infected cells for presentation of cytoplasmic antigens to class II–restricted T cells. J Exp Med 172:947–954.

Jardetzky TS, Lane WS, Robinson RA, Madden DR, Wiley DC (1991): Identification of self peptides bound to purified HLA-B27. Nature 353:326–329.

Jensen PE (1990): Regulation of antigen presentation by acidic pH. J Exp Med 171:1779–1784.

Jensen PE (1991a): Reduction of disulfide bonds during

antigen processing. Evidence from a thiol-dependent insulin determinant. J Exp Med 174:1121–1130.

Jensen PE (1991b): Enhanced binding of peptide antigen to purified class II major histocompatibility glycoproteins at acidic pH. J Exp Med 174:1111–1120.

Jin Y, Shih JWK, Berkower I (1988): Endosomal and nonendosomal processing pathways are accessible to both endogenous and exogenous antigen. J Exp Med 168:293.

Jones PP, Murphy DB, Hewgill D, McDevitt HO (1978): Detection of a common polypeptide chain in I-A and I-E sub-region immunoprecipitates. Immunochemistry 16:51–60.

Kelly A, Powis SH, Derr L-A, Mockridge I, Elliott T, Bastin J, Uchanska-Ziegler B, Ziegler A, Trowsdale J, Townsend A (1992): Assembly and function of the two ABC transporter proteins encoded in the human major histocompatibility complex. Nature 355:641–644.

Kelly A, Powis SH, Glynne R, Radley E, Beck S, Trowsdale J (1991): Second proteasome-related gene in the human MHC class II region. Nature 353:667–668.

Kleijmeer MJ, Kelly A, Geuze JJ, Slot JW, Townsend A, Trowsdale J (1992): Location of MHC-encoded transporters in the endoplasmic reticulum and cis-Golgi. Nature 357:342–344.

Kovac S, Schwartz RH (1985): The molecular basis of the requirement for antigen processing of pigeon cytochrome c prior to T cell activation. J Immunol 134:3233–3240.

Kurlander RJ, Shawar SM, Brown ML, Rich RR (1992): Specialized role for a murine class I-b MHC molecule in prokaryotic host defenses. Science 257:678–679.

Kvist S, Wiman K, Claesson L, Peterson PA, Dobberstein B (1982): Membrane insertion and oligomeric assembly of HLA-DR histocompatibility antigens. Cell 29:61–69.

Lamb CA, Cresswell P (1992): Assembly and transport properties of invariant chain trimers and HLA-DR–invariant chain complexes. J Immunol 148:3478–3482.

Lamb CA, Yewdell JW, Bennink JR, Cresswell P (1991): Invariant chain targets HLA class II molecules to acidic endosomes containing internalized influenza virus. Proc Natl Acad Sci USA 88:5998–6002.

Lanzavecchia A (1987): Antigen uptake and accumulation in antigen-specific B cells. Immunol Rev 99:39–51.

Lanzavecchia A (1990): Receptor-mediated antigen uptake and its effect on antigen presentation to class II–restricted T lymphocytes. Annu Rev Immunol 8:773–793.

Lanzavecchia A, Reid PA, Watts C (1992): Irreversible association of peptides with class II MHC molecules in living cells. Nature 357:249–252.

Lee P, Matsueda GR, Allen PM (1988): T cell recognition of fibrinogen. A determinant on the Aα-chain does not require processing. J Immunol 140:1063–1068.

Levy F, Gabathuler R, Larsson R, Kvist S (1991): ATP is required for in vitro assembly of MHC class I antigens but not for transfer of peptides across the ER membrane. Cell 67:265–274.

Ljunggren H-G, Stam NJ, Öhlén C, Neefjes JJ, Höglund P, Heemels M-T, Bastin J, Schumacher TNM, Townsend A, Kärre K, Pleogh HL (1990): Nature 346:476–480.

Lloyd JB (1986): Disulfide bond reduction in lysosomes: The role of cysteine. Biochem J 237:271.

Lloyd JB (1992): Lysosomal handling of cystine residues: Stoichiometry of cysteine involvement. Biochem J 286:979–980.

Long EO (1992): Antigen processing for presentation of $CD4^+$ T cells. The New Biologist 4:274–282.

Lorenz RG, Blum JS, Allen PM (1990): Constitutive competition by self proteins for antigen presentation can be overcome by receptor-enhanced uptake. J Immunol 144:1600–1606.

Lotteau V, Teyton L, Peleraux A, Nilsson T, Karisson L, Schmid SL, Quaranta V, Peterson PA (1990): Intracellular transport of class II MHC molecules directed by invariant chain. Nature 348:600–605.

Machy P, Bizozzero JP, Reggio H, Leserman L (1990): Endocytosis and recycling of MHC-encoded class II molecules by mouse B lymphocytes. J Immunol 145:1350–1355.

Madden DR, Gorga JC, Strominger JL, Wiley DC (1992): The three-dimensional structure of HLA-B27 at 2.1 Å resolution suggests a general mechanisms for tight peptide binding to MHC. Cell 70:1035–1048.

Malnati MS, Marti M, LaVaute T, Jaraquemada D, Biddison W, DeMars R, Long EO (1992): Processing pathways for presentation of cytosolic antigen to MHC class II–restricted T cells. Nature 357:702–704.

Martinez CK, Monaco JJ (1991): Homology of proteasome subunits to a major histocompatibility complex–linked LMP gene. Nature 353:664–667.

Marzella L, Glaumann H (1987): Autophagy, microautophagy and crinophagy as mechanisms for protein degradation. In Glaumann H, Ballard FJ (eds): Lysosomes: Their Role in Protein Breakdown. New York: Academic Press, pp 319–367.

Matsumura M, Fremont DH, Peterson PA, Wilson IA (1992): Emerging principles for the recognition of peptide antigens by MHC class I molecules. Science 257:927–934.

McCoy KL, Miller J, Jenkins M, Ronchese F, Germain RN, Schwartz RH (1989): Diminished antigen processing by endosomal acidification mutant antigen-presenting cells. J Immunol 143:29–38.

Mego JL (1984): Role of thiols, pH and cathepsin D in the lysosomal catabolism of serum albumin. Biochem J 218:775.

Mellins E, Smith L, Arp B, Cotner T, Celis E, Pious D (1990): Defective processing and presentation of exogenous antigens in mutants with normal HLA class II genes. Nature 343:71–74.

Michalek MT, Benacerraf B, Rock KL (1989): Two genetically identical antigen-presenting cell clones display heterogeneity in antigen processing. Proc Natl Acad Sci USA 86:3316–3320.

Momburg F, Ortiz-Navarrete V, Goulmy E, Van de Wal Y, Spits J, Powis SJ, Butcher GW, Howard JC, Walden P, Hammerling GJ (1992): Proteasome subunits encoded by the major histocompatibility complex are not essential for antigen presentation. Nature 360:174–177.

Moore MW, Carbone FR, Bevan MJ (1988): Introduction of soluble protein into the class I pathway of antigen processing and presentation. Cell 54:777–785.

Morrison LA, Lukacher AE, Braciale VL, Fan DP, Braciale TJ (1986): Differences in antigen presentation to MHC class I– and class II–restricted influenza virus-specific cytolytic T lymphocyte clones. J Exp Med 163:903–921.

Mouritsen S, Hansen AS, Petersen BL, Buus S (1992a): pH dependence of the interaction between immunogenic peptides and MHC class II molecules. Evidence for an acidic intracellular compartment being the organelle of interaction. J Immunol 148:1438–1444.

Mouritsen S, Meldal M, Werdelin O, Hansen AS, Buus S (1992b): MHC molecules protect T cell epitopes against proteolytic destruction. J Immunol 149: 1987–1993.

Nadimi F, Moreno J, Momburg F, Heuser A, Fuchs S, Adorini L, Hammerling GJ (1991): Antigen presentation of hen egg-white lysozyme but not of ribonuclease A is augmented by the major histocompatibility complex class II–associated invariant chain. Eur J Immunol 21:1255–1263.

Nag B, Passmore D, Deshpande SV, Clark BR (1992): In vitro maximum binding of antigenic peptides to murine MHC class II molecules does not always take place at the acidic pH of the in vivo endosomal compartment. J Immunol 148:369–372.

Neefjes JJ, Momburg F, Hämmerling GJ (1993): Selective and ATP-dependent translocation of peptides by the MHC-encoded transporter. Science 261:769–771.

Neefjes JJ, Pleogh HL (1992): Inhibition of endosomal proteolytic activity by leupeptin blocks surface expression of MHC class II molecules and their conversion to SDS resistant alpha beta heterodimers in endosomes. EMBO J 11:411–416.

Neefjes JJ, Stollorz V, Peters PJ, Geuze HJ, Pleogh HL (1990): The biosynthetic pathway of MHC class II but not class I molecules intersects the endocytic route. Cell 61:171–183.

Nelson CA, Roof RW, McCourt DW, Unanue ER (1992): Identification of the naturally processed form of hen egg white lysozyme bound to the murine major histocompatibility complex class II molecule I-A^k. Proc Natl Acad Sci USA 89:7380–7383.

Newcomb JR, Cresswell P (1993): Characterization of endogenous peptides bound to purified HLA-DR molecules and their absence from invariant chain-associated $\alpha\beta$ dimers. J Immunol 150:499–507.

Nguyen QV, Humphreys RE (1989): Time course of intracellular associations, processing and cleavages of Ii forms and class II major histocompatibility complex molecules. J Biol Chem 264:1631–1637.

Nuchtern JG, Biddison WE, Klausner RD (1990): Class II MHC molecules can use the endogenous pathway of antigen presentation. Nature 343:74–76.

Pamer EG, Harty JT, Bevan MJ (1991): Precise prediction of a dominant class I MHC-restricted epitope of *Listeria monocytogenes*. Nature 353:852–855.

Pamer EG, Wang C-R, Flaherty L, Fischer-Lindahl K, Bevan MJ (1992): H-2M3 presents a *Listeria monocytogenes* peptide to cytotoxic T lymphocytes. Cell 70:215–223.

Peters PJ, Neefjes JJ, Oorschot V, Ploegh HL, Geuze HJ (1991): Segregation of MHC class II molecules from MHC class I molecules in the Golgi complex for transport to lysosomal compartments. Nature 349:669–656.

Peterson M, Miller J (1992): Antigen presentation enhanced by the alternatively spliced invariant chain gene product p41. Nature 357:596–598.

Pfeifer JD, Wick MJ, Roberts RL, Findlay K, Normark SJ, Harding CV (1993): Phagocytic processing of bacterial antigens for class I MHC presentation to T cells. Nature 361:359–362.

Pfeifer JD, Wick MJ, Russell DG, Normark SJ, Harding CV (1992): Recombinant *E. coli* express a defined cytoplasmic epitope that is efficiently processed in macrophage phagolysosomes for class II MHC presentation to T lymphocytes. J Immunol 145:2576–2584.

Pierce SK, De Nagel DC, Van Buskirk AM (1991): A role for heat shock proteins in antigen processing and presentation. Curr Top Microbiol Immunol 167:83–92.

Pieters J, Horstmann H, Bakke O, Griffiths G, Lipp J (1991): Intracellular transport and localization of major histocompatibility complex class II molecules and associated invariant chain. J Cell Biol 115:1213–1223.

Pisoni RL, Acker TL, Lisowski KM, Lemons RM, Thoene JG (1990): A cysteine-specific lysosomal transport system provides a major route for the delivery of thiol to human fibroblast lysosomes: Possible role in supporting glycosomal proteolysis. J Cell Biol 110:327.

Porcelli S, Morita CT, Brenner MB (1992): CD1b restricts the response of human CD4$^-$8$^-$ T lymphocytes to a microbial antigen. Nature 360:593–597.

Powis SJ, Deverson EV, Coadwell WJ, Ciruela A, Huskisson NS, Smith H, Butcher GW, Howard JC (1992): Effect of polymorphism of an MHC-linked transporter on the peptides assembled in a class I molecule. Nature 357:211–215.

Powis SJ, Townsend AR, Deverson EV, Bastin J, Butcher GW, Howard JC (1991): Restoration of antigen presentation to the mutant cell line RMA-S by an MHC-linked transporter. Nature 354:528–531.

Puri J, Factorovich Y (1988): Selective inhibition of antigen presentation to cloned T cells by protease inhibitors. J Immunol 141:3313.

Rabinowitz S, Horstmann H, Gordon S, Griffiths G (1992): Immunocytochemical characterization of the endocytic and phagolysosomal compartments in peritoneal macrophages. J Cell Biol 116:95–112.

Reay PA, Wettstein DA, Davis MM (1992): pH dependence and exchange of high and low responder pep-

tides binding to a class II MHC molecule. EMBO J 11:2829–2839.
Reddy R, Zhou F, Nair S, Huang L, Rouse BT (1991): In vivo cytotoxic T lymphocyte induction with soluble proteins administered in liposomes. J Immunol 148:1585–1589.
Reid PA, Watts C (1990): Cell surface MHC glycoproteins cycle through primaquine sensitive intracellular compartments. Nature 346:655–657.
Riberdy JM, Cresswell P (1992): The antigen-processing mutant T2 suggests a role for MHC-linked genes in class II antigen presentation. J Immunol 148:2586–2590.
Riberdy JM, Newcomb JR, Surman MJ, Barbosa JA, Cresswell P (1992): HLA-DR molecules from an antigen-processing mutant cell line are associated with invariant chain peptides. Nature 360:474–477.
Roche PA, Cresswell P (1990): Invariant chain association with HLA-DR molecules inhibits immunogenic peptide binding. Nature 345:615–618.
Roche PA, Cresswell P (1991): Proteolysis of the class II–associated invariant chain generates a peptide binding site in intracellular HLA-DR molecules. Proc Natl Acad Sci USA 88:3150–3154.
Roche PA, Marks MS, Cresswell P (1991): Formation of a nine-subunit complex by HLA class II glycoproteins and the invariant chain. Nature 354:392–394.
Rock KL, Gamble S, Rothstein L (1990): Presentation of exogenous antigen with class I major histocompatibility complex molecules. Science 249:918.
Rock KL, Rothstein L, Gamble S, Fleischacker C (1993): Characterization of antigen-presenting cells that present exogenous antigens in association with class I MHC molecules. J Immunol 150:438–446.
Rodriquez GM, Diment S (1992): Role of cathepsin D in antigen presentation of ovalbumin. J Immunol 149:2894–2898.
Rotzschke O, Falk K, Deres K, Schild H, Norda M, Metzger J, Jung G, Rammensee HG (1990): Isolation and analysis of naturally processed viral peptides as recognized by cytotoxic T cells. Nature 353:326.
Rudensky AY, Preston-Hurlburt P, Al-Ramadi BK, Rothbard J, Janeway CA Jr (1992): Truncation variants of peptides isolated from MHC class II molecules suggest sequence motifs. Nature 359:429–431.
Rudensky AY, Preston-Hurlburt P, Hong S-C, Barlow A, Janeway CA Jr (1991): Sequence analysis of peptides bound to MHC class II molecules. Nature 353:622–627.
Sadegh-Nasseri S, Germain RN (1991): A role for peptide in determining MHC class II structure. Nature 353:167–170.
Salamero J, Humbert M, Cosson P, Davoust J (1990): Mouse B lymphocyte specific endocytosis and recycling of MHC class II molecules. EMBO J 9:3489–3496.
Schumacher TNM, Heemels M-T, Neefjes JJ, Kast WM, Melief CJM, Pleogh HL (1990): Direct binding of peptide to empty MHC class I molecules on intact cells and in vitro. Cell 62:563–567.
Sette A, Adorini L, Colon SM, Buus S, Grey HM (1989): Capacity of intact proteins to bind to MHC class II molecules. J Immunol 143:1265.
Sette A, Southwood S, O'Sullivan D, Gaeta FC, Sidney J, Grey HM (1992): Effect of pH on MHC class II–peptide interactions. J Immunol 148:844–851.
Shepherd JC, Schumacher TNM, Ashton-Rickardt PG, Imaeda S, Pleogh HL, Janeway CA, Tonegawa S (1993): TAP1-dependent peptide translocation in vitro is ATP dependent and peptide selective. Cell 74:577–584.
Shimonkevitz R, Colon S, Kappler JW, Marrack P, Grey HM (1984b): Antigen recognition by H-2 restricted T cells. II. A tryptic ovalbumin peptide that substitutes for processed antigen. J Immunol 133:2067–2074.
Shimonkevitz R, Kappler J, Marrack P, Grey H (1984b): Antigen recognition by H-2 restricted T-cells. I. Cell-free antigen processing. J Exp Med 158:303–316.
Silver JL, Guo H-C, Strominger JL, Wiley DC (1992): Atomic structure of a human MHC molecule presenting an influenza virus peptide. Nature 360:367–369.
Spies T, Bresnahan M, Bahram S, Arnold D, Blanck B, Mellins E, Pious D, DeMars R (1990): A gene in the human major histocompatibility complex class II region controlling the class I antigen presentation pathway. Nature 348:744–747.
Spies T, Cerundolo V, Colonna M, Cresswell P, Townsend A, DeMars R (1992): Presentation of viral antigen by MHC class I molecules is dependent on a putative peptide transporter heterodimer. Nature 355:644–646.
Spies T, DeMars R (1991): Restored expression of major histocompatibility class I molecules by gene transfer of a putative peptide transporter. Nature 351:323–324.
Stern LJ, Wiley DC (1992): The human class II MHC protein HLA-DR1 assembles as empty alpha beta heterodimers in the absence of antigenic peptide. Cell 68:465–477.
Stockinger B, Pessara U, Lin RH, Habicht J, Grez M, Koch N (1989): A role of Ia-associated invariant chains in antigen processing and presentation. Cell 56:683–689.
Streicher HA, Berkower IJ, Buch M, Guard FRN, Berzofsky JA (1984): Antigen conformation determines processing requirements for T-cell activation. Proc Natl Acad Sci USA 81:6831.
Swanson J, Bushnell A, Silverstein SC (1987): Tubular lysosomes morphology and distribution within macrophages depend on the integrity of cytoplasmic microtubules. Proc Natl Acad Sci USA 84:1921–1925.
Takahashi J, Cease KB, Berzofsky JA (1989): Identification of proteases that process distinct epitopes on the same protein. J Immunol 142:2221.
Teyton L, O'Sullivan D, Dickson PW, Lotteau V, Sette A, Fink P, Peterson PA (1990): Invariant chain distinguishes between the exogenous and endogenous antigen presentation pathways. Nature 348:39–44.
Townsend A, Bastin J, Gould K, Brownlee G, Andrew M, Coupar B, Boyle D, Chan S, Smith G (1988): Defective presentation of class I–restricted cytotoxic T lymphocytes in vaccinia infected cells is overcome by enhanced degradation of antigen. J Exp Med 168:1211–1224.

Townsend ARM, Gotch FM, Davey J (1985): Cytotoxic T cells recognize fragments of the influenza nucleoprotein. Cell 42:457–467.

Townsend ARM, Rothbard J, Gotch FM, Bahadur G, Wraith D, McMichael AJ (1986): The epitopes of influenza nucleoprotein recognized by cytotoxic T lymphocytes can be defined with short synthetic peptides. Cell 44:949–968.

Trowsdale J, Hanson I, Mockridge I, Beck S, Townsend A, Kelly A (1990): Sequences encoded in the class II region of the MHC related to the "ABC" superfamily of transporters. Nature 348:741–743.

Van Bleek BM, Nathenson SG (1990): Isolation of an endogenously processed immunodominant viral peptide from the class I H-$2K^b$ molecule. Nature 348:213.

Van Kaer L, Aston-Rickardt PG, Ploegh HL, Tonegawa S (1992): TAP1 mutant mice are deficient in antigen presentation, surface class I molecules and $CD4^-8^+$ T cells. Cell 71:1205–1214.

Van der Drift ACM, van Noort JM, Kruse J (1990): Catheptic processing of protein antigens: Enzymic and molecular aspects. Semin Immunol 2:255–271.

Van Noort JM, Boon J, Van der Drift AC, Wagenaar JP, Boots AM, Boog CJ (1991): Antigen processing by endosomal proteases determines which sites of spermwhale myoglobin are eventually recognized by T cells. Eur J Immunol 21:1989–1996.

Van Noort JM, Van der Drift ACM (1989): The selectivity of cathepsin D suggests an involvement of the enzyme in the generation of T cell epitopes. J Biol Chem 264:14159.

Vidard L, Rock KL, Benacerraf B (1992): Diversity in MHC class II ovalbumin T cell epitopes generated by distinct proteases. J Immunol 149:498–504.

Weber DA, Buck LB, Delohery TM, Agostine N, Pernis B (1991): Class II MHC molecules are spontaneously internalized in acidic endosomes by activated B cells. J Mol Cell Immunol 4:255–268.

Wei ML, Cresswell P (1992): HLA-A2 molecules in an antigen-processing mutant cell contain signal sequence–derived peptides. Nature 356:443–446.

Weiss S, Bogen B (1989): B lymphoma cells process and present their endogenous immunoglobulin to major histocompatibility complex–restricted T cells. Proc Natl Acad Sci USA 86:282–289.

Weiss S, Bogen B (1991): MHC class-II–restricted presentation of intracellular antigen. Cell 64:767–776.

Werdelin O, Mouritsen S, Petersen BL, Sette A, Buus S (1988): Facts on the association of antigen fragments with MHC molecules in cell-free systems, and speculation on the cell biology of antigen processing. Immunol Rev 106:181.

Wettstein DA, Boniface JJ, Reay PA, Schild H, Davis MM (1991): Expression of a class II major histocompatibility complex (MHC) heterodimer in a lipid-linked form with enhanced peptide/soluble MHC complex formation at low pH. J Exp Med 174:219–228.

Wick MJ, Pfeifer JD, Findlay KF, Harding CV, Normark SJ (1994): Compartmentalization of defined epitopes expressed in E. coli has only a minor influence on the efficiency of phagocytic processing for class I and class II major histocompatibility complex presentation to T cells. Infect Immun 61:4848–4856.

Yang Y, Fruh K, Chambers J, Waters JB, Wu L, Spies T, Peterson PA (1992a): Major histocompatibility complex (MHC)–encoded HAM2 is necessary for antigenic peptide loading onto class I MHC molecules. J Biol Chem 267:11669–11672.

Yang Y, Waters JB, Fruh K, Peterson PA (1992b): Proteasomes are regulated by interferon γ: Implications for antigen processing. Proc Natl Acad Sci USA 89:4928–4932.

Yoshikawa M, Watanabe M, Hozumi N (1987): Analysis of proteolytic processing during specific antigen presentation. Cell Immunol 110:431–435.

Yurin VL, Rudensky AY, Mazel SM, Blechman JM (1989): Immunoglobulin-specific T–B interaction. II. T cell clones recognize the processed form of B cells' own surface immunoglobulin in the context of the major histocompatibility complex class II molecule. Eur J Immunol 19:1685–1691.

Zhang W, Young ACM, Imarai M, Nathenson SG, Sacchettini JC (1992): Crystal structure of the major histocompatibility complex class I H-2Kb molecule containing a single viral peptide: Implications for peptide binding and T-cell receptor recognition. Proc Natl Acad Sci USA 89:8403–8407.

Ziegler K, Unanue ER (1982): Decrease in macrophage antigen catabolism by ammonia and chloroquine is associated with inhibition of antigen presentation to T cells. Proc Natl Acad Sci USA 78:175–178.

ABOUT THE AUTHOR

CLIFFORD V. HARDING is Assistant Professor of Pathology at Case Western Reserve University in Cleveland, OH, where he teaches and performs research in immunology, cell biology, and pathology. He received his A.B. *magna cum laude* with highest honors in Biology from Harvard College and subsequently earned his M.D. and Ph.D. in Cell Biology from Washington University, St. Louis, where he worked with Philip Stahl and John Heuser on endocytic transport and intracellular processing of transferrin. He remained at Washington University for postgraduate training in anatomic pathology and postdoctoral research in antigen processing with Emil Unanue. He then joined the faculty in Pathology at Washington University before moving to Case Western Reserve University. He is the recipient of a Pfizer Scholar Award and an American Cancer Society Junior Faculty Research Award. His recent research has addressed the cell biology of antigen processing, processing of bacteria by macrophages, processing of liposome-encapsulated antigens, and the intracellular localization, transport, and function of major histocompatibility molecules.

VIRAL-RELATED/HIV PROTEASE

Viral Proteases: Structure and Function

Tudor I. Oprea, Chris L. Waller, and Garland R. Marshall

INTRODUCTION

Post-translational modification of viral proteins is an essential part of the life cycle of many, if not all, classes of virus. Dependent on replication within the host cell, viruses subvert many cellular functions for their own purposes, often limiting their own genetic information to that required for synthesis of their characteristic capsid proteins plus those necessary for takeover and control of the host cell. Evolutionary pressure to minimize the size of the viral genome has led to production of polyproteins that are subsequently cleaved by virally encoded proteases or by constitutive enzymes of the infected cells [Kemp et al., 1992]. The fact that the virus encodes a protease is testimony to the critical role played by proteolysis in the viral life cycle. Not all viruses encode a protease, but an evolutionary advantage may be the ability to infect cells other than those expressing a particular proteolytic pathway for those that do. As our knowledge of these proteases in viral maturation and infectivity has increased, development of inhibitors of these enzymes has become a logical therapeutic approach. Inhibitors of HIV protease are undergoing clinical evaluation for treatment of AIDS, and inhibitors of other viral proteases as novel antivirals are under development.

In this review, a compilation of current knowledge on characterized viral proteases has been made. Several positive-stranded viruses (the Picornaviridae, Togaviridae, and Flaviviridae families) are discussed, with emphasis on the most studied members of each family.

The best studied viral protease, from the human immunodeficiency virus (HIV), of the Retroviridae family is discussed in more detail. Finally, a short summary of the role of cellular proteolytic enzymes in viral maturation is presented. Throughout this text, the following convention is used: When citing mutation studies, a one-letter code for amino acid residues was used (e.g., $D^{134}N$ denotes that aspartate^{134} was mutated to asparagine), while a three-letter code was used for sequence residues (e.g., Asp^{134} denotes the Aspartate^{134} in the primary sequence).

PICORNAVIRIDAE

Picorna is the acronym [Melnick, 1993] for *p*oliovirus, *i*nsensitivity to ether (a distinguishing property), *c*oxsackievirus, *o*rphan virus, *r*hinovirus, and *RNA* (its genomic characteristic), which are subgroups and properties considered important for naming the picornaviridae family. It includes the following genera [Melnick, 1991]: 1) Enterovirus (among which poliovirus, coxsackie A and B, and echovirus that infect humans), 2) cardiovirus (the encephalomyocarditis virus group), 3) rhinovirus (the common cold viruses), 4) aphthovirus (foot-and-mouth disease virus), 5) heparnavirus (hepatitis A virus of humans), and 6) other picornaviruses not yet assigned to genera (equine and invertebrate viruses). Picornaviruses are small, nonenveloped, positive-strand RNA viruses [Fenger, 1991b], and their viral growth cycle is dependent on the processing of a single

RNA open-reading frame (ORF) with the aid of viral proteases acting on the polyprotein product. For details concerning picornavirus protein processing, the reader is referred to the specific reviews of Lawson and Semler [1990] and Palmenberg [1990] and to the more general reviews of Wellink and van Kammen [1988], Krausslich and Wimmer [1988], and Kay and Dunn [1990].

Role of Viral Proteases During Replication of Poliovirus

Viral RNA is single stranded and covalently linked [Fenger, 1991b] to a viral protein, VPg, which is removed prior to translation by an enzyme detected in both infected and uninfected cells. Poliovirus mRNA is translated into a continuous polyprotein, P123 (Fig. 1). The P1–P2 cleavage site is located between Tyr-Gly and is mediated by the viral protease 2A ($2A^{pro}$) by an autocatalytic process [Yu and Lloyd, 1991]. The P2–P3 connection is cleaved by the viral protease 3C ($3C^{pro}$) at a Gln-Gly site [Lawson and Semler, 1990]. The resulting P1, P2, and P3 polyprotein subunits are cleaved by $3C^{pro}$ and yield 10 products [Fenger, 1991b], as noted above. $3C^{pro}$ flanking sequences are cleaved at Gln-Gly sites by autocatalysis [Wellink and vanKammen, 1988], while 3CD (the $3C^{pro}$ precursor) cleaves Gln-Gly bonds to release poliovirus capsid proteins [Lawson and Semler, 1990]. The VP_0 to VP_4 and VP_2 cleavage is not mediated [Fenger, 1991b] by $2A^{pro}$ or $3C^{pro}$, but is an autocatalytic process [Harber et al., 1991] required for viral infectivity (see text for further details).

Models of Picornavirus Proteases

Cysteine–active-center viral proteases were identified in the genomic sequences of four genera of the picornavirus family: rhinoviruses, enteroviruses, cardioviruses, and aphthoviruses [Bazan and Fletterick, 1988]. Two plant viruses also encode cysteine proteases that are homologous to picornaviral proteases: cowpea mosaic virus (a comovirus) and two potyviruses [Gorbalenya et al., 1989b]. These viruses contain positive-strand RNA molecules that are translated into single large polyproteins that are proteolytically processed at preferred Gln-Gly or related Gln-Ser, Glu-Ser, or Glu-Gly sites by the $3C^{pro}$ protease and at Tyr-Gly sites by the $2A^{pro}$ protease, respectively [Bazan and Fletterick, 1988; Wellink and van Kammen, 1988].

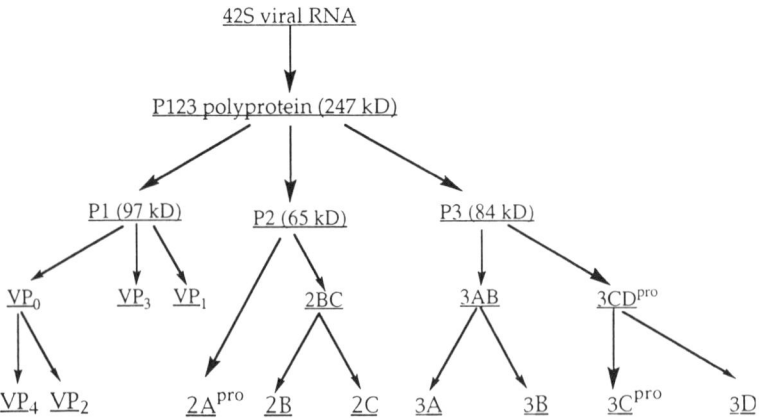

Fig. 1. *Translation and cleavage of poliovirus. Primary translation of genomic RNA (42S) yields a polyprotein (P123) that is cleaved in P1, P2, and P3. Subsequently, P1 is cleaved into structural proteins VP_0, VP_1, and VP_3. VP_0 is cleaved during viral maturation into VP_2 and VP_4. P2 is cleaved into several fragments, including the protease 2A. P3 is cleaved in two fragments, 3AB and 3CD. The latter, a protease itself, is cleaved in 3C (a protease) and 3D.*

The Bazan model. Based on computer-assisted sequence alignment and crystallographic data, the following model [Bazan and Fletterick, 1988] for viral 3C^{pro} and 2A^{pro} proteases was proposed. The picornavirus protease were suggested to be structurally related to serine proteases, not to cysteine proteases, with the active-site nucleophile being the sulfhydryl of cysteine instead of the hydroxyl of serine. The residues His$^{20/40}$, Asp$^{38/85}$, Cys$^{109/147}$ (numbers correspond to poliovirus 2A/3C proteases) were proposed to be absolutely conserved and were successfully superimposed on His^{57}, Asp^{102}, and Ser^{195} of the trypsin-like serine protease catalytic triad (residues numbered as in human trypsin). The proposed topology was 12 β-strands and one C-terminal α-helix [Bazan and Fletterick, 1988]. The alignment of the same residues was not possible with the catalytic triad (Asp^{32}, His^{64}, Ser^{220}) of the subtilisin-like Ser proteases.

The 3C^{pro} specificity pocket S1 is formed by Ala(Pro)-Thr, similar to the Ser-Thr pair of *Staphylococcus aureus* (Sa) protease residues located at the bottom of S1. In the cellular serine protease sequences, these residues are Asp-Ser (trypsin), Ser-Ser (chymotrypsin), and Ser-Gly (elastase). At the side of the S1 pocket, the following differences are observed: A hydrogen-bond pattern is proposed between His^{213}, Thr^{190} and the substrate-bound Gln for the 3C^{pro} viral protease (similar to the Sa protease, which binds Glu in S1); this sequence pattern is not matched by the cellular proteases, where residue 213 is typically a small hydrophobic amino acid residue (Val in trypsin and chymotrypsin, Thr in elastase). At the top of the S1 pocket, residue 215 is a small hydrophobic amino acid (Ala-Gly in the 3C^{pro} proteases, Gly in the Sa protease), as opposed to the large aromatic amino acid in cellular proteases (Trp in trypsin and chymotrypsin, Phe in elastase).

The 2A^{pro} picornaviral proteases were aligned with three small trypsin-like bacterial proteases (Fig. 2). Identical residues were noted at the interface between β-barrel domains, and they are essential to the enzymatic structure and function. In addition to the proposed catalytic triad (His^{57}, Asp^{102}, and Cys^{195}), other structural (Gly^{211} and Gly^{216}, which possibly play a role in folding) and functional (Tyr^{171}, which hydrogen bonds to Thr^{214}) determinants are markers for significant similarity. For a detailed discussion on 2A^{pro} see The 2A Protease, below.

The Gorbalenya model. A different model for 3C proteases was proposed based on sequence alignment (Finkelstein algorithm) of 3C proteases from picornaviruses and similar proteins from plant viruses (como-, nepo-, and potyviruses) to the x-ray crystallographically determined structures for three cellular proteases (trypsin, chymotrypsin, and elastase) and two bacterial proteases [Gorbalenya et al., 1989a] (see Fig. 3).

As in the Bazan model, a cysteine replaced the serine as the key residue in the catalytic triad. The topology of the viral proteases is proposed to consist of 12 β-strands and two terminal α-helices (at the C and N termini). This is somewhat different than that of chymotrypsin-like proteases (12 β-strands and a C-terminal α-helix), with the hypothesis that the 3C^{pro} molecule consists of two twisted antiparallel β-barrels (six β-strands each in the hydrophobic core) connected by a long loop (see Fig. 4).

The proposed catalytic triad His^{40}Glu(Asp)^{71} Cys^{147} differs from that of the Bazan model [Bazan and Fletterick, 1988] in that Glu(Asp)^{71} replaces Asp^{85}. The latter Cys^{147} residue is not considered important in the Gorbalenya model, since it lies in the loop connecting the two domains. This is supported by experimental evidence, discussed below [Kean et al., 1991]. The conserved Gly residues in the putative strands K and L are the same as in the previous model [Bazan and Fletterick, 1988]. In the 3C^{pro} molecules [Gorbalenya et al., 1989a], a partially conserved dipeptide Thr-Arg(Lys) is reportedly the determinant for substrate specificity (Thr is the same as Thr^{190} above, but the dipeptide is Ala(Pro)-Thr in the Bazan model).

The 3C Protease

In poliovirus-infected cells, autocatalytic cleavage of the P3 polyprotein results in the

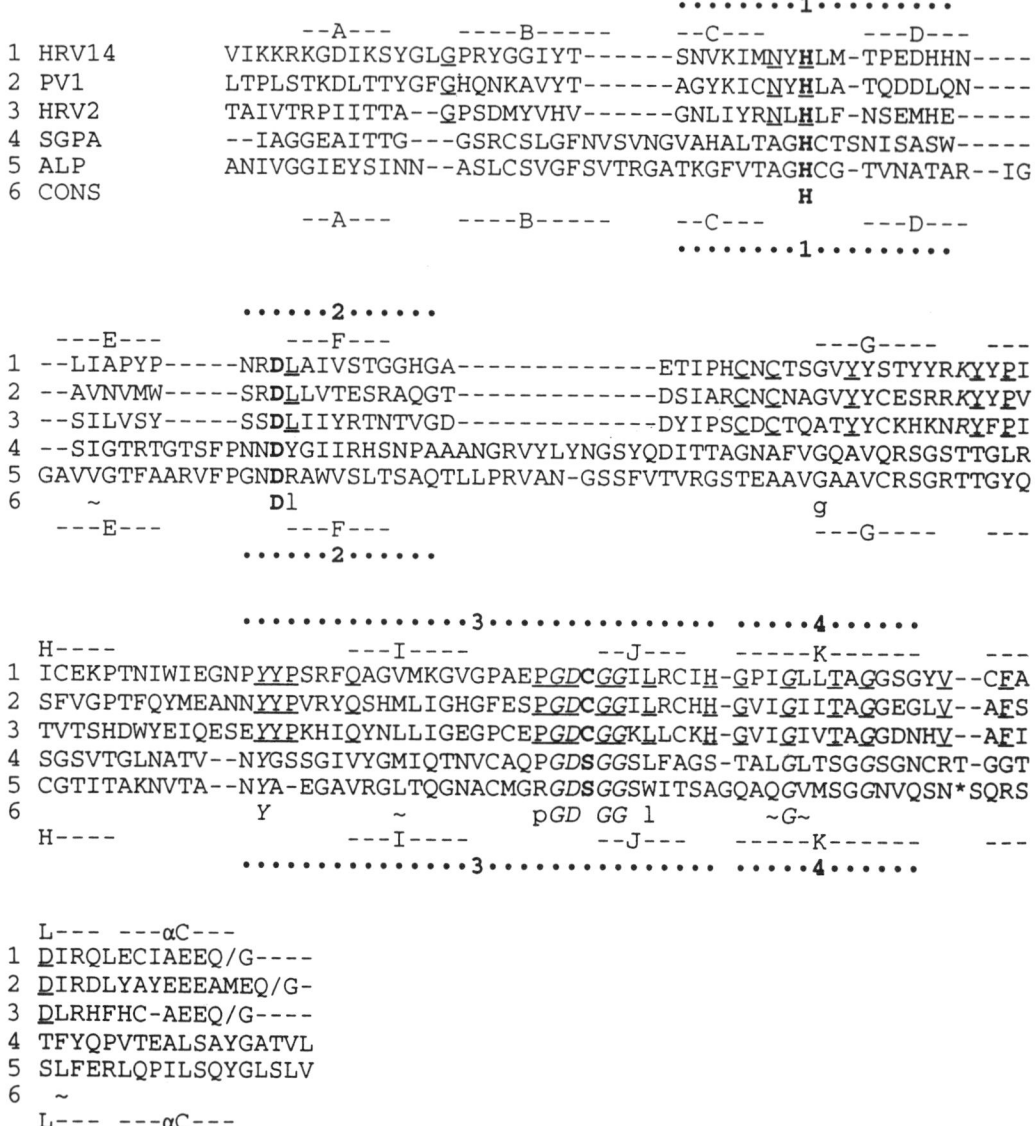

Fig. 2. Alignment of the picornaviral 2A proteases with small bacterial serine proteases. The similarity sequences (labeled 1–4) are marked (•). The secondary structure of the bacterial proteases (as observed in crystals) is shown: 12 β-strands (designated A–L) and the C-terminal α-helix (αC). Bold, putative catalytic residues; underlined, identical picornaviral residues (five sequences); ~, hydrophobic residues (V, L, I, M, F) in the original eight sequences; upper case, identical residues in all eight sequences; lower case, identical residues in six out of eight residues. The four picornaviruses are the same as in Figure 1; the bacterial proteases are SGPA, Streptomyces griseus protease A; ALP, Lysobacter enzymogenes α-lytic protease. CONS, consensus. Not shown: the bovine enterovirus and the coxsackievirus B3 strain Nancy 2A proteases and the S. griseus protease B sequence. [Modified from Bazan and Fletterick, 1988, with permission of the publisher.]

Fig. 3. Alignment of amino acid sequences of 3C-like proteases using 18 sequences, from which the consensus sequence was derived (residues that occurred in 14 out of 18 sequences were included in consensus). ~, Hydrophobic residues (V, L, I, M, F); bold, putative catalytic residues; italic, proposed binding site; un-

Viral Proteases

```
1 HRV14      ----------GPNTEFALSLLRKNIMTIT-----TSKGEFTGLGI-HDRVCVIPTH
2 PV1        ----------GPGFDYAVAMAKRNIVTAT-----TSKGEFTMLGV-HDNVAILPTH
3 HRV2       ----------GPEEEFGMSLIKHNSCVIT-----TENGKFTGLGV-YDRFVVVPTH
4 EMV        --------GPNPVDMFEKYVAKHVTAPIGFVYP-TGVSTQTCLLV-RGRTLVVNRH
5 FMDV       --------SGAPPTDLQKMVMGN-TKPVELNLDGKTVAICCATGV-FGTAYLVPRH
6 HAV        ---------SQSTLEIAGLVRNKLVQFGVGEKNGCVRWVMNALGV-KDDWLLVPSH
7 TEV        ----------GESLFKGPRDYNPISSTICHLTNESDGHTTSLYGIGFGPFIITNKH
8 CONS               ~      K     ~                ~G~        ~~  H
                           R

1 ----AQPGD--DVLV-----NGQKIRVKDKYKL--VDPENIN--LELTVLTLDRN---------
2 ----ASPGE--SIVI-----DGKEVEILDAKAL--EDQAGTN--LEITIITLKRN---------
3 ----ADPGK--EIQV-----DGITTKVIDSYDL--YSKNGIK--LEITVLKLDRN---------
4 ----MAESDWTSIVV-----RGVTHARSTVKIL-AIAKAGKE--TDVSFIRLSSG---------
5 ----LFAEKYDKIMLD---GRAMTDSDYRVFEF-EIKVKGQDMLSDAALMVLHRG---------
6 ----AYKFEKDYEMMEFYFNRGGTYYSISAGNV-VIQSLDGV-FQDVVLMKVPTI---------
7 ----LFRRNNGTLLV-----QSLH-GVFKVKNTTTLQQHLIDG-RDMIIIRMPKD---------
8      ~ ~                                    E~  ~~ ~
                                              D

1 EKF-RDIRGFIS-E-DLEGVD-ATLVVHSNNFT--NT--ILEVGPV---TMAGLIN--LSSTPT
2 EKF-RDIRPHIPTQ-ITETND-GVLIVNTSKYP--NM--YVPVGAV---TEQGYLN--LGGRQT
3 EKF-RDIRRYIPNN-EDDYPN-CNLALLANQPE--PT--IINVGDV---VSYGNIL--LSGNQT
4 PLF-RDNTSKFVKA-GDVLPT-GAAPVTGIMNT--DIP-MMYTGTF---LKAGVSVPVETGQTF
5 NCV-RDITKHF-RD-TARMKK-GTPVVGVVNNA--DVGRLIFSGEA---LTYKDIVVCMDGDTM
6 PKF-RDITQHFIKK-GDVPRALNRLATLVTTVN--GTPMLISEGPLKMEEKATYVHKKNDGTTV
7 --F-PPFPQKLKFR-EPQREE--RICLVTTNFQ------TKSMSSM---VSDTSC---TFPSSD
8     F   D~    ~                                  ~              ~

1 NRMIRYDYATK----TGQCGG-VLCAT-G---KIFGIH-VGG-NRQGFSAQLKK-QYFV----
2 ARTLMYNFPTR----AGQCGG-VITCT-G---KVIGMH-VGG-NGSHGFAAALKR-SYFT----
3 ARMLKYSYPTK----SGYCGG-VLYKI-G---QVLGIH-VGG-NGRDGFSAMLLR-SYFT----
4 NHCIHYKANTR----KGWCGSALLADL-GGSKKILGIH-SAG-SMGIAAASIVSQ-EMIRAV--
5 PGLFAYKAATR----AGYCGGAVLAKD-GADTFIVGTH-SAG-GNGVGYCSCVSR-SMLQKM-K
6 DLTVDQAWRGKGEGLPGMCGGALVSSNQSIQNAILGIH-VAG-GNSILVAKLVTQ-EMFQNI--
7 GIFWKHWIQTK----DGQCGSPLVSTRDG---FIVGIHSASNFTNTNNYFTSVPK-NFMELL--
8         TK        G CG  ~~      G    ~~G~H    G              ~    ~
           R

1 -----EKQ--
2 -----QSQ--
3 -----DVQ--
4 -VNAFEPQ--
5 AHVDPEPHHE
6 ---DKKIE--
7 ----TNQE--
8          Q
           E
```

derlined, residues that are invariant in picornaviral 3C proteases. The following seven sequences are shown above: HRV14 and HRV2 (human rhinovirus types 14 and 2); PV1 (poliovirus type 1); EMV encephalomiocarditis virus); FMDV (foot-and-mouth disease virus); HAV (hepatitis A virus [picornaviruses]); TEV, tobacco etch virus (potyvirus). Not shown: the putative protease sequences from coxsackie virus type B3; echovirus type 9; bovine enterovirus; bovine enterovirus; Theiler murine encephalomyelitis virus; the HRV types 1a, 1b, and 89 (picornaviruses); the cowpea mosaic virus (comovirus); the tomato black ring virus (nepovirus); the tobacco vein mottling virus (potyvirus); the southern bean mosaic virus (sobemovirus). [Modified from Gorbalenya et al., 1989, with permission of the publisher.]

```
              --A--     ---B----    ---C---           --D-       - ---E-----
      SGPA    4-eAItT-1-GSrCsLGf-6-vahALtagHcT-2 Sasw-0-  ---SIgTRtgt
      CHT    13-wQVsL-5-FhFCGgsL-2-ENWVVtaaHcg-4-DVVV-13-QKLKIaKVFKn
      TRP    13-yQVsL-3-YhFCGgsL-2-sQWVVsaaHcY-3-iQVr-13-QfISaSKSiVh
      ELA    13-SQIsL-8-AhtCGgtL-2-QNWVMtaaHcv-5-FrVV-13-QyVGVqKIvVh
                :         :           :              :               :
      HRV14  14-imtiT-3-GeFtGLGI-1-DrvcViptHaq-3-DVLV-2-  QKIRVkdYKYl
      PV1    14-ivtaT-3-GeFtmLGV-1-DNvAIlptHaS-3-SIVI-2-  KeVeIldakal
      HRV2   14-ScViT-a-GkFtGLGV-1-DrFVVvptHad-3-EIqV-2-  iTTKVidSYdl
      EMV    16-TApig-7-STqtcLlV-1-grtLVvnrHma-5-SIVV-2-  vTharSTVkIl
      FMDV   16-Tkpve-8-AicCatGV-1-gtaYLvprHlF-5-kIML-4-  mTdSdyRVFef
      HAV    15-vQfgV-8-WvmnaLGV-1-DDWLLvpsHaY-5-YEMM-7-  gTyysiSagnv
              --A--     ---B----    ---C---           --D-       - ---E-----


              ---F----     ----G---      ---H---      -I--
      SGPA    - 4-NDygIIRh-28-GQAVqrSg -4-LrSgsvt-16-MIQtN
      CHT    - 9-NDITLLKL-24-GTtcVTTg-16-qaSLplL-16-MIcag
      TRP    - 9-NDIMLIKL-22-GTQcLISg-16-cLKapiL-16-MFcag
      ELA    -11-yDIALLRL-24-NSPcyITg-15-qaYLpTV-18-MVcag
                ::   ::::                              ::
      HRV14  - 7-lELTVLtL-21-aTlVVhSn-5-  ILeVgpV-15-MIRyD
      PV1    - 7-lEITIItL-22-GVlIVnTs-5-  yVpVgaV-15-TLmyN
      HRV2   - 7-lEITVLKL-22-cnlaLLan-5-  IInVgdV-15-MLKys
      EMV    - 8-tDVSfIRL-22-GaApVTgi-6-  MMYtgTf-17-CIhyk
      FMDV   -10-sDaALMvL-21-GTPVVgvv-7-  LIFsgea-17-LFayk
      HAV    - 9-QDVVLMKV-23-NrlaTLvt-7-  LISegpL-20-TVDqa
              ---F----     ----G---      ---H---      -I--


              ---J---    -----K-----   -L--   ---αC---
      SGPA    -4-PGDsGGsLfAg-1-taLGLtSGGSG-5-GttF-5-EaLsaYga - 3
      CHT    -7-mGDsGGpLVCK-5-tLVGIvSwGSs-5-TpGV-5-ALV-NWVQ - 6
      TRP    -9-qGDsGGpVVCS-1-KLqGIvSwGSG-5-KpGV-5-Nyw-SWIK - 6
      ELA    -8-4GDsGGpLhCl-5-AVhGVtSfvSr-7-KptV-5-AyI-SWIN - 6
                ::   :   :          ::::              :
      HRV14  -4-tGQcGG-VLCa-2-KIfGIhvGGnG-0-RqGF-2-QLkkQYfv - 3
      PV1    -4-aGQcGG-VITc-2-KVIGMhvGGnG-0-ShGF-2-ALkrSYft - 3
      HRV2   -4-sGycGG-VLyK-2-qVLGIhvGGnG-0-RdGF-2-mLLrSYft - 3
      EMV    -4-kGwcGsaLLAd-5-KILGIhSaGSm-0-Giaa-2-iVsqEmIR - 9
      FMDV   -4-aGYcGGaVLAK-5-fIVGthSaGgn-0-GvGY-2-cVsrSmLQ -13
      HAV    -8-PGmcGGaLVSS-6-AILGIhvaGgn-0-SilV-2-lVtqEmfQ - 7
              ---J---    -----K-----   -L--   ---αC---
```

Fig. 4. Sequence alignment of 3C proteases and chymotrypsin-like proteases based on secondary structure superposition, proposed by Gorbalenya et al. [1989a]. The 12 β-strands (designated A-L), the C-terminal α-helix (αC), and some adjacent conserved regions are shown. The number of residues in each secondary structure element is shown. For picornaviral 3C proteases, the alignment shown in Figure 1 was used. For chymotrypsin-like proteases, data are from x-ray analysis. Numbers stand for length of spacer and terminal extensions. Amino acid residues having at least one identical or homologous counterpart in the other sequence set are designated by capitals. Colons, positions occupied by identical or homologous residues in at least half of the sequences of each of the sets; bold, putative catalytic residues. The six picornaviruses are the same as in Figure 1; the chymotrypsin-like proteases are SGPA, Streptomyces griseus protease A; CHT, chymotrypsin; TRP, trypsin; ELA, elastase.

release of two different proteins, 3C and 3CD, which have been reported to have proteolytic activity. 3C is responsible for cleavage of the P1, P2, and P3 polyprotein [Fenger, 1991] precursors, while 3CD is a protease polymerase that releases capsid proteins [Ypma-Wong et al., 1988] from P1.

3C is a 20-kDa protein [Baum et al., 1991] originally classified as a member of the cysteine protease class. The inhibitor profile [Baum et al., 1991] of 3C^{pro} showed that serine protease inhibitors TPCK (tosyl phenylalanyl chloromethyl ketone) and TLCK (tosyl lysyl chloromethyl ketone)—are weaker, while PMSF (phenylmethylsulphonyl fluoride) and 3,4-dichloroisocumarin were effective inhibitors of the protease, as well as $ZnCl_2$ (known to complex surface cysteines) and cystatin—both cysteine protease inhibitors. Other inhibitors, such as E64, leupeptin (cysteine protease inhibitors) and pepstatin A (aspartyl protease inhibitor) had no effect. These results provide experimental support for the computer-based models of Bazan and Gorbalenya that 3C^{pro} is a serine-like protease with a crucial cysteine residue.

The 73-kDa 3CD polyprotein, composed of the 3C protease and the 3D polymerase regions of the poliovirus genome, was shown to be stable to autodigestion (only 25% 3D release) and 3C-mediated proteolysis, both in vitro (in trans) and in poliovirus-infected HeLa cells [see Baum et al., 1991, and references therein]. Because of the role in capsid protein release from the P1 precursor, it was suggested that resistance to complete autodigestion is a regulatory feature required for proper virion assembly.

3C^{pro} site-directed mutagenesis studies in human rhinovirus (HRV) protease [Cheah et al., 1990] were conducted using the following point mutations: $C^{146}S$, $C^{146}M$, $C^{146}T$, $H^{40}D$, $D^{85}A$, $T^{141}S$, $G^{158}D$, $H^{160}N$, and $G^{162}D$. Cys^{146} is the HRV equivalent of Cys^{147} (the active-site nucleophile residue in picornaviruses). His^{40} and Asp^{85} are the active-site residues proposed in the Bazan model [Bazan and Fletterick, 1988]. The other mutated residues were selected after alignment [Bazan and Fletterick, 1988] of HRV-14 3C^{pro} with trypsin (Fig. 4). Thr^{141}, His^{160}, and Gly^{162} are the 3C equivalents of Ser^{190}, Val^{213}, and Trp^{215} in human trypsin, which are known to be important for substrate binding and specificity. Gly^{158} was chosen as an example of a highly conserved residue occurring in the vicinity of the predicted [Bazan and Fletterick, 1988] specificity pocket. Proteolytic activity was completely abolished for all Cys^{146}, His^{40}, and Asp^{85} mutations. While this seemed to confirm the catalytic triad proposed in the Bazan model, the Glu^{71} mutations were not investigated, and the possible implication of Glu^{71} in the binding site [Gorbalenya et al., 1989a] could not be ruled out. Proteolysis was abolished also in the Gly^{158}, His^{160}, and Gly^{162} mutants, which supports the originally proposed [Bazan and Fletterick, 1988] role for these residues in cleavage specificity. Consistent with these results, the $H^{161}G$ mutant in poliovirus (residue 161 is the equivalent of the His^{160} in HRV-14) resulted in complete loss of enzymatic activity [Ivanoff et al,. 1986], while mutations on Cys^{153} (a nonconserved residue) had a negligible effect. The $T^{141}S$ mutation markedly reduced enzyme activity, which could be explained by a weaker hydrogen bond interaction since serine has a shorter side chain. This is supported by immunoprecipitation data [Cheah et al., 1990] that suggests that Thr^{141} lies in an accessible surface region.

The 3C^{pro} catalytic triad controversy was addressed by Kean et al. [1991], who mutated Glu^{71} of the poliovirus type 1 protease to Asp or Gln, and Asp^{85} to Glu. The $E^{71}Q$ and $E^{71}D$ mutants did not give rise to a virus, and secondary cleavage products VP_0, VP_1, VP_3, 2C, and 3CD were not detected. However, high levels of primary cleavage products were identified, thus confirming that the P2/P3 cleavage could be carried out by the 3C^{pro} mutant proteases. These results suggest that a different catalysis mechanism exists for the two cleavage steps.

The $D^{85}E$ mutant is viable. Infectious virus with 20-fold lower specific infectivity compared with wild type was obtained after transfection. A similar phenotypic defect was shown for other 3C^{pro} mutants ($V^{54}A$ [Dewalt and Semler, 1987] and $I^{74}T$ [Kean et al., 1988]).

During the experiment, a large number of plaque variants were spontaneously recovered from the sequence encoding Glu^{85} to Asp^{85}, suggesting strong selective pressure to maintain aspartate at position 85. This residue is conserved in all sequenced picornaviruses. The $D^{85}N$ mutant could not be separated due to small differences in the hybridization temperature between wild-type and mutant cDNA. The $K^{60}I$ mutant shows increased efficiency in processing the 3C/3D junction, and it is possible that the region is involved in the recognition of the Gln-Gly pair at this site. The $I^{74}T$ mutant [Kean et al., 1988] 3C protease autodigested its N terminus, but not its C terminus, suggesting an altered site selectivity for the enzyme.

In a different determination, $K^{52}R$, $K^{52}I$, and $K^{52}T$ substitutions resulted in viable mutant polioviruses [Dewalt and Semler, 1987], while $G^{51}V$ and $G^{51}A$ were not viable. Surprisingly, $G^{51}D$ (a less conservative change) yielded a viable mutant, since Gly^{51} is a highly conserved residue. $V^{54}A$ and $V^{54}G$ mutations led to viable virus, while complete deletion of the Val^{54} codon led to loss of infectivity. This may be an effect of radical changes in the tertiary structure of the enzyme that altered the catalytic mechanisms.

Based on the hypothesis that $3C^{pro}$ digests unidentified *E. coli* proteins, which affect colony size and growth, a different strategy to isolate $3C^{pro}$ mutants was employed [Baum et al., 1991]. Several mutations resulting in large-sized colonies were assessed for cleavage of the 2C3AB polyprotein. Impaired proteolytic activity was noted for mutations near the active-site Cys^{147}. $A^{144}V$ (a conservative mutation) and $Q^{146}L$ mutants displayed 50%–80% and 30%–40% activity relative to wild-type cleavage at three different temperatures, respectively. An $H^{168}L$ mutant, a residue located in or near the substrate binding pocket, expressed 90%–130% activity. Other mutants, $P^{38}T$, $Y^{138}N$, and $A^{172}P$ yielded inactive proteases (at all three temperatures). This methodology allows the hypothesis that the active mutant proteases were ineffective in cleaving the *E. coli* proteins and is useful to point out mutants with specific picornaviral proteolytic activity impairment.

Degradation of $3C^{pro}$ was studied for the encephalomyocarditis virus (EMV) 3C protease, which undergoes rapid degradation both in vivo in mouse cells and in vitro in reticulocyte lysate [Oberst et al., 1993]. It was observed that EMV-$3C^{pro}$ is not directly involved in its own proteolysis and that $3C^{pro}$ activity is not required for the process. The results also suggest that the reticulocyte proteolytic system is selective toward viral proteins and that the process is ATP dependent. The similarity of EMV $3C^{pro}$ with other proteases is outlined in Figure 3 (picornaviral alignment) and Figure 4 (secondary structure prediction).

The 2A Protease

The 2A protease was studied by Yu et al. [1991, 1992] and is a small protein with a molecular weight of 16,400, originally classified as cysteine protease because of its sensitivity to alkylating agents. A rapid cleavage assay (based on fluorescein-labeled peptides) combined with a plasmid construction/expression method [Alvey et al., 1991] was used to mutate and assess 2A proteases. The observed low trans-cleavage activity is explained by viral protein in situ cis-cleavage, but both assay methods were used (with the synthetic peptide substrate STKNLTTY*GFGHQN(K)A, where lysine is the fluorescein-labeled residue).

Human rhinovirus serotype 2 (HRV2) $2A^{pro}$ was studied for synthetic substrate specificity [Sommergruber et al., 1992]. Changes show an open binding site for the P1 position (which can accommodate Met, Leu, Tyr, Phe, Thr, and Arg but not Val, Pro, Gly, and Asp), while an important role in enzyme–substrate interaction is played by P2 and P1´. Thus, no cleavage was observed when P1´ Gly was substituted with Phe, Thr, Lys, or Asp, as well as when P2 Thr was substituted with Ala, Val, Leu, Ile, Phe, Gly, Tyr, Asp, or Glu. However, cleavage was detected when P2 Thr was replaced with Pro, Ser, Asn, Gln, Lys, Arg, or His. Threonine at the P3 position is essential for substrate recognition in the P3 protein, where $2A^{pro}$ produces 3C´

and 3D′ in an alternate inefficient reaction [Yu and Lloyd, 1991].

A rapid screening method (Liebig et al., 1991] for viral protease mutants (proteinase trapping) that allows identification of both active and inactive proteinase mutants was used to study HRV2 2A^{pro} mutants. While S^{3}C and S^{83}G did not affect 2A^{pro} function, F^{130}S and F^{130}L mutants yielded inactive enzymes. Phe^{130} is a conserved residue in all picornaviral 2A^{pro} (see Fig. 2).

Autocatalytic activity [Yu and Lloyd, 1991] was absent in mutants H^{20}D, H^{20}N, and C^{109}S. These mutants show a complete loss of the post-translational cleavage of p220 (post-translational effects are discussed in Post-Translational Effects on the Host Cell Protein Synthesis, below). Autocatalytic activity was present in mutants Y^{88}F, Y^{88}L, Y^{88}S, Y^{89}F, and Y^{89}L, with some activity conserved in T^{124}S (better than T^{124}N) and D^{38}E (better than L^{39}G). During the post-translational p220 cleavage test, only Y^{88}F, Y^{89}F, and T^{124}S maintained wild-type activity levels. These results suggest that Asp^{38} may not be part of the catalytic triad, but rather part of a substrate-binding region, while His^{20} and Cys^{109} are essential for 2A^{pro} activity. In the Bazan model, Tyr^{88} and Thr^{124} were proposed to be stabilizing Asp^{38} via a putative hydrogen-bond network; however, since replacement of Tyr with Phe did not alter 2A^{pro} activity, this hypothesis was not validated.

The 2A^{pro} inhibitor profile [Yu and Lloyd, 1991] shows that a serine-protease inhibitor (PMSF) and two cysteine-protease inhibitors (N-ethylmaleimide and iodoacetamide) are effective, while other metallo-protease, aspartyl-protease, serine-protease, and specific cysteine-protease (e.g., E64) inhibitors had no effect. This profile categorizes the enzyme as a trypsin-like protease with cysteine as the nucleophile. In a similar study [Sommergruber et al., 1992] on HRV2 2A^{pro}, aspartyl-protease and metallo-protease inhibitors were found to have no effect (with the exception of EDTA; see below). The thiol protease inhibitors iodoacetamide and N-ethylmaleimide were strong inhibitors, whereas E64 was ineffective.

Antipain (a peptide aldehyde inhibitor of cysteine and serine proteases), chymostatin, and elastatinal (peptide aldehyde inhibitors of chymotrypsin-like and elastase-like serine proteases, respectively) were effective, whereas leupeptin (also a peptide aldehyde inhibitor) was not. The serine protease inhibitor TPCK (which also nonspecifically alkylates thiol groups) was significantly active, whereas TLCK (having similar mechanism) was not. These inhibitor studies preclude classification of HRV2 2A^{pro} as a conventional serine or cysteine protease.

Highly conserved residues in 2A^{pro} in picornaviruses are Cys^{55}, Cys^{57}, Cys^{115}, and His^{117}, which are a distinct feature [Yu and Lloyd, 1992] from 3C^{pro}. The following mutants had abolished proteolytic activity—C^{55}S, C^{57}S, C^{57}T, C^{115}P, C^{115}Y, C^{115}H, and H^{117}Q—while C^{64}N and C^{64}S did not abolish cleavage. Since these residues are located at the surface, two possibilities were examined: 1) S–S bridges and 2) cation-binding sites. Disulfide bridges could not be experimentally identified, and a divalent cation binding site could not be detected, although it is not excluded since EGTA and EDTA, which did not impair 2A^{pro} activity, may not be able to chelate internally buried complexed cations. However, the spacing between the Cys-His residues is different than the zinc finger-like binding motif. In a recent study [Sommergruber et al., 1992] EDTA at high concentrations was reported to inhibit HRV2 2A^{pro}, along with other inhibitors, as discussed above.

Post-Translational Effects on the Host Cell Protein Synthesis

The p220 protein is part of the cap-binding protein complex that mediates [Fenger, 1991b; Lloyd et al., 1988] association of cellular capped mRNA with ribosomes. p220 is probably cleaved by cellular proteases, a process that is induced by 2A^{pro}. Thus rapid inhibition of host–cell protein synthesis occurs. Poliovirus mRNA lacks the cap structure. Therefore, it is not affected by p220 cleavage and is able to direct ribosomal translation. Among other picornaviruses, the entero- and rhinoviruses, but not the cardioviruses, can induce p220 cleavage, while hepatitis A virus (HAV) produces persistent infection without inhibiting host–cell protein synthesis. Foot-and-mouth disease virus (FMDV) lacks 2A^{pro} sequences, but induces p220 cleavage. At least three FMDV proteinase activities have been described (see FMDV Protease Studies, below).

p68, which plays a critical role in the regulation of host–cell translation (like p220), is a 68-kDa interferon-induced protein degraded by a protease in poliovirus-infected cells [Black et al., 1993]. The protease is suggested to be of cellular origin, since any direct action of 2A^{pro}, 3C^{pro}, and 3CD^{pro} was excluded. Poliovirus-induced inhibition in all three host–cell polymerase (Pol) systems is correlated with the inactivation of specific transcription factors [Clark et al., 1993]. While the mechanism for the inhibition of Pol I–mediated transcription has not been elucidated, some of the mechanisms concerning Pol II– and Pol III–mediated transcription inhibition are discussed below.

Inhibition of Pol II–mediated transcription. TFIID [Roeder, 1991] is one of the five known transcription factors (TF) required [Greenblatt, 1991], in addition to RNA Pol II, for the specific transcription of a Pol II gene to take place in a reconstituted system. One of the first steps in the assembly of active Pol II transcription complexes is binding of TFIID to the TATA box sequence. Recently, the DNA-binding component of TFIID, TATA-binding protein (TBP), was cloned and characterized [see Clark et al., 1993, for references]. Wild-type 3C^{pro} (but not the C^{147}S mutant) directly cleaves [Clark et al., 1993] TBP and decreases TFIID activity in poliovirus-infected cells. However, yeast TBP or bacterially expressed human TBP only partially restores Pol II–mediated transcription in poliovirus-infected cell extracts, whereas purified TFIID completely restores transcription (S. Kliwer, M.E. Clark, and A. Dasgupta, unpublished data). Therefore, other activities in TFIID may be inactivated during poliovirus infection.

Inhibition of Pol III–mediated transcription. TFIIIC and TFIIIB are two transcription factors required for RNA Pol III activity. TFIIIC binds in a specific manner to the internal promoter of Pol III genes. In mock infected cells, two forms of TFIIIC could be detected by gel retardation assay: complex I (active form) and complex II (inactive). However, in poliovirus-infected cells, a new form, complex III, was detected [Clark et al., 1991]. Since treatment of complexes I and II resulted in complex III-like products, it was suggested that a limited proteolysis step is involved. Wild-type 3C^{pro} (but not the mutant V^{54}S 3C^{pro}) was able to produce complex TFIIIC cleavage products [Clark et al., 1991] (complex III) in both HeLa cells and in vitro. These data suggest that 3C^{pro}-induced cleavage of TFIIIC is a mechanism by which poliovirus infection results in inhibition of RNA Pol III transcription.

Maturation cleavage: Its role in infectivity. One of the final steps in picornavirus assembly is autocleavage of VP$_0$ to yield VP$_2$ and VP$_4$. Recently, it was shown [Lee et al., 1993] that some of the mutations introduced at the maturation cleavage site, Asn^{68} of VP$_4$ (N^{68}P and N^{68}T), but not at the proposed catalytic site, Ser^{10} of VP$_2$ (S^{10}A and S^{10}C) of the HRV14, yielded noninfectious virion particles. Thus, the maturation cleavage [Lee et al., 1993] is required for infectivity, while provirions lack infectivity. It has been earlier suggested that Ser^{10} of VP$_2$ might be the catalytic residue responsible for maturation cleavage, based on its hydrogen-bonding connection to Asn^{68} of VP$_4$ (at the C terminus) and its resemblance to serine proteases such as trypsin of the N-terminal sequence of VP$_2$. Maturation cleavage, the VP$_2$/VP$_4$ sequence, and the proposed VP$_2$/VP$_4$ structure are reviewed by Palmenberg [1990]. However, the results of Lee et al. [1993] show that the mutants S^{10}A and S^{10}C did not block the maturation cleavage of HRV14. A similar result was reported earlier for poliovirus maturation cleavage [Harber et al., 1991]. The slowing effect on maturation cleavage suggests that mutations of Ser^{10} induce modifications in the catalytic center.

Other 3C Related Proteases

The tobacco etch virus (TEV) 49-kDa protease. TEV, a member of the potato virus Y group (potyvirus) expresses an RNA genome from a single translation unit that encodes a 346-kDa protein that is co- and post-translationally processed to individual gene products by two TEV-encoded (87 kDa and 49 kDa) autocatalytically released proteins. The 49-kDa TEV protease is similar to other positive-stranded proteinases [Krausslich and Wimmer, 1988; Wellink and van Kammen, 1988] in several respects. Primarily, it cleaves the polyprotein between Gln-Gly or Gln-Ser dipeptides. Second, proteolytic activity is enhanced by dithiothreitiol. Third, the protein gene is adjacent to the RNA-dependent RNA polymerase gene. Finally, the C terminus contains a conserved Cys-rich motif. The TEV 49-kDa protease (see alignment in Fig. 3 and secondary structure model in Fig. 4) was proposed to be part of the trypsin-like cysteine protease family [Bazan and Fletterick, 1988] and of the chymotrypsin-like cysteine protease family [Gorbalenya et al., 1989a].

The catalytic residues and protease activity of the TEV 49-kDa protease were characterized [Dougherty et al., 1989] using inhibitor studies, site-directed mutagenesis, and computer-assisted molecular modeling. The inhibitor profile shows that aspartyl- and metalloprotease specific inhibitors (pepstatin A and EDTA, respectively), as well as some serine and cysteine protease inhibitors (PMSF, TLCK, and E64 respectively), had no effect. Of some effect were leupeptin (serine and cysteine protease inhibitor), aprotinin, and TPCK (serine protease inhibitors). The only effective inhibitors were iodoacetamide and N-ethylmaleimide (alkylating agents that attack thiol groups exposed by surface Cys residues) and $ZnSO_4$ (which was suggested to be coordinating Cys^{339}, His^{234}, and His^{335}).

Site-directed mutagenesis studies were consistent with the trypsin-like structural model [Bazan and Fletterick, 1988] with the following three exceptions: 1) His^{234}, not His^{216}, was found to be crucial for protease activity (the $H^{216}Y$ mutation did not impair proteolysis, while $H^{234}Y$ yielded an inactive protease), and the residue spacing between two of the catalytic residues, His^{234} and Asp^{269}, suggests a similar pattern with the flavivirus serine protease group [Bazan and Fletterick, 1989]; 2) $D^{269}V$ and $D^{269}N$ mutations abolished proteolysis, while $D^{269}E$ yielded altered cleavage activity; and 3) 10 different substitutions of Cys^{339} resulted in proteinases with no detected activity after either 1 or 12 hours. A $C^{339}S$ mutant yielded minor activity after 12 hours incubation, suggesting impaired substrate binding and/or reduced catalytic activity.

The TEV 49-kDa protease trypsin-like model [Bazan and Fletterick, 1989] proposes the two β-barrel-lobed trypsin fold, where the putative catalytic triad (His^{234}, Asp^{269} and Cis^{339}) is located on the loops that approach the active-site crevice. Additionally, His^{356} may play an important role in the P1 (Gln) substrate recognition, while Ser^{357} may stabilize the catalytic Asp^{269} by forming a specific hydrogen bond.

FMDV protease studies. Foot-and-mouth disease virus (FMDV) was crystallized [Acharya et al., 1989] at 2.9 Å resolution, and the tertiary structure of the capsid proteins (VP_1, VP_2, VP_3, and VP_4) was observed, together with the major antigenic site and other architectural details. The autocatalytic cleavage hypothesis of VP_0 to yield VP_4 and VP_2, proposed by Arnold et al. [1987] could not be confirmed in the FMDV particle crystal, since the incriminated residues are more than 8 Å apart. FMDV was proposed to encode cysteine–active-center viral proteases in both analyzed models [Bazan and Fletterick, 1988; Gorbalenya et al., 1989a], and primary sequences were aligned for $3C^{pro}$ models (see Fig. 3 for alignment with other picornaviruses, and Fig. 4 for the proposed secondary structure). Unlike other picornaviruses, the 2A fragment of FMDV is a 16-residue sequence [Lloyd et al., 1988] that does not contain the putative protease active site and is homologous to the C terminus of the cardiovirus 2A sequence. It was suggested that an alternate FMDV viral protease is being implicated in the p220 cleavage.

The viral genome encodes [Kleina and Grubman, 1992] a leader protease (L) located at the N terminus of the viral polyprotein that autocatalytically cleaves itself from the structural protein precursor P1 to expose a myristoylation site on VP$_4$, as well as the 3C^{pro} (which processes most of the other viral proteins, including capsid proteins). While p220 cleavage is mediated by 2A^{pro} in most picornaviruses, it was determined (using cDNA clones) that the leader protein of FMDV is required to initiate p220 cleavage [Devaney et al., 1988]. The leader protein lacks homology with 3C^{pro}. Recent [Kleina and Grubman, 1992] inhibition studies using E64 and analogs showed that the thiol protease inhibitors effectively block the autocatalytic cleavage of the leader protein. This yields a series of events that effectively reduces viral assembly (since P1 processing is blocked) and allows competitive translation for both host and viral mRNA (since p220 cleavage is delayed). Yet the precise role of L during p220 degradation remains unclear.

Antipicornavirus Therapy

Drug design studies of antipicornaviral agents, e.g., related to disoxaril 1 [Diana et al., 1993], are focused on the viral capsid-binding agents. Potential protease inhibitors have been investigated by Korant et al. [1986], but to our knowledge none are currently being tested for therapeutic purposes.

TOGAVIRIDAE

Togaviruses are small-enveloped (*toga* = cloak in Latin) vertebrate viruses containing a single-stranded RNA genome of positive sense. The Togaviridae family consists of a large number of pathogens classified [Melnick, 1991] as *Alphavirus* (arbovirus group A), *Rubivirus*, *Pestivirus* (mucosal disease virus group), and *Arterivirus*. Most members of this family replicate and are transmitted by arthropods, causing various frequently fatal, diseases. Representative members [Fenger, 1991b] of the family are Sindbis virus (SIN) and Semliki Forest virus (SFV), both from the *Alphavirus* genus.

Role of Viral Proteases During Replication of Togavirus

Genomic RNA (49S and 42S for SIN and SFV, respectively [Fenger, 1991b], is single stranded and of positive polarity. Upon infection of cells, it serves as mRNA that is translated to produce nonstructural (NS) viral proteins required for viral replication (early translation). RNA replication yields a full-length, negative-polarity RNA. The minus strand serves as template for the synthesis of nascent plus-strand genomic RNAs and for the transcription of a 26S subgenomic RNA [Fenger, 1991b].

Early translation. NS proteins are translated from the genomic RNA as two polyproteins, NS1234 (Fig. 5). Translation of NS123 in SIN terminates at a stop codon that can be bypassed, possibly by suppression, to yield [Li and Rice, 1989] the longer transla-

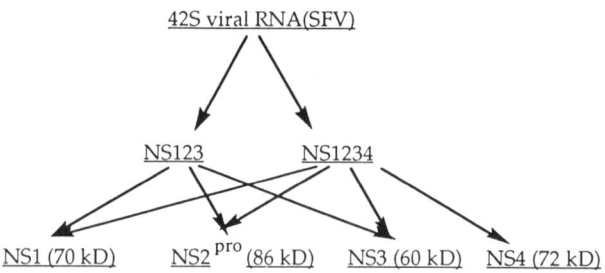

Fig. 5 *Early translation of togavirus (SFV). Genomic RNA (42S) is translated into a polyprotein that is cleaved into proteins NS1, NS2, NS3, and NS4. Protein NS4 in combination with other viral proteins (perhaps NS2 and NS1) forms the RNA polymerase. RNA replication yields a negative-strand RNA that serves as template for synthesis of full-length viral RNA and a 26S mRNA, the latter being translated in the late phase (see Fig. 7).*

tion product NS1234. NS1234 is the major SFV translation product, since the termination codon between NS3 and NS4 is absent [Takkinen, 1986]. Both NS123 and NS34 need to be coexpressed [Lemm and Rice, 1993b] for functional SIN RNA replication. Structural (a Gly-Asp-Asp consensus sequence [Fenger, 1991]) and biochemical [Lemm and Rice, 1993a,b] data suggest that NS4 is the viral polymerase, and undergoing studies [Lemm and Rice, 1993a] are focused on the elucidation of the alphavirus replication process and the role of the NS4–NS2 association.

The nonstructural NS2 protein is a papain-like thiol protease. SIN NS2 was reported to be an autoprotease that cleaves the nonstructural viral polyprotein [Ding and Schlesinger, 1989] at Ala-Ala bonds located between NS1–NS2 and NS3–NS4, respectively. The protease was localized at the C-terminal half of NS2 [Hardy and Strauss, 1989] and is a thiol protease related to papain [Strauss et al., 1992]. Mutation studies [Strauss et al., 1992] identified Cys^{481} and His^{558} as residues of the catalytic site. While an asparagine residue has been implicated in the active-site of papain, none of the conserved asparagines were proven to be essential for protease activity. Papain-like proteases have an Ala(Gly) succeeding the catalytic His residue, while in NS2 the residue is Trp. However, the abolished proteolytic activity in the W^{559}A mutant suggests that Trp^{559} is required for enzyme activity. All mutations with abolished activity were lethal, indicating that proteolysis of the early translation products is essential for SIN replication.

Sequence alignment studies [Koonin et al., 1992] of the functional domains in the nonstructural polyproteins for the hepatitis E virus (HEV) with the "alpha-like" group (including SIN and SFV, among other alphaviruses; rubella virus; and the beet necrotic yellow vein virus, a plant furovirus) allowed identification of a putative papain-like protease ("X") domain, approximately 300 residues long, showing striking similarity between the HEV sequence and the rubella virus sequence. The conserved Cys^{483} and His^{590} (HEV) and Cys^{1151} and His^{1272} (rubella virus) were proposed to be the catalytic residues (Fig. 6). The protease block was completely absent from the plant furovirus, suggesting different expression strategies between animal and plant viruses.

Late translation. The virion structural proteins are translated as a single 130-kDa polyprotein from the 26S subgenomic positive-strand mRNA. This polyprotein is cleaved into four structural proteins: C (core protein), E1, E2, and E3 (Fig. 7) [Fenger, 1991b].

The SIN core protein (SCP) can autocatalytically cleave itself [Aliperti and Schlesinger, 1978] from the nascent polyprotein, exposing a hydrophobic signal sequence (6K) at the N terminus of the adjacent region (E1). In a similar process, the SFV core protein autocatalytically cleaves itself [Melancon and Garoff, 1987] from the 62-kDa precursor of the capsid proteins E2 and E3, at a Trp-Ser bond [Wellink and van Kammen, 1988]. Cleavages between the 6K signal peptide and E1 (at an Ala-Tyr site) between 6K and the 62-kDa precursor (at a Ala-Glu site) and within the precursor to release E3 and E2 (at an Arg-Ser site) are processes not mediated by viral proteases [Wellink and van Kammen, 1988].

SIN and SFV Core Proteins Are Serine Proteases

The crystallographic structure of the SCP dimer has been reported [Choi et al., 1991] at a resolution between 6 and 3 Å. Each SCP subunit consists of 264 amino acid residues, of which the N-terminal 113 residues are variable, while the C-terminal 151 residues are conserved among alphaviruses (see Fig. 8). The N-terminal fragment contains no aromatic residues between positions 22 and 115, while the C-terminal region is exceptionally rich in basic residues and prolines. Out of 163 C-terminal amino acid residues, 113 were identical between SIN and SFV core proteins [Boege et al., 1981]. The longest identical peptide in this region comprises 18 residues (189–206 in SIN and 193–210 in SFV, respectively). The longest identical peptide in the N terminus region has four residues (21–24 in SIN and 14–17 in

```
HEV    (433-592)   QCRRWLSAGFHLDPRVLVFDESAPCHCRTAI-RKALSKFCCFMKWLGQE

RubV  (1109-1274)  RCRGW--HGMP-QVRCTPSNAHAA-LCRTGVPPRASTR--G-GE-LDPN


HEV    CTCFLQPAEGAVGDQGHDNEAYEGSD-VDPA-ESAISDISGSYVVPGTALQ-PLYQALDLP

RubV   -TCWLRAA-ANVAQAARACGAYTSAGCPKCAYGRALSEARTHEDFAALSQRWSASHADASP


HEV    AEIVARAGRLTATV--KVSQV-DGR-IDC--ETLLGNKTFRTSFVD-GAVLETNG-PE---

RubV   DGTGDPLDPLMETVGCACSRWVGSEHEAPPDHLLVS-LHRAPNGPWGVVLEVRARPEGGN


HEV    R--HNL

RubV   PTGHFV
```

Fig. 6. *Sequence alignment of the putative papain-like proteases of HEV (Burma strain) and rubella virus (RubV), based on the functional mapping of the nonstructural polyprotein "X" domains [Koonin et al., 1992]. Bold, putative catalytic residues; italics, identical residues; underlined, similar residues. [Adapted from Koonin et al., 1992, with permission of the publisher.]*

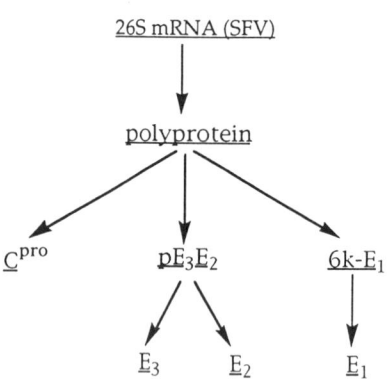

Fig. 7. *Late translation of togavirus (SFV). A 26S messenger RNA is translated into a polyprotein from which a capsid protein (C) and two precursor envelope proteins, pE_3E_2 and $6k$-E_1, are cleaved. Cleavage of pE_3E_2 to form E_3 and E_2 glycoproteins occurs in Golgi membranes, probably mediated by a cellular protease. A cellular signalase removes the 6k N-terminal peptide from E_1.*

SFV, respectively), but physicochemical properties are more conserved in the N-terminal region than the primary structures [Boege et al., 1981].

The SCP tertiary structure [Choi et al., 1991] shows two similar β-barrel domains, each composed of six β-strands (denoted A–F) and one α-helix (denoted $α_A$ and $α_B$ for each barrel, respectively) with the substrate-binding site situated between. The tertiary structure of SCP was compared with human chymotrypsin and with bacterial α-lytic protease (Fig. 9). In the first domain, the Ser(Thr)[54] of chymotrypsin is replaced by Val[136] in SCP (causing a hydrogen bond to be absent), while the "calcium-binding loop" (residues 65–84 in chymotrypsin) and the $βE_1$–$βF_1$ loop are missing in both SCP and α-lytic protease. The essential His[141] (SCP) is located in an α-helix, as is the corresponding His[57] in chymotrypsin. The connection between the first and second domains is shorter and different in SCP from that in chymotrypsin and α-lytic protease. In the second domain, the connection between $βA_2$ and $βB_2$, the "autolysis

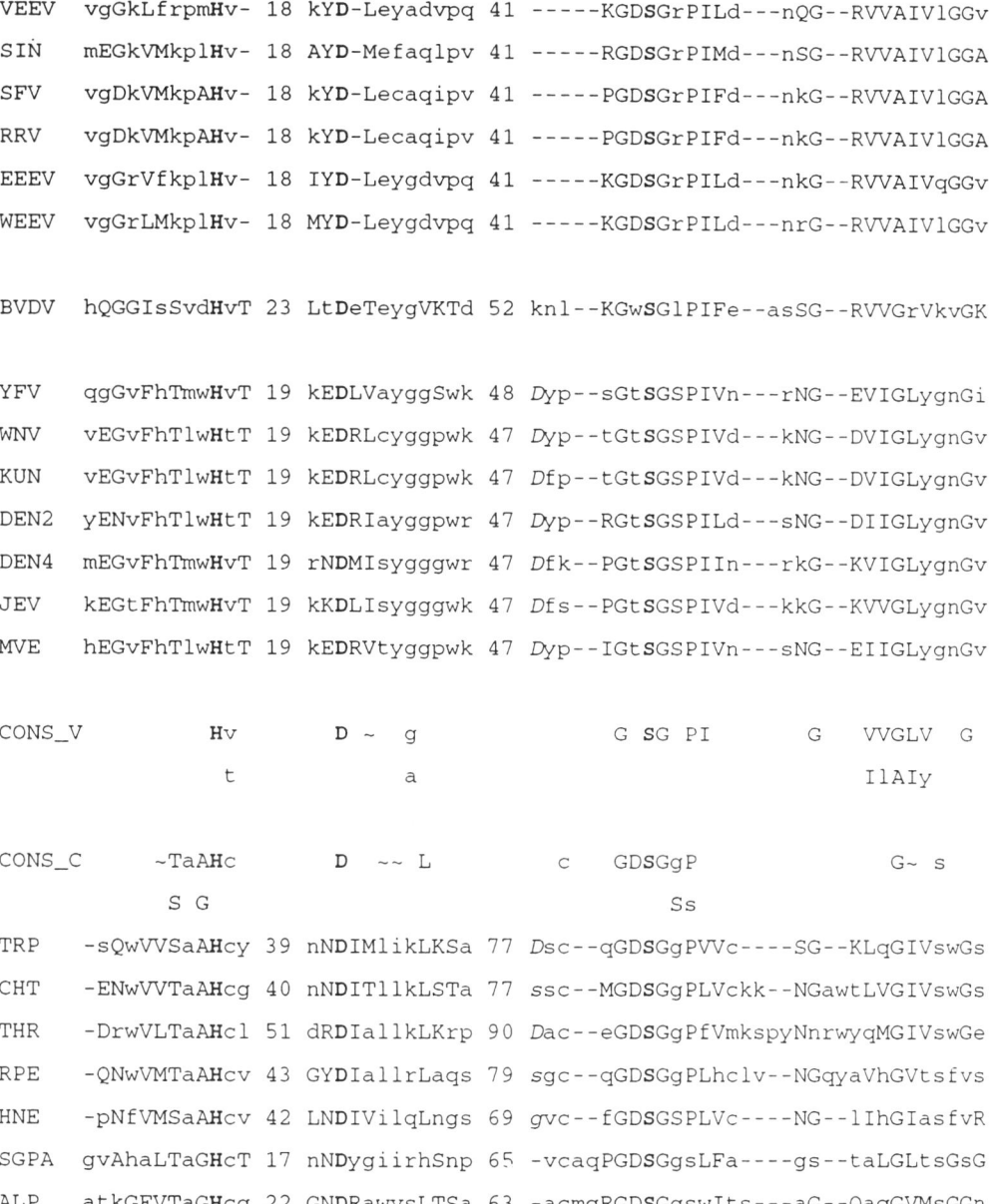

Fig. 8. Sequence alignment of putative catalytic residues of flavivirus and pestivirus NS3 protein N-terminal domains with other cellular and viral proteases (capsid proteins), as proposed by Gorbalenya et al. [1989b]. Bold, putative catalytic residues; capitals, identical or similar residues in both viral and cellular proteases; italics, possible S1 specificity binding site residue; ~, hydrophobic residues (L, V, I, M); numbers indicate the length of spacer. The flaviviruses are YFV, yellow fever; WNV, West Nile; KUN, Kunjin; DEN 2 and DEN4, Dengue types 2 and 4; JEV, Japanese encephalitis; MVE, Murray Valley encephalitis. The pestivirus is BVDV, bovine viral diarrhea virus. The alphaviruses are VEEV, Venezuelan equine encephalitis; SIN, Sindbis; SFV, Semliki Forest virus; RRV, Ross river; EEEV, Eastern equine encephalitis; WEEV, Western equine encephalitis. The cellular proteases are TRP, trypsin; CHT, chymotrypsin; THR, thrombin; RPE, rat pancreatic elastase; HNE, human neutrophil elastase. The bacterial proteases SGPA and ALP are the same as in Figure 3. CONS_V is the consensus pattern for viral proteases; CONS_C, the consensus pattern for cellular proteases, both patterns derived allowing one exception [Gorbalenya et al., 1989b].

Fig. 9. Alignment of Sindbis virus core protein (SCP) with chymotrypsin (CHT), α-lytic protease (ALP), on the basis of structural comparison, and poliovirus 3C protease (PV1), based on the model of Gorbalenya et al. [1989]. Secondary structure of CHT: A_1-F_1, A_2-F_2, β-strands; α_A, first domain-helices; α_B, second domain α-helices; αC, C-terminal α-helix; the functional loops of CHT are also marked. Bold, catalytic residues; underlined, conserved residues; numbers indicate the length of the spacer; dotted underlined residues mark additional comments: one residue should be inserted between KK in SCP, and the APV sequence in ALP is a modified inhibitor of ALP. [Modified from Choi et al., 1991, with permission of the publisher.]

loop" (residues 145–154 in chymotrypsin) is much shorter in SCP and α-lytic protease, while the "methionine loop" (residues 164-182) is completely absent in SCP, but is present in α-lytic protease. The highly conserved region Gly^{213}–Gly^{216} is similarly arranged at the βC_2–βD_2 loop (two residues) and in βD_2 (two residues) in all three proteinases. Instead of the final α-helix (16 residues in both chymotrypsin and α-lytic protease), the SCP C-terminal fragment (βF_2) points the last 5 residues into the active site.

No disulfide bonds were observed in SCP, although these are essential [Anthony et al.,

1992] for other SIN capsid protein stability (e.g., E2 and E3). However, both α-lytic protease and chymotrypsin have several S–S bonds (two and five, respectively). There is a higher degree of resemblance of SCP to the α-lytic bacterial serine protease than to chymotrypsin [Choi et al., 1991]. The Gly-Asp-Ser-Gly sequence (residues 213–216) of the SCP protein, located in the C-terminal region, corresponds to the highly conserved sequence around the catalytic Ser^{195} in the chymotrypsin-like serine proteases [Boege et al., 1981]. Hahn et al. [1985] suggested that His^{141}, Asp^{147}, and Ser^{215} in SCP could constitute the required catalytic triad for a serine protease based on temperature-sensitive mutant studies. Thus, the mutants $K^{138}I$ (close to His^{141}) and $P^{218}S$ (close to Ser^{215}), both modified close to, or within, the active site, block capsid protein formation [Hahn et al., 1985].

Melancon and Garoff [1987] determined, via site-directed mutagenesis, that Ser^{219} in the SFV core protein (the residue homologous to Ser^{215} in SCP) is essential for catalysis. In their study [Melancon and Garoff, 1987], the sequence Gly-Asp-Ser^{219}-Gly was replaced with Gly-Asp-Arg-Ser-Thr and was shown to completely abolish in vitro cleavage. Based on sequence comparison of alphavirus core proteins (SCP and SFV, among others) and the N-terminal domain of flavivirus nonstructural NS3 proteins, it was suggested [Gorbalenya et al., 1989b] that Asp^{163} (which was conserved in the alignment), not Asp^{147}, is implicated in catalysis since its location relative to the conserved His residue corresponds better to that of chymotrypsin-like proteases (Fig. 8). Hahn and Strauss [1990] confirmed that His^{141} and Ser^{215} are part of the SCP catalytic triad and provided evidence to support Asp^{163}, not Asp^{147}, as essential for catalysis. The mutants $H^{141}R$, $D^{163}H$, $S^{215}T$, and $S^{215}C$ did not inhibit polyprotein processing, yet viral assembly was abolished. Proteolysis was inhibited for the mutants $H^{141}P$, $H^{141}A$, $S^{215}I$, and $S^{215}A$ (the infectivity was not tested), but not for $D^{163}Q$ (which yielded infectious virus) and for $D^{163}H$ (noninfectious virus). The Ser^{215} mutants support the serine-protease known mechanism [Warshel et al., 1989] that the –OH of Ser makes a nucleophilic attack at the scissile bond. $S^{215}C$ mutant was 60% active (compared with wild type), which suggests that the –SH function of Cys may act in a similar fashion as the wild-type –OH [Hahn and Strauss, 1990]. This result is not confirmed by either experimental simulations of the papain triad in a trypsin mutant (rat $S^{195}C$ trypsin enzymatic activity is 10^5 times lower than wild type [Hiagaki et al., 1989]) or replacement of cysteine with serine at the catalytic site of a picornaviral protease [Yu and Lloyd, 1991] ($C^{109}S$ mutation yields an inactive 2A protease).

The active center residue superposition of both SCP and α-lytic protease gave an rms deviation for C_α atoms of 0.8 Å. The superposed residues were H^{141}/H^{57}, Asp^{163}/Asp^{102}, the conserved sequence Gly^{213}-Asp^{214}-Ser^{215}-Gly^{216} to the corresponding Gly^{193}-Asp^{194}-Ser^{195}-Gly^{196}, and Leu^{231}/Ser^{214}. Gly^{213} forms an oxyanion hole for the substrate carbonyl preceding the scissile bond. Asp^{194} in chymotrypsin (Asp^{214} in SCP) forms an ion pair with the N terminus (Arg^{138} in bacterial enzymes) when the mammalian zymogen is activated by cleavage at residue 16. This rearranges the specificity pocket and permits binding of the substrate. In SCP, Asp^{214} is associated with His^{193} and Arg^{217}. However, SCP does not require any activation step, being active immediately after synthesis. Other conserved residues in SCP are Leu^{169}, Gly^{187}, and Gly^{23}. These residues possibly serve a role in folding. Ser^{214} is a highly conserved residue in chymotrypsin-like proteinases and is active in the catalytic site, forming a hydrogen bond with Asp^{102}. However, the corresponding residue in SCP is Leu^{231} (matching a similar Val^{219} in the 3C poliovirus protease [Gorbalenya et al., 1989b]). In the alignment of SCP, chymotrypsin, α-lytic protease, and 3C poliovirus protease, several conserved residues including the catalytic triad (with the proposed Ser/Cys substitution in 3C) are observed.

SCP performs only one cleavage, removing itself from the polyprotein, and, thereafter, it no longer acts as a protease [Kemp et al., 1992]. After cleavage, its new C-terminal Trp is left blocking the active site [as shown by Choi et al., 1991]. Additionally, the C-terminal Trp^{264} occupies the hydrophobic pocket defined by Val^{20} (which corresponds to Asp^{189} in trypsin, the major determinant for P1 specificity), Gly^{210}, Gly^{211}, Gly^{213}, Gly^{232}, and Gly^{233}. In Figure 9, the APV inhibitor sequence of α-lytic protease is matched to the EEW sequence in SCP. In the SCP crystal, the catalytic residue Asp^{163} is exposed to the solvent, thus being less able to cooperate with His^{141}. In other serine proteases, the catalytic aspartate is buried, stabilizing the imidazole proton of histidine and allowing the motion [Singer et al., 1993] of the hydrolitic water during the nucleophilic attack. It was suggested [Choi et al., 1991] that Asp^{163} is buried while the dissociation from the rest of the polyprotein occurs.

FLAVIVIRIDAE

The family Flaviviridae includes the genera *Flavivirus* (also known as *arbovirus group B*), which was originally included in the Togaviridae [Melnick, 1991], *Pestivirus* and *hepatitis C virus* [Lin et al., 1993]. However, there are structural and functional differences between alphavirus and flavivirus. Flaviviruses are enveloped small, positive, single-strand RNA viruses [Fenger, 1991b], and the genomic RNA, containing a single ORF, is surrounded by icosahedral capsids composed of a single 14-kDa protein, C. Among species with a pathogenic effect in humans, yellow fever virus (YFV) and hepatitis C virus (HCV) will be discussed.

Role of Viral Proteases During Replication of Flavivirus

Viral RNA (44S), which is translated into a single polyprotein (Fig. 10) with the order anchC-prM-E-NS1-NS2A-NS2B-NS3-NS4A-NS4B-NS5, is the only viral mRNA detected in infected cells [Krausslich and Wimmer, 1988]. The structural proteins, derived from the N-terminal region of the polyprotein, are the capsid protein C precursor (anchC), the membrane protein precursor, prM, and the envelope (E) protein. The remaining 75% of the polyprotein is processed to yield at least seven nonstructural proteins (NS1 to NS5) that are implicated in replication.

Polyprotein proteolytic cleavage requires both cellular and viral proteases [Wellink and van Kammen, 1988]. Thus, viral proteases cleave NS2A-NS2B and NS2B-NS3 at Arg-Ser bonds, NS3-NS4 and NS4-NS5 at Arg-Gly bonds, and NS4A-NS4B at an Arg-Val site [Wellink and van Kammen, 1988], while the other sites are cleaved by host proteases.

Three types of mechanisms are employed in processing of the viral polyprotein: 1) The sequences preceding the N termini of prM, E, NS1, and NS4B [Lin et al., 1993] are rich in aromatic residues, and it is believed that they

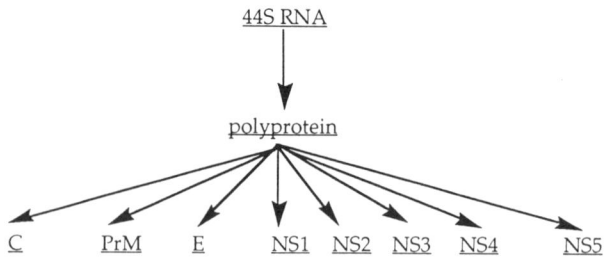

Fig. 10. *Protein translation of flavivirus. Genomic RNA (44S) is translated into a polyprotein that is cleaved into three membrane-associated polypeptides (PrM, E, and NS1), a capsid protein (C), and proteins NS2, NS3, NS4, and NS5. The cleavage process is mediated by cellular and viral proteases.*

act as hydrophobic signals for translocation of these proteins into the lumen of ER. These signal sequences are then removed to separate the proteins, a process that is mediated by cellular signalases. 2) prM is cleaved to M, which is incorporated into viral particles. A Cys-Trp dipeptide (characteristic for thiol proteases) is located within the prM primary sequence [Dalgarno et al., 1986], suggesting that prM (21 kDa) autocleaves into M (8.5 kDa, nonglycosylated) and a polypeptide (12.5 kDa). 3) Cleavage at sites flanked by two basic amino acids (e.g., Arg-Arg or Arg-Lys) at the P2 and P1 site and a polar residue (Ser or Gly) at the P1´ position is catalyzed by a viral serine protease, which requires NS2B [Chambers et al., 1991] and NS3 [Chambers et al., 1990]. This process is employed during the formation of several viral polypeptides associated with the ER or Golgi.

The Flavivirus NS3 N-Terminal Domain Serine Proteases

The N-terminal domains of the NS3 regions of flavi- and pestiviruses were identified [Gorbalenya et al., 1989b], by sequence comparison, as serine proteinases. These regions of the viral polyprotein contain conserved segments similar to those around the catalytic residues of serine chymotrypsin-like proteases and have statistically significant sequence similarity to the protease domains of alphavirus capsid proteins. The sequence GxSGxP (x—any residue), which resembles the conserved motif around the catalytic serine residue of chymotrypsin-like serine proteases, and the invariant His and Asp residues were detected [Gorbalenya et al., 1989b] when N-terminal domains of flavivirus NS3 proteins (viruses: YFV, MVE [Murray Valley encephalitis], WNV [West Nile], Dengue types 2 and 4, Japanese encephalitis, and Kunjin) of a pestivirus p125 protein (bovine viral diarrhea) and the capsid proteins of several alphaviruses (SFV, SIN, among others) were aligned using computer-assisted sequence comparison (see Fig. 8). The invariant residues Gly^{167} and Ser^{179} were proposed to be part of the substrate-binding pocket. However, the specificity is different for alphavirus capsid proteins (Trp in P1, chymotrypsin-like) compared with flaviviruses (Arg/Lys-Arg in P2-P1, trypsin-like). The inhibition by TPCK and the preferential cleavage site are consistent with the proposal that the NS3 region encodes a 180-residue serine protease.

Using an independent alignment, Bazan and Fletterick [1989] showed the conservation of nucleophilic residues identical to Ser^{195} (and the corresponding Cys in picornaviral 3C proteases) and other amino acid equivalents of the catalytic triad (His^{57} and Asp^{102}) and substrate-binding pocket. The best similarity sequence region is mapped to the C-terminal third of the trypsin fold, within the second β-barrel domain, at the area that contributes the catalytic serine and forms the substrate-binding pocket. It was proposed [Bazan and Fletterick, 1989] that the flavivirus serine protease pocket mirrors the pancreatic and bacterial (*Streptomyces griseus*) trypsin enzymes (specific for Arg/Lys residues) in positioning an electrostatically complementary Asp^{189} residue (Asp^{132} in YFV) at the bottom of the binding pocket.

The positions of the three conserved amino acid residues of YFV NS3 corresponding to the trypsin catalytic triad (His^{53}, Asp^{77}, Ser^{138}) were mutated [Chambers et al., 1990] using the following substituents: $H^{53}A$, $D^{77}N$, $D^{77}A$, $S^{138}A$, and $S^{138}C$. In the in vitro assay, RNA translation of mutants $H^{53}A$, $D^{77}N$, $D^{77}A$, and $S^{138}A$ failed to yield detectable cleavage products, while the $S^{138}C$ mutant yielded diminished quantities. Two mutations that either abolish ($S^{138}A$) or significantly reduce ($S^{138}C$) cleavage activity in vitro were analyzed in vivo after transfection of the full-length YFV cDNA templates encoding the mutations to BHK-21 cells. No infectivity was observed compared with wild-type virus. These results suggest that the studied residues are important for catalytic activity. The modulating function of NS2B was reported elsewhere [Chambers et al., 1991].

The HCV-Encoded Serine Protease

Recently, nine cleavage products were identified after HCV polyprotein expression

[Grakoui et al., 1993b] and the 70-kDa HCV NS3 protein was shown to be nearly identical in size with the NS3 protein of flaviviruses (80 kDa in some pestiviruses). The N-terminal one-third of HCV NS3 protein functions [Grakoui et al., 1993a] as a serine protease mediating viral polyprotein cleavage at several sites. Cleavages between NS3 and NS4A, NS4A and NS4B, NS4B and NS5A, and NS5A and NS5B were abolished for the $H^{1083}A$ and $S^{1165}A$ mutants, while the mutations $C^{1078}A$ and $S^{1164}A$ had no effect on polyprotein cleavage. However, the HCV NS3 does not cleave [Grakoui et al., 1993b] the NS2A–NS2B and NS2B–NS3 sites, contrary to other flavivirus NS3 proteases. These data support the serine protease model for NS3, with His^{1083} and Ser^{1165} as part of the catalytic triad.

Examination of conserved features at HCV NS3 cleavage sites made possible several observations after alignment [Grakoui et al., 1993b] of the sequences flanking all four potential cleavage sites (Fig. 11). A Cys at P1 is common, with the exception of the NS3–NS4A, where Thr is found. Additionally, a Ser is common in P1′, with the exception of two HCV isolates, where Ala is present at the NS4A–NS4B site. All four cleavage sites contain an acidic residue at the P6 position. Asp is constantly present at the NS3–NS4A and NS4A–NS4B cleavage sites and sometimes at the NS4B–NS5A and NS5A–NS5B, where it can be replaced by Glu (more frequently at the NS5A–NS5B site). A possible cleavage recognition motif, Asp/Glu (P6)-X-X-X-X-Cys/Thr-Ser/Ala, is present at five additional sites in the HCV H strain polyprotein, but cleavage products of these alternative sites were not identified, suggesting that additional factors play a role in the cleavage site preferences of HCV NS3 protease. Compared with other flavivirus NS3 proteases, the HCV NS3 protease has the functional key residues in the catalytic triad (His^{1083} and Ser^{1165}), Asp^{1107} (proposed to be part of the catalytic triad), and Gly^{1163}, which can be superimposed on His^{57}, Asp^{102}, Ser^{195} (catalytic triad), and Gly^{193} of the trypsin sequence. However, cleavage site specificity is different, as discussed above.

RETROVIRIDAE

Retroviridae is a family of enveloped RNA viruses that contain much of their genetic information in three distinct genomic elements arranged from 5′ to 3′ as follows: 1) the *gag* (group-specific antigen) precursor, which is translated from 35S mRNA as a 55-kDa polyprotein and encodes the structural proteins of the virus including the matrix protein p17 (MA), capsid protein p24 (CA), and nucleocapsid p7 (NP); 2) the *pol* precursor, which is also translated from 35S mRNA initially as part of a 160-kDa *gag–pol* fusion product and is

P6	P5	P4	P3	P2	P1	P1′	P2′	P3′	P4′	P5′	cleavage site
Asp	Leu	Glu	Val	Val	**Thr**	**Ser**	Thr	Trp	Val	Leu	NS3-NS4A
			Ile	Met			Ser				
Asp	Glu	Met	Glu	Glu	**Cys**	**Ala**	Ser	His	Leu	Pro	NS4A-NS4B
						Ser	Gln	R/K	Ala	Ala	
Asp	Cys	Ser	Thr	Pro	**Cys**	**Ser**	Gly	Ser	Trp	Leu	NS4B-NS5A
Glu		S/P	I/V								
Glu	Asp	Val	Val	Cys	**Cys**	**Ser**	Met	Ser	Tyr	Thr	NS5A-NS5B
Asp	S/G	Ile	Ile					Ser			

Fig. 11. *Alignment of sequences flanking the potential cleavage sites for the HCV proteins. See explanations in text. [Modified from Grakoui, 1993b, with permission of the publisher.]*

autocatalytically processed into the viral replication enzymes protease p11 (PR), reverse transcriptase p51 (RT), RNaseH p15 (RNH), and integrase p34 (IN); and 3) the *env* precursor, which encodes the envelop glycoproteins that are processed by cellular enzymes, i.e., gp160 [Ding and Schlesinger, 1989; Krausslich and Wimmer, 1988; Wagner et al., 1992; Wellink and van Kammen, 1988].

The location of the protease (PR) between the structural and nonstructural units of the *gag–pol* polyprotein facilitates cleavage and further processing of the *pol* precursor. Consequently, viral infectivity is dependent on the proteolytic activity of the protease. This has been confirmed by the production of noninfectious virions containing unprocessed polyprotein by murine leukemia virus [Crawford and Goff, 1985] and human immunodeficiency virus type 1 deficient in protease [Kohl et al., 1988]. Particle formation is not, however, hindred by the lack of protease. These observations form the basis of drug therapy based on retroviral protease inhibitors (discussed in detail below).

It has been suggested that viruses initially express their gene products as large polyproteins in order to economize the amount of genetic material that must be carried and subsequently transcribed and to maximize the diversity of viral proteins needed for replication in the host [Krausslich and Wimmer, 1988]. Furthermore, proteolysis provides a mechanism by which viruses can coordinately regulate several different proteins from an mRNA from which only one protein can be synthesized—monocistronic mRNA. Additionally, viruses encode for a proteolytic enzyme that regulates the conversion of these large precursor polyproteins into biologically active structural and functional products that are required for the successful replication of viral progeny [Robins, 1993].

Several benefits of this strategy are noted: 1) encoding a protease provides a functionally active enzyme for use in the host cell's cytoplasm—not necessarily available normally—that 2) eliminates the dependence of the virus on host cell enzymes that have numerous other substrates and 3) allows virus to replicate in many divergent cell types [Robins, 1993].

Structural Characteristics of Retroviral (Aspartic) Proteases

Human immunodeficiency virus type 1 (HIV-1) is an aspartic protease. Early examination of the amino acid sequence of HIV-1 afforded the discovery of a conserved Asp-Thr(Ser)-Gly sequence that is homologous to the active-site of mammalian cellular aspartic proteases [Kohl et al., 1988] (Fig. 12). Aspartic proteases are large (>300 residues; molecular mass, ~35 kDa) monomeric enzymes composed of two homologous domains proposed to have evolved by the processes of gene duplication and fusion [Billich and Winkler, 1991]. Retroviral proteases, however, are smaller (~130 residues), with the conserved Asp-Thr(Ser)-Gly found only once in the sequence. Using pattern recognition, structure prediction, and molecular modeling techniques, sequences of aspartic and retroviral proteases were examined, and the HIV-1 protease was suggested to be a dimer (molecular mass of 11.5 kDa) comprised of two identical 99 residue subunits, each corresponding to a single domain of the aspartic protease [Pearl and Taylor, 1987]. Further evidence of the characterization of HIV-1 protease as an aspartic protease resulted from its inhibition by the protypical aspartic protease inhibitor pepstatin A [Katoh et al., 1987; Nutt et al., 1988; Seelmeier et al., 1988].

Homology modeling studies with Rous sarcoma virus (RSV). In 1989, the three-dimensional structure of the RSV protease was solved. The RSV protease monomer consists of 124 amino acids and exists as a C2-symmetric homodimer with an overall structure similar to aspartic proteases. The N-terminal (1-7) and C-terminal (119-124) residues of each monomer constitute β-strands that are part of the dimer interface (analogous to the inter-domain junction in typical aspartic proteases). The central cores of the monomers are located on both sides of interface region. These core re-

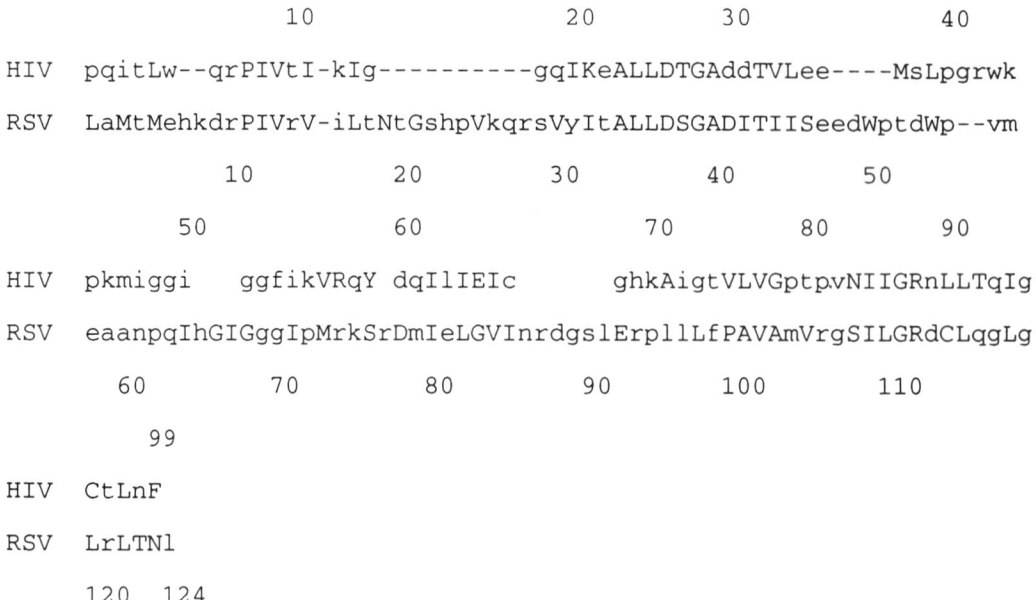

Fig. 12. *Sequence alignment of human immunodeficiency virus (HIV) protease and Rous sarcoma virus (RSV) protease. Upper case, solvent-inaccessible residues; lower case, solvent-accessible residues. [Modified from Blundell et al., 1991, with permission of the publisher.]*

gions exhibit a characteristic topological feature of a sandwich of two four-stranded β-sheets with perpendicular chain directions. One sheet is constituted by antiparallel β-chains formed by residues 13-21, 28-34, 80-89, and 92-98. The other sheet is described as two superimposed ψ-structures. Residues 35-39 and 99-104 form central two antiparallel β-chains, while the two outer b-chains are parallel to their neighbors. The conserved active-site residues Asp^{37}-Thr^{38}-Gly^{39} are located between two β-chains, with proximal residues hydrogen bonded such as to form the so-called fireman's grip configuration—a common trait of aspartic proteases. A substrate-binding cleft is formed by flaplike processes (one from each monomer) extending over the active-site residues [Miller et al., 1989a; Hellen et al., 1989].

The confirmation of the RSV protease as an aspartic protease, the implication of HIV-1 protease as belonging to the aspartic protease family, and the overall similarity in the sizes of the two formed the basis of homology modeling prior to the actual crystallization of the HIV-1 protease. In homology modeling, a fundamental assumption is that identical residues will be similar in structure, while modified conformations may be required for dissimilar sequences. The two sequences were initially aligned via superimposition of the conserved active-site residues and a second conserved sequence, Gly-Arg-Asn/Asp, located near the C terminus. Where sequence similarity was noted, the amino acids of the HIV-1 protease were substituted for the residues in the RSV protease with side chains in either identical or sterically accessible conformations. At only 99 residues in length, the HIV-1 protease is a bit shorter than the RSV protease. By deleting residues from the surface loops, a structural alignment of the proteases was made possible [Miller et al., 1989a]. A comparison of the modeled HIV-1 protease with the experimentally determined structure of a synthetic HIV-1 protease at 2.8 Å resolution [Blumenstein et al., 1989] revealed an rms deviation of 1.4 Å for 86 superimposed

Cα atoms. The residues of the flap region were mispredicted possibly as a result of the lack of experimental evidence for this region in the RSV protease structure. The residues making contact with the substrate were correctly modeled, suggesting, in the absence of experimental evidence, that the modeled protease could be used to aid in the design of novel substrates (inhibitors) [Weber, 1990]. In a more general sense, the method of homology modeling was supported by these results. For homology modeling studies of the HIV-1 protease based on other known aspartic protease structures, including endothiapepsin, penicillopepsin, and/or rhizopuspepsin, the reader is referred to works by Pearl and Taylor [1987] and Pechik et al. [1989].

Crystal structure of HIV-1 protease. X-ray crystallography confirmed that the HIV-1 protease belongs to the same family as the mammalian cellular aspartic proteases. The general description of the structure of HIV-1 protease will, therefore, be addressed in the context of typical cellular aspartic proteases, specifically pepsins. While pepsin-like enzymes (i.e., endothiapepsin, penicillopepsin, rhizopuspepsin, hexagonal porcine pepsin, and chymosin) exhibit a C2 axis of symmetry as the monomer, retroviral proteases including RSV protease and HIV-1 protease exhibit this form of symmetry only as the dimeric structure. Retroviral and pepsin-like aspartic proteases possess a common super-secondary structure in that each subunit of the retroviral protease corresponding to each lobe of the pepsin-like protease comprised of two similar motifs formed from four antiparallel strands. In retroviral proteases, each subunit possesses four antiparallel β-strands (a,b,c,d and a´,b´, c´,d´ from N to C termini). Taken together, the strands from both subunits form a disorganized sheet. Two pairs of parallel strands are formed by strands c and d´ and strands c´ and d. A second sheet folded over and orthogonal to the first is formed by a pair of β-hairpins given by strands b and c and strands b´ and c´, which are hydrogen bonded together around a C2 axis of symmetry. A third interdomain sheet, composed of antiparallel β-strands, forms the base of the active-site cleft. In retroviral proteases strand a and the C-terminal β-strand, and the corresponding strands in the other subunit, form a sheet of four strands, while in pepsin-like enzymes strand a and two C-terminal β-strands from each lobe contribute to a six-stranded, interdomain sheet [Blundell et al., 1991].

The binding domain of the enzymes may be classified according to substrate. HIV-1 protease has an extended binding region where as many as eight consecutive amino acids of the substrate are in contact with the cleft. According to the convention of Schechter and Berger [1970], substrate residues are designated P1 to PN extending distally from the scissile amide bond to the N terminus of the substrate and P1´ to PN´ in the C-terminal direction. This scheme may be extended to binding sites in the protease that are labeled S1 to Sn and S1´ to SN´, respectively. The physical characteristics of each binding site are discussed below in the context of substrate-specificity studies and inhibitor design.

Active-Site Residues, Transition State Models, and Mechanisms

Comparisons of the x-ray structures of complexes of aspartic proteases (both monomeric pepsin-like proteases and dimeric retroviral proteases) with inhibitors containing transition-state isosteres initially suggested a common mode of interaction of the inhibitor with the enzymes [Blundell et al., 1991]. This was confirmed by x-ray analysis of protease–inhibitor complexes [Miller et al., 1989a]. In HIV-1 protease, the carbonyls of Gly^{27} and Gly^{127} accept hydrogen bonds from the substrate NH functions of P1 and P2´. The carbonyl functions of the substrate in P2 and P1´ are hydrogen bonded to a conserved hydrolytic water molecule that in turn is stabilized by hydrogen bonds to Ile^{50} and Ile^{150}. Gly^{48} and Gly^{148} serve analogous functions in P2 and P3´. In P3 and P2´, Asp^{29} and Asp^{129} donate hydrogen bonds to the substrate backbone (Fig. 13) [Blundell et al., 1991].

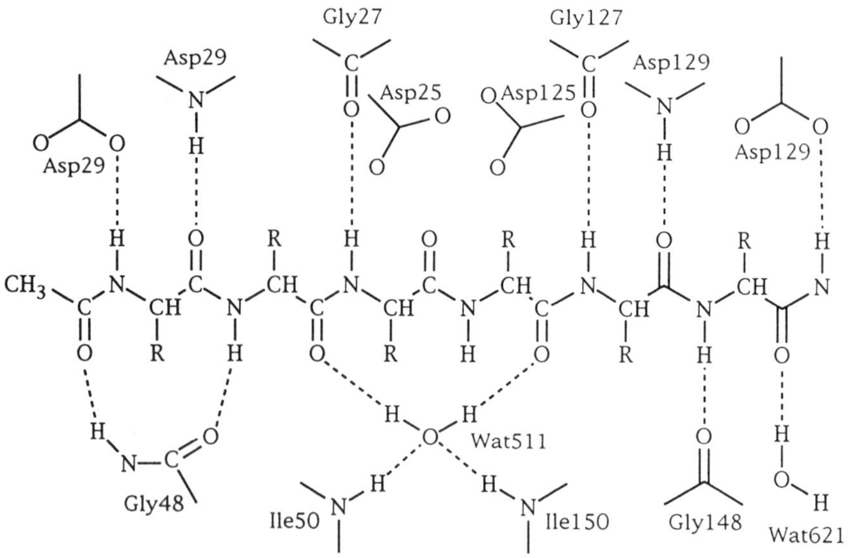

Fig. 13. *Schematic representation of hydrogen bonding interactions between HIV-1 protease and inhibitor (substrate). [Modified from Miller et al., 1989b, with permission of the publisher.]*

The amino acid sequences around the active site are equivalent in retroviral and pepsin-like proteases. Most amino acids that are conserved, or conservatively varied, between retroviral and pepsin-like proteases are solvent inaccessible or participate in hydrogen bonds to mainchain carbonyl or NH functions. A significant conservation is the sequence hydrophobe-hydrophobe-glycine (Ile[84]-Ile[85]-Gly[86] in HIV-1 protease and Leu[84]-Leu[85]-Gly[86] in endothiapepsin). The residue in position 84 is conserved for purposes of binding the substrate and contributes to the formation of specificity subsites S1 and S1′. Residue 85 contributes to the protein core, and residue 86 (Gly in both cases) is packed close to the active-site residues, and for this reason a side chain is not tolerated [Blundell, et al., 1991].

The conserved active-site residues in pepsin-like proteases (Asp[32]-Thr[33]-Gly[34] and Asp[215]-Thr[216]-Gly[217]) and retroviral proteases (Asp[25]-Thr[26]-Gly[27] and Asp[125]-Thr[126]-Gly[127]) of pepsin possess buried solvent-inaccessible polar residues (Thr) that are hydrogen bonded to main chain carbonyls and NHs of the other lobe or subunit. These residues are important for the maintenance of the three-dimensional structure. It is postulated that the glycines are conserved between sequences so as not to disrupt the spatial positioning of the active-site aspartates [Blundell et al., 1991].

The flap regions in the uncomplexed protease are seen to make intermolecular contacts in crystal structures. Upon binding, a conformational change in the body of the enzyme corresponds to a hinge motion as flaps close (corresponds to movement of one lobe of pepsin-like proteases on binding). In the complexed state, the flaps interact with substrates through direct hydrogen bonding [Blundell et al., 1991] and through coordination with a conserved water molecule that is tetrahedrally coordinated to the main-chain nitrogen atoms from Ile[50] and Ile[150] of the flaps and to the carbonyl oxygens from the P2 and P1 substituents of the inhibitors [Swain et al., 1991]. Compared with nonviral aspartic proteases, HIV-1 protease shows more interactions with a longer substrate. This along with large flap movement restricts the conformation of the protease cleavage sites in the retroviral polyprotein precursor [Gustchina and Weber, 1990].

The overall scheme of the hydrolytic action of HIV-1 protease is analogous to that of the pepsin-like proteases [Blundell et al., 1991]. The hydrolytic water molecule is polarized by coordination with the anionic residues Asp^{25} and Asp^{125}. Nucleophilic attack at the carbonyl of the scissile bond of the substrate results in the formation of a gem diol intermediate that is stabilized by hydrogen bonds to the negatively charged Asp^{25} and from the neutralized (protonated) Asp^{125}. The tetrahedral intermediate is destabilized upon protonation of the nitrogen of the scissile bond, which is suggested to be mediated by the protonated Asp^{125} residue. This is facilitated by nitrogen inversion and rotation about the $C(OH)_2$-N bond, resulting in a staggered conformation about the P1 and P1´ Cαs of the substrate. The nitrogen is then within hydrogen-bonding distance of Asp^{125}, and, as protonation occurs, the hydrogen bond from the statine-like hydroxyl to Asp^{25} is removed. Products are formed as the hydrogen of the statine-like hydroxyl is moved to Asp^{25}, with the resulting N-terminal half existing as a carboxylic acid dimer in conjunction with Asp^{25} (Fig. 14).

Substrate-Specificity Studies

With the exception of renin [Henrikson and Poorman, 1990], most aspartic proteases hydrolyze a wide variety of substrates; that is, they express ill-defined substrate specificities. To assess the substrate specificity of the HIV-1 protease, an expanded data base consisting of 40 regions of viral polyproteins and nonviral proteins known to be sensitive to hydrolysis by HIV-1 protease was subjected to statistical analysis. Rationally designed synthetic peptide substrates were excluded from the data set to avoid biasing the results. The frequency of occurrence of the amino acids in each substrate-binding pocket (P4–P4´) was tabulated and compared with the natural abundance of each particular amino acid in a random data set of globular proteins (Fig. 15) [Poorman et al., 1991].

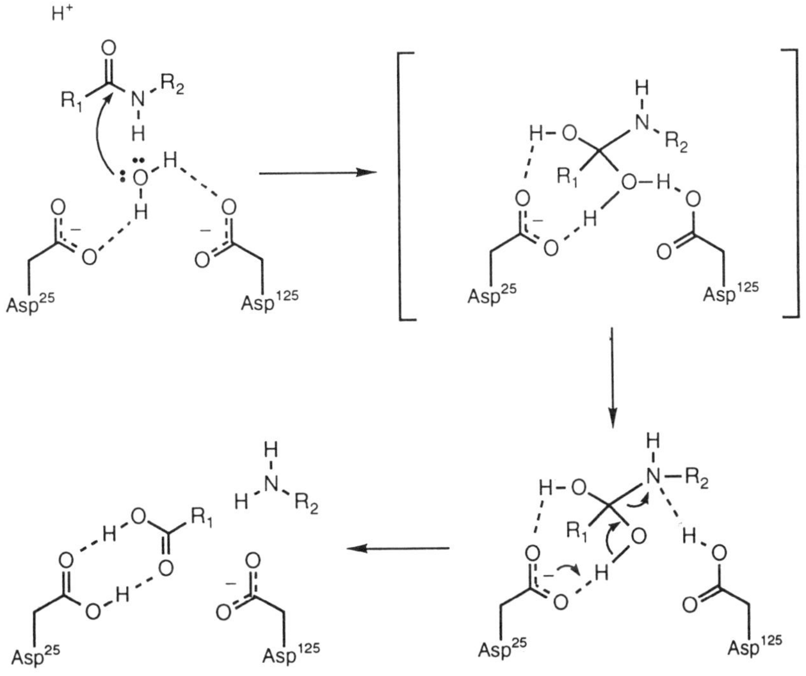

Fig. 14. *Proposed hydrolytic mechanism of HIV-1 protease.*

P4 - Pro(2.8), Ala(2.4), Ser(2.2)	0.8
P3 - Gln(6.5), Glu(4.7)	1.2
P2 - Val(6.2), Asn(5.3), Ile(5.3)	1.8
P1 - Phe(13.7), Leu(6.8), Met(4.2), Asn(2.8), Tyr(2.2)	3.3
P1' - Phe(4.5), Pro(4.0), Tyr(3.5), Met(2.2)	1.2
P2' - Glu(15.0), Gln(6.5)	4.1
P3' - Phe(3.2), Thr(2.9), Glu(2.1), Arg(2.2)	1.1
P4' - Phe(3.2), Pro(2.8), Met(2.2), Ser(2.2)	0.8

Fig. 15. *Substrate specificity table.*

For each subsite, the abundance of each residue is listed with respect to the number of standard deviations above what was expected for a globular protein. Only surabundant residues (greater than two standard deviations) are listed. The overall selectivity of each particular subsite is then expressed in terms of the ratio of the number of surabundant residues to that expected from a random distribution multiplied by the ratio of the number of surabundant residues to the total number of residues at a given subsite [Poorman et al., 1991].

From the results above, it can be noted that the contribution of the subsites to specificity is not uniform. The contribution of P4, P3´, and P4´ to the specificity of the HIV-1 protease is minimal, while the P2, P1, and P2´ subsites exhibit the highest stringency for particular amino acids. One might expect that, due to the symmetric nature of the active site, equivalent contributions to specificity would be attained at the equivalent sites P1 and P1´ surrounding the scissile bond. This is not the case, as evidenced by the high degree of selectivity noted for the P1 subsite, with Phe being the most surabundant residue—while the P1´ subsite expresses little selectivity. Additionally, from the surabundance of Gln and Glu at subsites P3 and P2´, it is suggested that P1 is the center of symmetry in the enzyme, which is a reflection of the structural properties of residues. Yet, the distribution of symmetry would indicate that P1´ is the center of symmetry, reflecting possibly the diversity of mechanistic roles served by the subsites and the directionality of the intrinsically asymmetrical substrates [Poorman et al., 1991].

In general, it would appear that only a few positions of the substrate need to be optimized in order to exhibit productive binding. Direct experimental evidence of subsite specificity was obtained from the effect of the systematic replacement of the P3, P2, P1, and P1´ residues of the heptapeptide corresponding to the matrix/capsid protein junction in the *gag* protein (Ser-Gln-Asn-Tyr-Pro-Ile-Val) upon hydrolysis rates by HIV-1 protease [Billich and Winkler, 1991]. In agreement with the above results, hydrophobic residues (Phe, Met) are preferred in S1 and S1´, although proline and β-branched amino acids are not tolerated in S1. Furthermore, in S2, small apolar amino acids (Ala and Val) as well as polar amino acids (Asn, Asp, Glu, and Cys) are equally accommodated. The presence of Ser or Gln in the S2 subsite prevents hydrolysis. As indicated previously, the S3 subsite shows only minimal specificity, and this is supported experimentally, since all amino acids with the exception of Pro can be tolerated.

Approaches to Inhibitor Design

Substrate-based inhibitors. Previous studies with inhibitors of other aspartic proteases, i.e., renin [Greenlee, 1990], provided insight into the design of HIV-1 protease inhibitors. These studies indicated that peptide-like substrates incorporating transition-state mimetics of the scissile amide bond (i.e., hydroxy or statine) should be highly potent and selective. HIV-1 protease cleaves the polyprotein at several sites in order to liberate the *gag* polyproteins itself and reverse transcriptase (RT). To facilitate hydrolysis, it has been suggested that heptapeptide (i.e., Ser-Gln-Asn-Tyr-Pro-Ile-Val) [Darke et al., 1988, 1989] or hexapeptide (i.e., Ac-Thr-Ile-Met-Met-Gln-Arg-NH$_2$) [Toth et al., 1990] substrates were necessary. These peptides have served as templates that have been modified to yield HIV-1 protease inhibitors successfully. To date, several transition-state mimetic inhibitors have

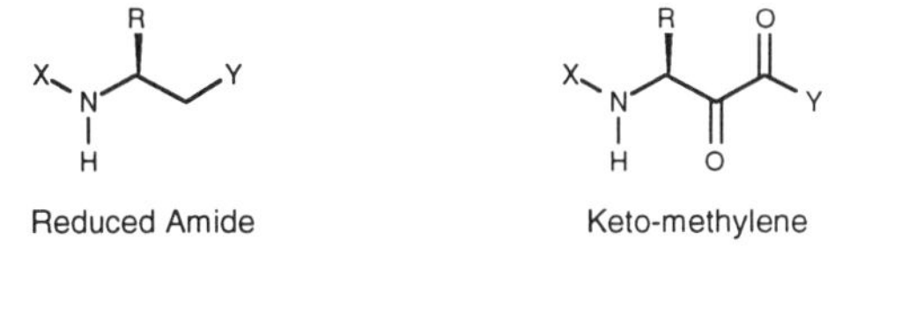

Fig. 16. *Examples of transition-state mimetics used in substrate-based inhibitors.*

been developed, including reduced amide [Toth et al., 1990], ketomethylene [Marinier et al., 1994], hydroxyethylamine [Krohn et al., 1991; Rich et al., 1991; Roberts et al., 1990; Tucker et al., 1992], statine, norstatine [Tam et al., 1992], dihydroxyethylene [Thaisrivongs et al., 1991], and hydroxyethylurea [Getman et al., 1993].

Examination of the crystallographic data for several HIV-1 protease–inhibitor complexes reveals that the inhibitors bind pseudosymmetrically in the active-site cleft using topologically-equivalent hydrogen-bonding functions [Blundell et al., 1991]. Inhibitors bind in extended conformation as evidenced by the structures of MVT101, JG365, and U85548E. For hexa- and heptapeptides (JG365), the preferred configuration about the tetrahedral carbon of the isostere is S. However, for shorter inhibitors (Roche), those possessing R alcohols exhibit the greatest activity. A conformational alteration allows for the alcohol functions, in both cases, to be oriented similarly, with the major intermolecular differences occurring in the P2 position.

A comparison of the crystal structures for three structurally diverse inhibitors (MVT-101, JG365, and U85548E) of HIV-1 protease allowed for the following observations to be made. Subsites S1–S1´ are composed mainly of hydrophobic residues, but polar contacts between active-site aspartates and inhibitor backbone hydroxyls are noted. The residues of inhibitors in these areas are either aliphatic or aromatic, usually hydrophobic, and can be moderately large (i.e., Phe-Pro, Leu-Val, Nle-Nle). Side chains of the inhibitors in the P1 subsite extend furthest away from backbone, implying that the P1 pocket is larger than the P1´ pocket. However, as the overall length of the inhibitor is decreased, larger aromatic or aliphatic hydrophobic groups are required in P1´ in order to maintain potency [Moore and Dreyer, 1993; Swain et al., 1991].

The side-chain residues forming S2- and S2´-binding pockets are mostly hydrophobic, but both hydrophobic and hydrophilic, often β-branched, residues may occupy these sites. A substitution of Asn in P2 and increased hydrophilicity in P2´ are associated with diminished inhibitory activity. An isoleucine forms hydrophobic contacts with different groups in the binding pocket, with the amides being held in place by hydrogen bonds. Asparagine amides point toward S4 and glutamine amides point toward the extremity and make polar contact with the main chain of Gly48. The amide side chains in S2–S2´ subsites are stabilized by polar contact with carbonyl oxygens of P3–P3´ inhibitor residues and main-chain atoms of Asp^{29} and Asp^{30} [Moore and Dreyer, 1993; Swain et al., 1991].

The S3 and S3´ subsites are distinct, with the outer walls of both being formed by protease loops containing $Phe^{53(153)}$ and $Pro^{181(81)}$. Aliphatic, polar, and ionic side chains are found in P3 and P3´ of inhibitors. In S3´, the residue Trp^{106} is inserted between Phe^{153} and Pro^{8}, which makes Arg[206] of MVT assume a compact (strained) conformation. In S3, Trp[6] is too far away to influence binding of P3 residue [Swain et al., 1991].

Subsites S4–S4′ are poorly defined and not really considered to be true binding pockets, since only a portion of the atoms in the P4 residue of the inhibitor are surrounded by atoms of the protease. In JG365 and U85548E, the P4 residue is a Ser. Additional subsites will not be discussed, since only a limited number of HIV-1 protease inhibitors (i.e., U85548E) extend into these regions that are not classified as binding pockets [Swain et al., 1991].

Although peptides and peptide-like compounds can be used in vitro to inhibit protease activity, the limited bioavailability due to poor oral absorption, instability due to amidases and peptidases, and rapid biliary metabolism limit their usage as practical therapeutic agents. Currently, a goal of many research laboratories is to minimize the peptide-like character of compounds [Lyle et al., 1991; Ghosh et al., 1993].

C2-symmetric inhibitors. An alternative to transition-state mimetics based on natural asymmetric substrates is the utilization of C2 symmetric inhibitors. It was suggested that inhibitors with a C2 axis of symmetry may exploit the symmetry of protease itself and may bind more tightly than the native peptide, which is not symmetric [Blundell and Pearl, 1989]. Since active-site residues (Asp-Thr/Ser-Gly) are conserved between mammalian cellular and retroviral aspartic proteases, the design of inhibitors specific for one or the other is made difficult [Rayan et al., 1991]. In this respect, additional advantages of C2 symmetric inhibitors are that they are inherently less peptidic in nature and as such are less likely to interact with normal mammalian cellular aspartic proteases that are less symmetric than HIV-1 protease thus increasing specificity and potentially minimizing side effects of therapy.

Diol functionalities with either RR or SS stereochemistry as transition-state mimetics represent a true class of C2-symmetric inhibitors. Although both hydroxyl groups do not interact with both active-site aspartates, these inhibitors are generally more potent than the corresponding pseudosymmetric mono alcohols [Moore and Dryer, 1993].

Caveat No. 1. While some C2-symmetric inhibitors bind the enzyme symmetrically [Erickson et al., 1990], substrate symmetry does not ensure a symmetric mode of interaction with the protease. This was recently confirmed by x-ray analysis of Cbz-Val-(Ph[CH_2]$_3$)C-(OH)C(OH)Ph[CH_2]$_3$)-Val-Cbz complexed with HIV-1 protease [Dreyer et al., 1993].

Caveat No. 2. The current generation of C2-symmetric compounds does not possess favorable water solubility characteristics. Most often, efforts to improve the water solubility of a given compound have decreased the inhibitory activity. To date, the C2 symmetrics that have been subjected to pharmacokinetic testing have not proven to be suitable for oral administration (the preferred route) and only lend themselves to administration via intravenous injection [Kageyama et al., 1992].

Mechanism-based inhibitors. The earlier studies of renin inhibitors indicated that irreversible inhibitors would be difficult to design [Greenlee, 1990]. However, a tripeptide inhibitor incorporating an epoxide ring system has been developed (Fig. 17). Attack of the epoxide by an active-site aspartate results in an ester complex. The protease inhibition by this compound is irreversible and competitive (activity is blocked by a potent transition-state inhibitor). This compound represents the first potent example of a mechanism-based inhibitor of the HIV-1 protease [Moore et al., 1992].

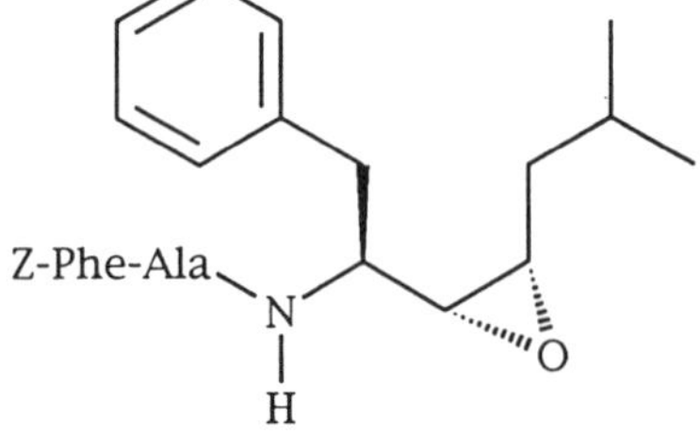

Fig. 17. *Structure of mechanism-based inhibitor. [Reproduced from Moore et al., 1992, with permission of the publisher.]*

Other inhibitors. Direct, rapid, and irreversible inhibition of HIV-1 protease by stoichiometric amounts of Cu^{2+} has been observed. It is known that copper interacts with Cys residues of proteins. A possible mechanism for this inhibition is the disruption of the dimer interface in which Cys^{95} is intimately involved. This hypothesis is supported by experiments in which the proteolytic activity of a mutant strain of the enzyme, lacking cysteine residues, was not inhibited by copper. However, upon addition of dithiothreitol (DTT) to the incubation mixture containing protease, substrate, and copper, inhibition was once again noted. Consequently, it has also been postulated that Cu^{2+} may indirectly inhibit proteolytic activity mediated by the formation of a copper–DTT complex. Regardless of the mechanism, these studies may stimulate the development of novel copper-containing HIV-1 protease inhibitors [Karlstrom and Levine, 1991].

Reversible and competitive inhibition of the hydrolytic activity of aspartic proteases (including HIV-1 protease, renin, and pepsin) has been observed to be characteristic of zinc. Contrary to inhibition by copper, zinc inhibition is not due to the induction of dimer instability. Presently, it is suggested that Zn^{2+} binds at or near the active-site aspartates. These results provide limited rationale for the use of Zn^{2+} supplementation in AIDS therapy [Zhang et al., 1991].

Most recently, certain metal ions (i.e., uranium [UO_2^{2+}], vanadium (V^{5+}), lead (Pb^{2+}), titanium [Ti^{4+}], and gold [AuIII]) have been found to be inhibitors ($IC_{50} \sim 1$–$20\,\mu M$) of HIV-1 protease. Inhibition is not believed to be mediated through nonspecific charge interactions, yet via coordination to the carboxylate side chains of Asp^{25} and Asp^{125} of the active site. Additionally, inhibition appears to be a function of the hardness (high charge:size ratio), with hard/acidic metals displaying the greatest inhibitory activity [Woon et al., 1992].

QSAR Studies: De Novo Design

An essential feature of any successful drug development campaign is the discovery of the relationship between molecular structure and activity. A qualitative assessment of the structure–activity relationship (SAR) for inhibitors of the HIV-1 protease has been presented above. While one may be able to design an inhibitor based on such information, predictions of activity for the resultant compound are not possible. If, however, one is able to assign a numerical value to the contribution of a given molecular feature to the overall activity, then a quantitative forecast of the activity, based on the principle of additivity, may be made.

Various approaches have been developed to aid in the formulation of quantitative structure–activity relationships (QSAR) [Martin, 1978]. The traditional methods utilize an overall measure of the hydrophobicity of the molecule usually described as the partitioning of the substance between octanol and water (P). Physical descriptors representative of steric bulk (Es) and electronic character (σ) are also typically considered in conventional QSAR. The relationship between molecular physicochemical characteristics and biological activity are then expressed as follows:

$$1/\log IC_{50} = a + \beta_1 \log P + \beta_2 (\log P)^2 + \beta_3 Es + \beta_4 \sigma$$

where IC_{50} is the concentration that is required to suppress proteolytic activity to 50% of the normal level, a is the intercept value, β_n are the regression coefficients, P is the octanol/water partition coefficient, E_s is the Taft steric constant, and σ is the electronic constant.

Traditional techniques are typically indiscriminant of the conformational and configurational characteristics of the molecules under study. The vast amount of crystallographic data available for the HIV-1 protease, both uncomplexed and complexed with a wide variety of inhibitor structures, has provided for an unprecedented description of the molecular environment of the active site of HIV-1 protease. From these crystal structures, the receptor-bound conformation of a variety of inhibitors has been determined. In addition, the relative orientation of one ligand in the active site may be directly compared with another via superposition of the protease–inhibitor structures. This information has been utilized recently in the development of a three-dimensional QSAR for HIV-1 protease inhibitors. By sampling the steric and electrostatic fields at regularly spaced intersections on a grid encompassing the molecular volumes of a series of molecules in a common orientation (receptor-bound geometries) using the technique of comparative molecular field analysis (CoMFA) [Cramer et al., 1988], a correlation was established between variations in field values with respect to inhibitory potency. The use of molecular fields as descriptors allowed for the accurate prediction of activity for molecules that were substantially structurally disparate from those on which the regression model was based [Waller et al., 1993].

An alternative approach to the design of novel inhibitors based on information derived from crystallographic analysis is known as *de novo design*. Beginning only with the enzyme structure, it is possible to fabricate a ligand that is complementary to the binding site [DesJarlais et al., 1990]. Several methods are currently under development that construct new inhibitors based on fragments taken from user-generated or crystallographic data bases [Bohm, 1991; Ho and Marshall, 1993a,b]. De novo construction on an atom-by-atom basis is also being explored. Still in the development phase, it is anticipated that de novo design will facilitate the development of a new generation of HIV-1 protease inhibitors.

Mutation Studies

It is known that disabling the HIV-1 protease via mutation results in the production of noninfectious progeny containing unprocessed *gag* and *gag/pol* protein precursors. To identify functionally important regions of the protease, a simple mutagenesis procedure capable of introducing missense mutations at each single residue of the protease was employed [Loeb et al., 1989a,b]. The results suggest that, in general, the number of sensitive amino acids is large relative to the size of the protein, on the order of 40%. Specifically, there appear to be three mutationally sensitive areas: 1) Ala^{22}-Leu^{33} (runs along binding cleft and into core of protein), 2) Ile^{47}-Gly^{52} (flap region), and 3) Thr^{74}-Arg^{87} (C-terminal region, a basic structural feature of aspartic proteases) [Blundell et al., 1991].

The region Ala^{22}-Leu^{33} is highly conserved between the sequences of the active sites of the aspartic proteases, including retroviral proteases [Toh et al., 1985]. Mutations in this area resulted in three types of mutants: 1) indistinguishable from wild type, 2) displaying some appropriately processed and some larger unprocessed and partially processed products, and 3) displaying no normal processing. Of 36 independent mutations, only 11 (31%) resulted in mutants that displayed any levels of normal processing (groups 1 and 2). Of the 11, only 2 ($E^{21}V$ and $D^{30}E$) yielded mutants that displayed activity comparable with the wild type. The active-site residues Asp^{25}-Thr^{26}-Gly^{27}-Ala^{28}-Asp^{29} are extremely sensitive to mutations. All mutations in this region resulted in no, or extremely low, levels of proteolytic processing. Additionally, only conservative changes were allowed at Ala^{22}, Leu^{23}, Leu^{24}, Thr^{31}, and Val^{32} [Loeb et al., 1989a,b].

In contrast to the highly conserved regions, the adjacent sequence Glu^{34}-Met^{46} is much more tolerant to mutations. Of 27 substitutions tested, only 19% resulted in phenotypes incapable of normal levels of processing. Nineteen of the remaining 22 mutants expressed activity similar to the wild type. The conservation of active-site residues between the aspartic and retroviral proteases and the sensitivity of these residues of HIV-1 protease to mutation reported in these early studies supported the classification of retroviral proteases as aspartic proteases [Loeb et al., 1989a,b].

Extensive studies have been performed on three HIV-1 protease mutants in an effort to explore the energetic importance of amino acid side chains in substrate and inhibitor binding and to qualitatively evaluate the individual effects upon kinetic parameters. The importance of hydrophobic interactions was explored with the use of a $V^{82}A$ mutant, while examination of hydrogen-bonding interactions was the primary objective of the $R^{8}E$ and $D^{29}A$ mutant studies. The $V^{82}A$ mutation was discovered to affect substrate-hydrolysis rates only moderately, with K_m only slightly higher and K_{cat} approximately fourfold less. Much more disabling were the R^8E and $D^{29}A$ mutants, with both exhibiting K_m values 10-fold above and K_{cat} values 50-fold below native enzyme values. Using two inhibitors varying only in the side chain (benzyl versus hydrogen) at the P1´ position (near Val^{82} in the binding site), it was noted that a reduction in the amount of hydrophobic surface area in the S1´-binding pocket reduced the activity of the benzyl-containing inhibitor, while the activity of the less hydrophobic inhibitor was moderately increased. The proteolytic inhibitory activity of the benzyl-containing inhibitor (the most potent of the two) in the R^8E and $D^{29}A$ mutants was reduced 10-fold and 50-fold, respectively. These results suggest that while Arg^8 and Asp^{29} are not necessarily required for the protease to exhibit activity, a charge–charge interaction between the two may contribute to stabilization of the tertiary structure. Additionally, it is suggested by crystallographic data that Asp^{29} contributes a significant hydrogen bond to the substrates or inhibitors [Darke et al., 1991].

CELLULAR ENZYMES INVOLVED IN VIRAL MATURATION

Other possible therapeutic targets for antiviral therapy are the constitutive enzymes essential for processing polyprotein components of the virus. As an example, inhibition of HIV protease only blocks the maturation of one of the polyprotein gene products, that of *gag/pol*. The fate and possible pathological effects of the continued expression and maturation of the products of the *env* gene during therapeutic inhibition of HIV protease are not known. Inhibiting constitutive enzymes as an approach to viral therapy does offer the advantage that the target enzymes are not virally encoded, and, therefore, the problem of resistance developing as a result of the rapid mutational rate of the viral genome would not occur. The rapidly evolving area of cellular processing enzymes, the convertases, is reviewed.

Convertases

Over the years, the concept of a protein precursor that undergoes limited proteolysis by one or more specific protease—convertase—to release the active component(s) was found to be appropriate for prohormones such as insulin, hormonal receptors, viral surface glycoproteins, adhesion molecules, zymogens, and many other proteins [Kido et al., 1993; Barr, 1991; Steiner et al., 1992]. The identification and characterization of the convertases proved difficult until the molecular characterization of the subtilisin-like proteinase [Fuller et al., 1988] Kex2 involved in the biosynthesis of yeast pheromones led to the subsequent identification of the mammalian enzymes. Yeast Kex2 cleaved mammalian precursors both in vitro and in vivo, implying that Kex2 represents a prototype for a mammalian subtilisin-like proteinase(s) with specificity for pairs of basic residues. Identification of three mammalian homologs of Kex2 quickly appeared, furin [Fuller et al., 1989; Roebroek et al., 1986], PC1 [Seidah et al., 1990] and PC2 [Smeekens and Steiner, 1990; Seidah et al., 1990], followed by the elucidation of the cDNA sequences of mouse [Seidah et al., 1991b; Smeekens et al., 1991] and human PC1 [Seidah et al., 1992b].

Coexpression of furin, PC1, and PC2 with suspected proprotein substrates demonstrated that all three enzymes cleave these precursors at specific pairs of basic residues. While furin cleaves proproteins that in vivo are normally expressed in cells devoid of secretory granules (constitutively secreting cells), PC1 and PC2 demonstrated selectivity for activation of precursors synthesized in cells containing secretory granules [Seidah et al., 1991a]. Furin appears to be ubiquitously expressed in most cells [Schalken et al., 1987], whereas PC1 and PC2 are mostly found in endocrine and neuroendocrine cells and tissues that are endowed with a regulated secretory pathway [Smeekens and Steiner, 1990; Seidah et al., 1990; Seidah et al., 1991a,b, 1992b; Smeekens et al., 1991; Schalken et al., 1987]. PC1 and PC2 have been shown to represent major processing enzymes of prohormones within endocrine and neuroendocrine tissues. Recently, three new members were identified, namely, PC4 [Nakayama et al., 1992; Seidah et al., 1992a], PACE4 [Kiefer et al., 1991], and PC5 [Seidah et al., 1991a; Lusson et al., 1993].

Activation of the proconvertases. The alignment of the sequences of PC1, PC2, furin, PACE4, PC4, and PC5 with that of subtilisin suggested that each convertase possesses a unique prosegment that by analogy to prosubtilisin [Wells et al., 1983] might have to be excised in order to generate an active proteinase [Barr, 1991; Smeekens and Steiner, 1990; Seidah et al., 1990, 1991a,b, 1992a,b; Roebroek et al., 1986; Smeekens et al., 1991; Nakayama et al., 1992; Kiefer et al., 1991]. Recent studies with the convertases kexin [Wilcox and Fuller, 1991; Germain et al., 1992] and furin [Leduc et al., 1992] demonstrated that these two processing enzymes are able to excise their prosegments autocatalytically. In the case of the yeast kexin [Wilcox and Fuller, 1991; Germain et al., 1992], this reaction seems to take place within the endoplasmic reticulum (ER), resulting in an active enzyme [Chaudhuri et al., 1992].

Cleavage of virus envelope glycoproteins by convertases. Aside from the cleavage of endogenous prohormones, convertases are also involved in the activation of other precursors such as certain cell surface receptors (e.g., the insulin receptor, adhesion molecules, and viral membrane glycoproteins). In some of these cases, it has been shown that the consensus sequence Arg_4-X-Lys/Arg-Arg_{-1} is critical for cleavage activation [Stieneke-Grober et al., 1992]. Evidence was presented that the glycoprotein prM of the yellow fever virus [Randolph et al., 1990], the F-fusion protein of mumps [Yamada et al., 1988], New Castle disease virus F1 protein [Morrison et al., 1985], the hemagglutinin (HA) of fowl plague virus [Stieneke-Grober et al., 1992], and the surface glycoprotein gp160 of HIV-1 [Hallenberger et al., 1992] are activated in the Golgi apparatus, possibly within the trans-Golgi network. Furthermore, a Ca^{2+}-dependent proteinase similar, if not identical, to furin has been demonstrated to be the enzyme responsible for the activation of HA of fowl plague virus [Stieneke-Grober et al., 1992] and the envelope glycoprotein gp160 of the human immunodeficiency virus HIV-1 [Hallenberger et al., 1992]. Since in most cases cleavage of the prosurface glycoprotein by host cell enzyme(s) is necessary for the virion to become infective, it is possible that furin or other convertases (such as PC5 or PACE4) play an important role in the life cycles of various viruses and their interplay with host cells.

Proteolysis and Viral Maturation

A common feature of viral replication is the production of large polyproteins that contain several viral protein products linked together. Within the sequence are specific motifs that specify cleavage sites for endopeptidases, encoded by either the viral genome or the cellular processing enzymes. A variety of enzyme classes (Asp, Cys, and Ser proteinases) have all been shown to be encoded by various classes of viruses. For example, Grakoui et al. [1993a] have recently shown that hepatitis C virus encodes a serine protease with a special cleavage-site motif. In addition, the viruses often utilize normal cellular processing routes, including endoproteolysis, for membrane glycoproteins in order to assemble the mature viral outer membranes. Inhibition of virally encoded proteases as an approach to antiviral therapy has been demonstrated by Korant et al. [1986], who showed that an endogenous inhibitor of cysteine proteases, cystatin, inhibited replication of poliovirus in tissue culture as well as by the multiple studies on HIV protease inhibitors.

While the polyprotein coded by the *gag/pol* gene of HIV is processed by the virally encoded HIV protease, the other polyprotein encoded by *env* is processed by cellular proteases. Mutational analysis [Guo et al., 1990; Freed et al., 1989; Bosch and Pawlita, 1990] of gp160 in the region of the cleavage site has identified a crucial Arg residue at position 518, whose mutation to Thr blocks gp160 cleavage, syncytium formation, and viral infectivity. Based on the sequence of the cleavage site, the dibasic endopeptidases [Barr, 1991] of mammalian cells are the probable enzymes responsible. In particular, the signal sequence RXK/RR has been identified [Hosaka et al., 1991] as a cleavage signal for secretory and membrane proteins. Envelope glycoproteins of many viruses, including HIV (cleaved sequence = RVVQREKR-AV), are produced from polyprotein precursors by cleavage at RXK/RR sites.

One concern in the use of HIV protease inhibitors is the continued build-up of mature envelope glycoprotein, which may lead to syncytia formation. Selective inhibition of the proteolytic pathways leading to mature envelope proteins may be a significant adjunct therapy to HIV protease inhibition. At worst, selective inhibitors would prove useful pharmacological agents with which to dissect the roles of the different proteolytic pathways. Recently, Hallenberger et al. [1992] showed that chloromethylketone (cmk) peptide substrate analogs decanoyl-RAKR-cmk and decanoyl-REKR-cmk could inhibit the furin-based cleavage of HIV gp160 in CV-1 cells in which both gp160 and human furin were coexpressed. The REKR analog inhibited formation of infectious HIV virus in the MT-4 lymphocyte cell line at concentrations of 2 mM (approximately 50% inhibition), demonstrating the acute feasibility of this approach.

Recently, another enzyme from human T4 lymphocytes has been implicated by Kido et al. [1993] in the cleavage of gp160. This 26-kDa enzyme is inhibited by leupeptin, benzamidine, aprotinin, and HI-30, but is not affected by $CaCl_2$, $MgCl_2$, or chelating agents, which differentiates it from the convertase family. The relative importance of the different proteolytic enzymes expressed in T4 lymphocytes and macrophages in the processing of gp160 remains to be determined. Specific proteolytic inhibitors under development will assist in this task.

ACKNOWLEDGMENTS

The authors thank the National Institutes of Health (for Research Cooperative Agreement grant AI27302 and program project grant GM24483) for support during the writing of this review. C.L.W. acknowledges support from NIH Training grant T32HL07275. In addition, the critical input of Prof. Nabil Seidah (Clinical Research Institute of Montreal) and Dr. Gary Jacob (Monsanto) to the section on convertases is gratefully noted.

REFERENCES

Acharya R, Fry E, Stuart D, Fox G, Rowlands D, Brown F (1989): The three-dimensional structure of foot-and-mouth disease virus at 2.9 Å resolution. Nature 337:709–716.

Aliperti G, Schlesinger MJ (1978): Evidence for an autoprotease activity of Sindbis virus capsid protein. Virology 90:366–369.

Alvey JC, Wyckoff EE, Yu SF, Lloyd R, Ehrenfeld E (1991): Cis- and trans-cleavage activities of poliovirus 2A protease expressed in *Escherichia coli*. J Virol 65:6077–6083.

Anthony RP, Paredes AM, Brown DT (1992): Disulfide bonds are essential for the stability of the Sinbis virus envelope. Virology 190:330–336.

Arnold E, Luo M, Vriend G, Rossmann MG, Palmenberg AC, Parks GD, Nicklin MJ, Wimmer E (1987): Implications of the picornavirus capsid structure for polyprotein processing. Proc Natl Acad Sci USA 84:21–25.

Barr PJ (1991): Mammalian subtilisins: The long-sought dibasic processing endoproteases. Cell 66:1–3.

Baum EZ, Bebernitz GA, Palant O, Mueller T, Plotch SJ (1991): Purification, properties, and mutagenesis of poliovirus 3C protease, Virology 185:140–150.

Bazan JF, Fletterick RJ (1988): Viral cysteine proteases are homologous to the trypsin-like family of serine proteases: Structural and functional implications. Proc Natl Acad Sci USA 85:7872–7876.

Bazan JF, Fletterick RJ (1989): Detection of a trypsin-like serine protease domain in flaviviruses and pestiviruses. Virology 171:637–639.

Berger A, Schechter I (1970): Mapping the active site of papain with the aid of peptide substrates and inhibitors. Phil Trans R Soc Lond-Series B: Biol Sci 257:249–264.

Billich A, Winkler G (1991): Analysis of subsite preferences of HIV-1 proteinase using MA/CA junction peptides substituted at the P3-P1´ positions. Arch Biochem Biophys 290:186–190.

Black TL, Barber GN, Katze MG (1993): Degradation of the interferon-induced 68000-Mr protein kinase by poliovirus requires RNA. J Virol 67:791–800.

Blumenstein JJ, Copeland TD, Oroszlan S, Michejda CJ (1989): Synthetic non-peptide inhibitors of HIV protease. Biochem Biophys Res Commun 163:980–987.

Blundell T, Pearl L (1989): Retroviral proteinases. A second front against AIDS. Nature 337:596–597.

Blundell TL, Cooper JB, Sali A, Zhu Z-y (1991): Comparisons of the sequences, 3-D structures and mechanisms of pepsin-like aspartic proteinases. Adv Exp Med Biol 306:443–453.

Boege U, Wengler G, Wengler G, Wittmann-Liebold B (1981): Primary structure of the core proteins of the alphaviruses Semliki Forest virus and Sindbis virus. Virology 113:293–303.

Bohm H-J (1991): The computer program LUDI: A new method for the de novo design of enzyme inhibitors. J Comput Aided Mol Design 6:61–78.

Bosch V, Pawlita M (1990): Mutational analysis of the human immunodeficiency virus type 1 *env* gene product proteolytic cleavage site. J Virol 64:2337–2344.

Chambers TJ, Grakoui A, Rice CM (1991): Processing of the yellow fever virus nonstructural polyprotein: A catalytically active NS3 proteinase domain and NS2B are required for cleavage at dibasic sites. J Virol 65:6042–6050.

Chambers TJ, Weir RC, Grakoui A, McCourt DW, Bazan JF, Fletterick RJ, Rice CM (1990): Evidence that the N-terminal domain of nonstructural protein NS3 from yellow fever virus is a serine protease responsible for site-specific cleavages in the viral polyprotein. Proc Natl Acad Sci USA 87:8898–8902.

Chaudhuri B, Latham SE, Helliwell SB, Seeboth P (1992): A novel Kex2 enzyme can process the proregion of the yeast alphafactor in the endoplasmic reticulum instead of the Golgi. Biochem Biophys Res Commun 183:212–219.

Cheah K-C, Leong LE-C, Porter AG (1990): Site-directed mutagenesis suggests close functional relationship between a human rhinovirus 3C cysteine protease and cellular trypsin-like serine proteases. J Biol Chem 265:7180–7187.

Choi H-K, Tong L, Minor W, Dumas P, Boege U, Rossmann MG, Wengler G (1991): Structure of Sindbis virus core protein reveals a chymotrypsin-like serine proteinase and the organization of the virion. Nature 354:37–43.

Clark ME, Haemmerle T, Wimmer E, Dasgupta A (1991): Poliovirus proteinase 3C converts an active form of transcription factor IIIC to an inactive form: A mechanism for inhibition of host cell polymerase III transcription by Poliovirus. EMBO J 10:2941–2947.

Clark ME, Lieberman PM, Berk AJ, Dasgupta A (1993): Direct cleavage of human TATA-binding protein by poliovirus protease 3C in vivo and in vitro. Mol Cell Biol 13:1232–1237.

Cramer RD III, Patterson DE, Bunce JD (1988): Comparative molecular field analysis (CoMFA). 1. Effect of shape on binding of steroids to carrier proteins. J Am Chem Soc 110:5959–5967.

Crawford S, Goff SP (1985): A deletion mutation in the 5′ part of the *pol* gene of Moloney murine leukemia virus blocks proteolytic processing of the *gag* and *pol* polyproteins. J Virol 53:899–907.

Dalgarno L, Trent DW, Strauss JH, Rice CM (1986): Partial nucleotide sequence of the Murray valley encephalitis virus genome. J Mol Biol 187:309–323.

Darke PL, Kohl NE, Hanobik MG, Leu CT, Vacca JP, Guare JP, Heimbach JC, Dixon RAF (1991): Interaction of mutant forms of the HIV-1 protease with substrate and inhibitors. Adv Exp Med Biol 306:483–487.

Darke PL, Leu C-T, Davis LJ, Heimbach JC, Diehl RE, Hill WS, Dixon RAF, Sigal IS (1989): Human immunodeficiency virus protease: Bacterial expression and characterization of the purified aspartic protease. J Biol Chem 264:2307–2312.

Darke PL, Nutt RF, Brady SF, Garsky VM, Ciccarone TM, Leu C-T, Lumma PK, Freidinger RM, Veber DF, Sigal IS (1988): HIV-1 protease specificity of peptide cleavage is sufficient for processing of *gag* and *pol* polyproteins. Biochem Biophys Res Commun 156:297–303.

DesJarlais RL, Seibel GL, Kuntz ID, Furth PS, Alvarez JC, Montellano PROd, DeCamp DL, Babe LM, Craik CS (1990): Structure-based design of nonpeptide inhibitors specific for the human immunodeficiency virus 1 protease. Proc Natl Acad Sci USA 87:6644–6648.

Devaney MA, Vakharia VN, Lloyd RE, Ehrenfeld E, Grubman MJ (1988): Leader protein of foot-and-mouth disease virus is required for cleavage of the p220 component of the cap-binding protein complex. J Virol 62:4407–4409.

Dewalt PG, Semler BL (1987): Site-directed mutagenesis of proteinase 3C results in a poliovirus deficient in synthesis of viral RNA polymerase. J Virol 61:2162–2170.

Diana GD, Nitz TJ, Mallamo JP, Treasurywala A (1993): Antipicornavirus compounds: Use of rational drug design and molecular modelling. Antiviral Chem Chemother 4:1–10.

Ding M, Schlesinger MJ (1989): Evidence that Sindbis virus nsP2 is an autoprotease which processes the virus nonstructural polyprotein. Virology 171:280–284.

Dougherty WG, Parks TD, Cary SM, Bazan JF, Fletterick RJ (1989): Characterization of the catalytic residues of the tobacco etch virus 49-kDa proteinase. Virology 172:302–310.

Dreyer GB, Boehm JC, Chenera B, DesJarlais RL, Hassell AM, Meek TD, Tomaszak TA Jr, Lewis M (1993): A symmetric inhibitor binds HIV protease asymmetrically. Biochemistry 32:937–947.

Erickson J, Neidhart DJ, VanDrie J, Kempf DJ, Wang XC, Norbeck DW, Plattner JJ, Rittenhouse JW, Turon M, Wideburg N, Kohlbrenner WE, Simmer R, Helfrich R, Paul DA, Knigge M (1990): Design, activity, and 2.8 Å crystal structure of a C2 symmetric inhibitor complexed to HIV-1 protease. Science 249:527–533.

Fenger TW (1991a): Replication of DNA viruses. In

Belshe RB (eds): Textbook of Human Virology. St. Louis: Mosby Year Book, pp 24–73.

Fenger TW (1991b): Replication of RNA Viruses. In Belshe RB (eds): Textbook of Human Virology. St. Louis: Mosby Year Book, pp 74–115.

Freed EO, Myers DJ, Risser R (1989): Mutational analysis of the cleavage sequence of the human immunodeficiency virus type 1 envelope glycoprotein precursor gp160. J Virol 63:4670–4675.

Fuller RS, Brake AJ, Thorner J (1989): Intracellular targeting and structural conservation of a prohormone-processing protease. Science 246:482–486.

Fuller RS, Sterne RE, Thorner J (1988): Enzymes required for yeast prohormone processing. Annu Rev Biochem 50:345–362.

Germain D, Dumas F, Vernet T, Bourbonnais Y, Thomas DY, Boileau G (1992): The pro-region of the Kex2 endoprotease of *Saccharomyces cerevisiae* is removed by self-processing. FEBS Lett 299:283–286.

Getman DP, DeCrescenzo GA, Heintz RM, Reed KL, Talley JJ, Bryant ML, Clare M, Houseman KA, Marr JJ, Mueller RA, Vazquez ML, Shieh H-S, Stallings WC, Stegeman RA (1993): Discovery of a novel class of potent HIV-1 protease inhibitors containing the (R)-hydroxyethylurea isostere. J Med Chem 36:288–291.

Ghosh AK, Thompson WJ, McKee SP, Duong TT, Lyle TA, Chen JC, Darke PL, Zugay JA, Emini EA, Schleif WA, Huff JR, Anderson PS (1993): 3-Tetrahydrofuran and pyran urethanes as high-affinity P2-ligands for HIV-1 protease inhibitors. J Med Chem 36:292–294.

Gorbalenya AE, Donchenko AP, Blinov VM, Koonin EV (1989a): Cysteine proteases of positive strand RNA viruses and chymotrypsin-like serine proteases. FEBS Lett 243:103–114.

Gorbalenya AE, Donchenko AP, Koonin EV, Blinov VM (1989b): N-terminal domains of putative helicases of flavi- and pestiviruses may be serine proteases. Nucleic Acids Res 17:3889–3897.

Grakoui A, McCourt DW, Wychowski C, Feinstone SM, Rice CM (1993a): Characterization of the hepatitis C virus-encoded serine proteinase: Determination of proteinase-dependent polyprotein cleavage sites. J Virol 67:2832–2843.

Grakoui A, Wychowski C, Lin C, Feinstone SM, Rice CM (1993b): Expression and identification of hepatitis C virus polyprotein cleavage products. J Virol 67:1385–1395.

Greenblatt J (1991): RNA polymerase-associated transcription factors. Trends Biochem Sci 16:408–411.

Greenlee WJ (1990): Renin inhibitors. Med Res Rev 10:173–236.

Guo H-G, Veronese Fd, Tschachler E, Pal R, Kalyanaraman VS, Gallo RC, Reitz MS Jr (1990): Characterization of an HIV-1 point mutation blocked in envelope glycoprotein cleavage. Virology 174:217–224.

Gustchina A, Weber IT (1990): Comparison of inhibitor binding in HIV-1 protease and in non-viral aspartic proteases: The role of the flap. FEBS Lett 269:269–272.

Hahn CS, Strauss JH (1990): Site-directed mutagenesis of the proposed catalytic amino acids of the Sindbis virus capsid protein autoprotease. J Virol 64:3069–3073.

Hahn CS, Strauss EG, Strauss JH (1985): Sequence analysis of three Sindbis virus mutants temperature-sensitive in the capsid protein autoprotease. Proc Natl Acad Sci USA 82:4648–4652.

Hallenberger S, Bosch V, Angliker H, Shaw E, Klenk H-D, Garten W (1992): Inhibition of furin-mediated cleavage action of HIV-1 glycoprotein gp160. Nature 360:358–361.

Harber JJ, Bradley J, Anderson CW, Wimmer E (1991): Catalysis of poliovirus VP0 maturation cleavage is not mediated by serine 10 of VP2. J Virol 65:326–334.

Hardy WR, Strauss JH (1989): Processing the non-structural polyproteins of Sindbis virus: Nonstructural proteinase is in the C-terminal half of nsP2 and functions both in cis and in trans. J Virol 63:4653–4664.

Hellen CUT, Krausslich H-G, Wimmer E (1989): Proteolytic processing of polyproteins in the replication of RNA viruses. Biochemistry 28:9881–9890.

Henrikson RL, Poorman RA (1990): In Laragh JH, et al. (eds): Hypertension: Pathophysiology, Diagnosis, and Management. New York: Raven Press, pp 1179–1196.

Hiagaki JN, Evnin LB, Craik CS (1989): Introduction of a cysteine protease active site into trypsin. Biochemistry 28:9256–9263.

Ho CMW, Marshall GR (1993a): FOUNDATION: A program to retrieve subsets of query elements, including active site region accessibility, from three-dimensional databases. J Comput Aided Mol Des 7:3–22.

Ho CMW, Marshall GR (1993b): SPLICE: A program to assemble partial query solutions from three-dimensional database searches into novel ligands. J Comput Aided Mol Des 7:623–647.

Hosaka M, Nagahama M, Kim W-S, Watanabe T, Hatsuzawa K, Ikemizu J, Murakami K, Nakayama K (1991): Arg-X-Lys/Arg-Arg motif as a signal for precursor cleavage catalyzed by furin within the constitutive secretory pathway. J Biol Chem 266:12127–12130.

Ivanoff LA, Towatari T, Ray J, Korant BD, Petteway SR Jr (1986): Expression of a site-specific mutagenesis of the poliovirus 3C protease in *Escherichia coli*. Proc Natl Acad Sci USA 83:5392–5396.

Kageyama S, Weinstein JN, Shirasaka T, Kempf DJ, Norbeck DW, Plattner JJ, Erickson J, Mitsuya H (1992): In vitro inhibition of human immunodeficiency virus (HIV) type 1 replication by C2 symmetry-based HIV protease inhibitors as single agents or in combinations. Antimicrob Agents Chemother 36:926–933.

Karlstrom AR, Levine RL (1991): Copper inhibits the protease from human immunodeficiency virus 1 by both cysteine-dependent and cysteine-independent mechanisms. Proc Natl Acad Sci USA 88:5552–5556.

Katoh I, Yasunaga T, Ikawa Y, Yoshinaka Y (1987): Inhibition of retroviral protease activity by an aspartyl protease inhibitor. Nature 329:654–656.

Kay J, Dunn BM (1990): Viral proteinases: Weakness in strength. Biochim Biophys Acta 1048:1–18.

Kean KM, Agut H, Fichot O, Wimmer E, Girard M (1988): A Poliovirus mutant defective for self-cleavage at the COOH-terminus of the 3C protease exhibits secondary processing defects. Virology 163:330–340.

Kean KM, Teterina NL, Marc D, Girard M (1991): Analysis of putative active site residues of the poliovirus 3C protease. Virology 181:609–619.

Kemp G, Webster A, Russell WC (1992): Proteolysis is a key process in virus replication. Essays Biochem 27:1–16.

Kido H, Kamoshita K, Fukotomi A, Katunuma N (1993): Processing protease for gp160 human immunodeficiency virus type I envelope glycoprotein precursor in human T4 lymphocytes. J Biol Chem 268:13406–13413.

Kiefer MC, Tucker JE, Joh R, Landsberg KE, Saltman D, Barr BJ (1991): Identification of a second human subtilisin-like protease gene in the *fes/fps* region of chromosome 15. DNA Cell Biol 10:757–769.

Kleina LG, Grubman MJ (1992): Antiviral effects of a thiol protease inhibitor on foot-and-mouth disease virus. J Virol 66:7168–7175.

Kohl NE, Emini EA, Schleif WA, Davis LJ, Heimbach JC, Dixon RAF, Scolnick EM, Sigal IS (1988): Active Human Immunodeficiency Virus Protease is Required for Viral Infectivity. Proc Natl Acad Sci USA 85:4686–4690.

Koonin EV, Wlodawer AE, Purdy MA, Rozanov MN, Reyes GR, Bradley DW (1992): Computer-assisted assignment of functional domains in the nonstructural polyprotein of hepatitis E virus: delineation of an additional group of positive-strand RNA plant and animal viruses. Proc Natl Acad Sci USA 89:8259–8263.

Korant B, Towatari T, Ivanoff L, Petteway SJ, Brzin J, Lenarcic B, Turk V (1986): Viral therapy: Prospects for protease inhibitors. J Cell Biochem 32:91–95.

Krausslich H-G, Wimmer E (1988): Viral proteinases. Annu Rev Biochem 57:701–754.

Krohn A, Redshaw S, Ritchie JC, Graves BJ, Hatada MH (1991): Novel binding mode of highly potent HIV-proteinase inhibitors incorporating the (R)-hydroxyethylamine isostere. 34:3340–3342.

Lawson MA, Semler BL (1990): Picornavirus protein processing—enzymes, substrates, and genetic regulation. Curr Top Microbiol Immunol 161:49–87.

Leduc R, Molloy SS, Thomas BA, Thomas G (1992): Activation of human furin precursor processing endoprotease occurs by an intramolecular autoproteolytic cleavage. J Biol Chem 267:14304–14308.

Lee W-M, Monroe SS, Rueckert RR (1993): Role of maturation cleavage in infectivity of picornaviruses: Activation of an infectosome. J Virol 67:2110–2122.

Lemm JA, Rice CM (1993a): Assembly of functional Sindbis virus RNA replication complexes: Requirement for coexpression of P123 and P34. J Virol 67:1905–1915.

Lemm JA, Rice CM (1993b): Roles of nonstructural polyproteins and cleavage, products in regulating Sindbis virus RNA replication and transcription. J Virol 67:1916–1926.

Li G, Rice CM (1989): Mutagenesis of the in frame opal termination codon preceding ns P4 of Sindbis virus: Studies of translational readthrough and its effects on virus replication. J Virol 63:1326–1337.

Liebig H-D, Skern T, Luderer M, Sommergruber W, Blaas D, Kuechler E (1991): Proteinase trapping: Screening for viral proteinase mutants by alpha complementation. Proc Natl Acad Sci USA 88:5979–5983.

Lin C, Amberg SM, Chambers TJ, Rice CM (1993): Cleavage at a novel site in the NS4A region by the yellow fever virus NS2B-3 proteinase is a prerequisite for processing at the downstream 4A/4B signalase site. J Virol 67:2327–2335.

Lloyd RE, Grubman MJ, Ehrenfeld E (1988): Relationship of p220 cleavage during picornavirus infection to 2A proteinase sequencing. J Virol 62:4216–4223.

Loeb DD, Hutchinson CA III, Edgell MH, Farmerie WG, Swanstrom R (1989a): Mutational analysis of human immunodeficiency virus type 1 protease suggests functional homology with aspartic proteinases. J Virol 63:111–121.

Loebb D, Swanstrom R, Everitt L, Manchester M, Stamper SE, Hutchison CA III (1989b): Complete mutagenesis of the HIV-1 protease. Nature 340:397–400.

Lusson J, Viean D, Hamelin J, Day R, Chretien M, Seidah NG (1993): cDNA structure of the mouse and rat subtilisin/Kexin-like Pc5: A candidate proprotein convertase expressed in endocrine and nonendocrine cells. Proc Natl Acad Sci USA 90:6691–6695.

Lyle TA, Wiscount CM, Guare JP, Thompson WJ, Anderson PS, Darke PL, Zugay JA, Emini EA, Schleif WA, Quintero JC, Dixon RAF, Sigal IS, Huff JR (1991): Benzocycloalkyl amines as novel C-termini for HIV protease inhibitors. J Med Chem 34:1228–1230.

Marinier A, Toth MV, Houseman K, Mueller R, Marshall GR (1994): HIV-1 protease inhibitors: Ketomethylene isosteres with unusually high affinity compared with hydroxyethylene isostere analogs. Bioorg Med Chem (in press).

Martin YC (1978): Quantitative Drug Design. New York: Marcel Dekker, 402 pp.

Melancon P, Garoff H (1987): Processing of the Semliki Forest virus structural polyprotein: Role of the capsid protease. J Virol 61:1301–1309.

Melnick JL (1991): Structure and Classification of Viruses. In Belshe RB (eds): Textbook of Human Virology. St. Louis: Mosby Year Book, pp 1–23.

Miller M, Jaskolski M, Rao JKM, Leis J, Wlodawer A (1989a): Crystal structure of a retroviral protease proves relationship to aspartic protease family. Nature 337:576–579.

Miller M, Schneider J, Sathyanarayana BK, Toth MV, Marshall GR, Clawson L, Selk L, Kent SBH, Wlodawer A (1989b): Structure of complex of synthetic HIV-1 protease with a substrate-based inhibitor at 2.3 Å resolution. Science 246:1149–1152.

Moore ML, Dreyer GB (1993): Substrate-based inhibitors of HIV-1 protease. Perspect Drug Discovery Design 1:85–108.

Moore ML, Fakhoury SA, Bryan WM, Bryan HG, Thomaszek JTA, Grant SK, Meek TD, Huffman WF (1992): Peptidyl epoxides as potent, active site-directed irreversible inhibitors of HIV-1 protease. In Smith JA, Rivier JE (eds): Peptides, Chemistry, and Biology. Proceedings of the Twelfth American Peptide Symposium. Leiden: ESCOM, pp 781–782.

Morrison T, Ward LJ, Semerjian A (1985): Intracellular processing of the Newcastle disease virus fusion glycoprotein. J Virol 53:851–857.

Nakayama K, Kim W-S, Torij S, Hosaka M, Nakagawa T, Ikemizu J, Baba T, Murakami K (1992): Identification of the fourth member of the mammalian endoprotease family homologous to the yeast Kex2 protease: Its testis-specific expression. J Biol Chem 267:5897–5900.

Nutt RF, Brady SF, Darke PL (1988): Chemical synthesis and enzymatic activity of a 99-residue peptide with a sequence proposed for the human immunodeficiency virus protease. Proc Natl Acad Sci USA 85:7129–7133.

Oberst MD, Gollan TJ, Gupta M, Peura SR, Zydlewski JD, Sudarsanan P, Lawson TG (1993): The encephalomyocarditis virus 3C protease is rapidly degraded by an ATP-dependent proteolytic system in reticulocyte lysate. Virology 193:28–40.

Palmenberg AC (1990): Proteolytic processing of picornaviral polyprotein. Annu Rev Microbiol 44:603–623.

Pearl LH, Taylor WR (1987): A structural model for the retroviral proteases. Nature 329:351–354.

Pechek IV, Gustchina AE, Andreeva NS, Fedorov AA (1989): Possible role of some groups in the structure and function of HIV-1 protease as revealed by molecular modeling studies. FEBS Lett 247:118–122.

Poorman RA, Tomaselli AG, Heinrikson RL, Kezdy FJ (1991): A cumulative specificity model for proteases from human immunodeficiency virus types 1 and 2, inferred from statistical analysis of an extended substrate data base. J Biol Chem 266:14554–14561.

Randolph V, Winkler G, Stollar V (1990): Acidotropic amines inhibit proteolytic processing of flavivirus prM protein. Virology 174:450–458.

Rayan A, Fliess A, Kotler M, Chorev M, Goldblum A (1991): Theoretical models of aspartic proteases: Active site properties, dimer stability and interactions with model inhibitors. Adv Exp Med Biol 306:555–558.

Rich DH, Sun C-Q, Vara Prasad JVN (1991): Effect of hydroxyl group configuration in hydroxyethylamine dipeptide isosteres on HIV protease inhibition. Evidence for multiple binding modes. J Med Chem 34:1222–1225.

Roberts NA, Martin JA, Kinchington D, Broadhurst AV, Craig JC, Duncan IB, Galpin SA, Handa BK, Kay J, Krohn A, Lambert RW, Merrett JH, Mills JS, Parkes KEB, Redshaw S, Ritchie AJ, Taylor DL, Thomas GJ, Machin PJ (1990): Rational design of peptide-based HIV proteinase inhibitors. Science 248:358–361.

Robins TPJ (1993): HIV protease inhibitors: Their anti-HIV activity and potential role in treatment. J AIDS 6:162–170.

Roebroek AJ, Schalken JA, Leunissen JAM, Onnekink C, Gloemers HPJ, Ven WJMVd (1986): Evolutionary conserved close linkage of the c-*fes/fps* proto-oncogene and genetic sequences encoding a receptor-like protein. EMBO J 5:2197–2202.

Roeder RG (1991): The complexities of eukaryotic transcription initiation: Regulation of preinitiation complex assembly. Trends Biochem Sci 16:402–408.

Schalken JA, Roebroek AJ, Oomen PPCA, Wagenaar SSC, Debruyne FMJ, Ven WJMVd (1987): Fur gene expression as a discriminating marker for small cell and nonsmall cell lung carcinomas. J Clin Invest 80:1545–1549.

Seelmeier S, Schmidt H, Turk V, von der Helm K (1988): Human immunodeficiency virus has an aspartic type protease that can be inhibited by pepstatin A. Proc Natl Acad Sci USA 85:6612–6616.

Seidah NG, Day R, Hamelin J, Gaspar A, Collard MW, Chretien M (1992a): Testicular expression of PC4 in the rat: Molecular diversity of a novel germ cell-specific Kex2~subtilisin-like proprotein convertase. Mol Endocrinol 6:1559–1569.

Seidah NG, Day R, Marcinkiewicz M, Benjannet S, Chretien M (1991a): Mammalian neural and endocrine pro-protein and pro-hormone convertases belonging to the subtilisin family of serine proteinases. Enzyme 45:271–284.

Seidah NG, Gaspar L, Mion P, Marcinkiewicz M, Mbikay M, Chretien M (1990): cDNA sequence of two distinct pituitary proteins homologous to Kex2 and Furin gene products: Tissuespecific mRNAs encoding candidates for pro-hormone processing proteinases. DNA 9:415–424.

Seidah NG, Hamelin J, Gaspar A, Day R, Chretien M (1992b): The cDNA sequence of the human pro-hormone and pro-protein convertase PC1. DNA Cell Biol 11:283–289.

Seidah NG, Marcinkiewicz M, Benjannet S, Gaspar L, Beaubien G, Mattei MG, Lazure C, Mbikay M, Chretien M (1991b): Cloning and primary sequence of a mouse candidate pro-hormone convertase PC1 homologous to PC2, furin and Kex2: Distinct chromosomal localization and mRNA distribution in brain and pituitary as compared to PC2. Mol Endocrinol 5:122.

Singer PT, Smalås A, Carty RP, Mangel WF, Sweet RM (1993): The hydrolytic water molecule in trypsin, revealed by time-resolved Laue crystallography. Science 259:669–673.

Smeekens SP, Avruch AS, LaMendola J, Chan SJ, Steiner DF (1991): Identification of a cDNA encoding a second putative prohormone convertase related to PC2 in AtT20 cells and islets of Langerhans. Proc Natl Acad Sci USA 88:340–344.

Smeekens SP, Steiner DF (1990): Identification of a human insulinoma cDNA encoding a novel mammalian protein structurally related to the yeast dibasic processing protease Kex2. J Biol Chem 265:2997–3000.

Sommergruber W, Ahorn H, Zophel A, Maurer-Fogy I, Fessl F, Schnorrenberg G, Liebig H-D, Blaas D, Kuechler E, Skern T (1992): Cleavage specificity on synthetic peptide substrates of human rhinovirus 2 proteinase 2A. J Biol Chem 267:22639–22644.

Steiner DF, Smeekens SP, Ohagi S, Chan SJ (1992): The new enzymology of precursor processing endoproteases. J Biol Chem 267:23435–23438.

Stieneke-Grober A, Angliker MVH, Shaw E, Thomas G, Roberts C, Klenk H-D, Garten W (1992): Influenza virus hemagglutinin with multibasic cleavage site is activated by furin, a subtilisin-like endoprotease. EMBO J 11:2407–2414.

Strauss EG, Groot RJD, Levinson R, Strauss JH (1992): Identification of the active site residues in the nsP2 proteinase of Sindbis virus. Virology 191:932–940.

Swain AL, Gustchina A, Wlodawer A (1991): Comparison of three inhibitor complexes of human immunodeficiency virus protease. Adv Exp Med Biol 306:433–441.

Takkinen K (1986): Complete nucleotide sequence of the nonstructural protein genes of Semliki Forest virus. Nucleic Acids Res 14:5667–5682.

Tam TF, Carriere J, MacDonald ID, Castelhano AL, Pliura DH, Dewdney NJ, Thomas EM, Bach C (1992): Intriguing structure–activity relations underlie the potent inhibition of HIV protease by norstatine-based peptides. J Med Chem 35:1318–1320.

Thaisrivongs S, Tomasselli AG, Moon JB, Hui J, McQuade TJ, Turner SR, Strohbach JW, Howe WJ, Tarpley WG, Heinrikson RL (1991): Inhibitors of the protease from human immunodeficiency virus: Design and modeling of a compound containing a dihydroxyethylene isostere insert with high binding affinity and effective antiviral activity. J Med Chem 34:2344–2356.

Toh H, Kikuno R, Hayashida H, Miyata T, Kugimiya W, Inouye S, Yuki S, Saigo K (1985): Close structural resemblance between putative polymerase of a *Drosophila* transposable genetic element 17.6 and pol gene product of Moloney murine leukaemia virus. EMBO 4:1267–1272.

Toth MV, Chiu F, Glover G, Kent SBH, Ratner L, Heyden NV, Green J, Rich DH, Marshall GR (1990): Inhibitors of HIV protease based on modified peptide substrates. In Rivier JE, Marshall GR (eds): Peptides: Chemistry, Structure, and Biology. Proceedings of the 11th American Symposium. Leiden: ESCOM, pp 835–838.

Tucker TJ, Lumma WC Jr., Payne LS, Wai JM, de Solms SJ, Giuliani EA, Darke PL, Heimbach JC, Zugay JA, Schleif WA, Quintero JC, Emini EA, Huff JR, Anderson PS (1992): A series of potent HIV-1 protease inhibitors containing a hydroxyethyl secondary amine transition state isostere: Synthesis, enzyme inhibition, and antiviral activity. J Med Chem 35:2525–2533.

Wagner R, Fliessbach H, Wanner G, Motz M, Niedrig M, Deby G, Brunn Av, Wolf H (1992): Studies on processing, particle formation, and immunogenicity of the HIV-1 gag gene product: a possible component of a HIV vaccine. Arch Virol 127:117–137.

Waller CL, Oprea TI, Giolitti A, Marshall GR (1993): 3-D QSAR of human immunodeficiency virus (I) protease inhibitors. I. A CoMFA study employing experimentally-determined alignment rules. J Med Chem. 36:4152–4160.

Warshel A, Naray-Szabo G, Sussman F, Hwang J-K (1989): How do serine proteases really work? Biochemistry 28:3629–3637.

Weber IT (1990): Evaluation of homology modeling of HIV protease. Proteins Struct Funct Genet 7:172–184.

Wellink J, van Kammen A (1988): Proteases involved in the processing of viral polyproteins. Arch Virol 98:1–26.

Wells JA, Ferrari E, Henner DJ, Estell DA, Chen EY (1983): Nucleic Acids Res 18:7911–7925.

Wilcox CA, Fuller R (1991): Post-translational processing of the prohormone-cleaving Kex2 protease in the *Saccharomyces cerevisiae* pathway. J Cell Biol 115:297–307.

Woon TC, Brinkworth RI, Fairlie DP (1992): Inhibition of HIV-1 proteinase by metal ions. Int J Biochem 24:911–914.

Yamada A, Tackeuchi K, Hishiyama M (1988): Intracellular processing of mumps virus glycoproteins. Virology 165:268–273.

Ypma-Wong MF, Dewalt PG, Johnson VH, Lamb JG, Semler BL (1988): Protein 3CD is the major poliovirus proteinase responsible for cleavage of the P1 capsid precursor. Virology 166:265–270.

Yu SF, Lloyd RE (1991): Identification of essential amino acid residues in the functional activity of poliovirus 2A protease. Virology 182:615–625.

Yu SF, Lloyd RE (1992): Characterization of the roles of conserved cysteine and histidine residues in poliovirus 2A protease. Virology 186:725–735.

Zhang Z-Y, Reardon IM, Hui JO, O'Connell KL, Poorman RA, Tomasselli AG, Heinrikson RL (1991): Zinc inhibition of renin and the protease from human immunodeficiency virus type 1. Biochemistry 30:8717–8721.

NOTE ADDED IN PROOF

The X-ray structure at 2.3 Å resolution of a double mutant ($C^{24}S$ and $C^{172}A$) 3C protease from hepatitis A virus (HAV) was recently reported [Allaire et al., 1994]. The observed protein structure is consistent with earlier predictions (see Figs. 3 and 4), and confirms the role of Cys^{172} (nucleophile) and His^{44} (base) in the catalytic site. However, Asp^{84} does not interact with His^{44}, as predicted earlier [Gorbalenya et al., 1989a]. The role of Asp^{84} and structural features of HAV-3C are presented [Allaire et al., 1994].

Allaire M, Chernaia MM, Malcolm BA, James MNG (1994): Picornaviral 3C cysteine proteinases have a fold similar to chymotrypsin-like serine proteinases. Nature 369:72–76.

ABOUT THE AUTHORS

GARLAND ROSS MARSHALL was born in San Angelo, Texas. He attended the California Institute of Technology, receiving a B.S. in Biology in 1962. As the first graduate student of Prof. R.B. Merrifield, Nobel Laureate, at Rockefeller University, he participated in the development of solid-phase peptide chemistry. After receiving his Ph.D. in 1966, he joined the faculty of Washington University School of Medicine in St. Louis, Missouri, where he is Professor of Molecular Biology and Pharmacology and of Biochemistry and Molecular Biophysics. In recognition of his contributions to drug design, Prof. Marshall was named the 1988 recipient of the ACS Award in Medicinal Chemistry. In addition, he received the Medal XL-Lecia Politechniki Lodzkiej in 1987 and a D.Sc. (hon. causa) in 1993 from the Technical University, Lodz, Poland, in recognition of his contributions to peptide chemistry and design of peptidomimetics. Prof. Marshall serves as the Director of the Center of Molecular Design at Washington University, an interdisciplinary group focused on developing the technology of computational chemistry and computer-aided drug design.

TUDOR IONEL OPREA was born in Timisoara, Romania. He received an M.D. in 1990 and a Ph.D. in physiology in 1992 from the University of Timisoara, where he is an Assistant Professor of Physiology. He spent the 1990/1991 academic year as visiting scientist at the State University of Utrecht (The Netherlands) doing molecular pharmacology and computational chemistry. Since the fall of 1992, he has been a postdoctoral research associate at the Center for Molecular Design at Washington University, doing 3D-QSAR.

CHRISTOPHER LEE WALLER was born in Salisbury, NC, and educated at Davidson College. He received his Ph.D. in Medicinal Chemistry from the University of North Carolina in 1992 for studies on inhibitors of chloride transport in astrocytes. He joined the Center of Molecular Design in 1992, where his postdoctoral research centered on three-dimensional quantitative structure–activity studies. Dr. Waller has recently joined the scientific staff of the United States Environmental Protection Agency, Research Triangle Park, NC.

Index

2A proteases, overview of, 190–191
A1s9X gene, ubiquitin and, 28
A1s9Y-1 gene, ubiquitin and, 28
N-Acetyl-leucyl-leucyl-methioninal (ALLM), endoplasmic reticulum retained protein degradation and, 142, 144, 151
N-Acetyl-leucyl-leucyl-norleucinal (ALLN), endoplasmic reticulum retained protein degradation and, 142, 144, 151
Acetylcholine receptor subunits, endoplasmic reticulum retained protein degradation and, 138
Acetylcholinesterase, endoplasmic reticulum retained protein degradation and, 138
Acid hydrolases
　autophagy and, 66–67, 69–70
　hepatic endosomes and, 89
Acid phosphatase, hepatic endosomes and, 89, 91
Acridine orange, hepatic endosomes and, 94, 97
Adenosine monophosphate (AMP), ubiquitin and, 5
Adenosine triphosphate (ATP)
　autophagy and, 66, 68–69
　endoplasmic reticulum and, 144, 149–151
　hepatic endosomes and, 94, 97–100
　lysosomes and, 55, 61–62
　ubiquitin and, 3–5, 9–10, 13–14, 19, 23
Adenylate cyclase, hepatic endosomes and, 100
α-Adrenergic pathway autophagy and, 82
Alanine
　autophagy and, 69, 77–78, 80–82
　ubiquitin and, 7–8, 24, 35, 42
Aldehydes, ubiquitin and, 12
Alkaline phosphatase, yeast vacuole and, 115, 119–121, 128
ALLM. *See N*-Acetyl-leucyl-leucyl-methioninal
ALLN. *See N*-Acetyl-leucyl-leucyl-norleucinal
α-amino group, of protein substrate, ubiquitin-protein ligases and, 7–9
α-factor, yeast vacuole and, 122, 128
α-toxin, autophagy and, 68, 75–76
α_2-repressor protein, yeast vacuole and, 123–124
Amastatin, hepatic endosomes and, 101
Amines, ubiquitin and, 6, 11
Amino acids
　autophagy and, 65–83
　hepatic endosomes and, 94
　lysosomes and, 55–56, 58–59, 62
　ubiquitin and, 4–9, 11–14, 16, 24–27, 30, 35, 37, 41–44
　yeast vacuole and, 115, 123, 129
Aminooxyacetate, autophagy and, 78
Aminopeptidases
　hepatic endosomes and, 90
　yeast vacuole and, 115, 119
Ammonium chloride
　endoplasmic reticulum and, 141, 144
　hepatic endosomes and, 98
　lysosomes and, 61
AMP. *See* Adenosine monophosphate
Analog-containing proteins, yeast vacuole and, 126
Angiotensin-converting enzyme, hepatic endosomes and, 105
ANP. *See* Atrial natriuretic peptide
Antibodies
　autophagy and, 68
　monoclonal
　　antigen catabolism and, 170
　　hepatic endosomes and, 96
　　lysosomes and, 60, 62
　polyclonal, hepatic endosomes and, 96
Antigens, processing of, 163–173
α_1-Antitrypsin, endoplasmic reticulum retained protein degradation and, 138, 152
ApI precursor, yeast vacuole and, 121, 128
Apolipoprotein B, endoplasmic reticulum retained protein degradation and, 138–140, 145–146
Apolipoprotein B-100, hepatic endosomes and, 91–92
Apoptosis, ubiquitin and, 19
Arabidopsis thaliana, ubiquitin and, 26, 28, 31, 33–36, 38
Arg-βgal, ubiquitin and, 9
Arginine
　lysosomes and, 55–56
　ubiquitin and, 7–8, 16, 42
Arginyl-tRNA protein transferase, ubiquitin and, 8, 42
ASFV-UBC gene, ubiquitin and, 29
Asialoglycoprotein receptor, H2a subunit, endoplasmic reticulum retained protein degradation and, 138, 141, 150

Asparagine
  autophagy and, 77
  ubiquitin and, 8, 42
Aspartate, autophagy and, 78
Aspartate aminotransferase, selective proteolysis by lysosomes and, 57, 60
Aspartic acid, ubiquitin and, 7–8, 42
Aspartic proteases
  hepatic endosomes and, 90
  retroviral, 203
Aspartyl proteases, antigen catabolism and, 166
ATE1 gene, ubiquitin and, 42
ATP. *See* Adenosine triphosphate
ATPase, yeast vacuole and, 115, 122
Atrial natriuretic peptide (ANP), hepatic endosomes and, 103
Autophagocytosis. *See* Autophagy
Autophagy, amino acids and, 77–82
  β-antagonists and, 82–83
  deprivation-induced breakdown of protein and RNA, 70–72
  general, 126–127
  glucagon and, 82–83
  hormonal regulation and, 80–83
  insulin and, 80–82
  intralysosomal pools of degradable protein and RNA, 72–73
  introduction to, 65
  macroautophagy and, 67–73, 77
  mechanism of, 66–77
  microautophagy and, 73–75
  protein degradation and, 65–66
  regulation of, 77–83
  RNA degradation and, 66
  selective, 127
  unified mechanism for autophagic proteolysis, 75–77
  vacuole and, 67–71, 126–129
AXR1 gene, ubiquitin and, 26, 28

Bacitracin, hepatic endosomes and, 98, 101, 103
Baculovirus, ubiquitin and, 25
Bafilomycin, yeast vacuole and, 127
Bazan model, for viral proteases, 185–186
B cells, endoplasmic reticulum retained protein degradation and, 139–140
Benzamidine, hepatic endosomes and, 98
Bestatin, hepatic endosomes and, 101
β-antagonists, autophagy and, 82–83
BHK cells, ubiquitin and, 10
Bombyxin, hepatic endosomes and, 103, 105
Borohydride, ubiquitin and, 12
Brain, selective proteolysis by lysosomes and, 60
Brefeldin A, endoplasmic reticulum retained protein degradation and, 142–143, 147

3C proteases, overview of, 185, 188–190, 192–194
C1 gene, ubiquitin and, 33

C2 gene, ubiquitin and, 33–34
C2-symmetric inhibitors, viral proteases and, 210
C3 gene, ubiquitin and, 33–34
C5 gene, ubiquitin and, 33–34
C8 gene, ubiquitin and, 33–34
C9 gene, ubiquitin and, 33–34
Ca^{2+}, endoplasmic reticulum retained protein degradation and, 144, 149
Caenorhabditis elegans, ubiquitin and, 27, 29
Calpain inhibitors, endoplasmic reticulum retained protein degradation and, 142
Calreticulin, endoplasmic reticulum retained protein degradation and, 151
cAMP. *See* Cyclic adenosine monophosphate
N-Carbenzoxy-L-phenylalanine chloromethyl ketone (ZPCK), endoplasmic reticulum retained protein degradation and, 142
Carbodiimides, water-soluble, selective proteolysis by lysosomes and, 58
Carbon, yeast vacuole and, 124–125
Carboxykinases, yeast vacuole and, 128
Carboxypeptidases
  endoplasmic reticulum and, 152–153
  hepatic endosomes and, 90–92
  yeast vacuole and, 115, 119–121, 123
Cardiac muscle, autophagy and, 77
L-Carnitine, autophagy and, 79
Carrier protein, ubiquitin and, 5
Catabolite inactivation, yeast vacuole and, 124, 129
Cathepsins
  antigen catabolism and, 166, 168–169
  endoplasmic reticulum and, 152
  hepatic endosomes and, 89–92, 94
CCl_4, endoplasmic reticulum retained protein degradation and, 150
CD4 molecule, endoplasmic reticulum retained protein degradation and, 138, 148, 151
CDC34 gene, ubiquitin and, 6, 12, 28, 30–31
Cell cycle, ubiquitin and, 6, 19, 30–31, 43–44
Cell-free assays, endoplasmic reticulum retained protein degradation and, 149–151
Cell surface receptors, ubiquitin and, 17
Cellular proteins, in vivo degradation of, ubiquitin in, 13-18
CF1–3 conjugate-degrading factors, ubiquitin and, 10–11, 19
c-fos oncoprotein, ubiquitin and, 14–15
p-Chloromercuribenzoic acid, hepatic endosomes and, 94, 101, 103
Chloromethyl ketones, endoplasmic reticulum retained protein degradation and, 142
Chloroquine
  autophagy and, 83
  endoplasmic reticulum and, 141
  hepatic endosomes and, 93–95, 97–99
CHO cells
  autophagy and, 68, 70
  insulin in, 95, 96

Index

c-jun oncoprotein, ubiquitin and, 15
Clathrin, yeast vacuole and, 122
CLB genes, ubiquitin and, 44
CMP, autophagy and, 66
c-myb oncoprotein, ubiquitin and, 14
c-myc oncoprotein, ubiquitin and, 14
Co^{2+}, endoplasmic reticulum retained protein degradation and, 144
Convertases, viral maturation and, 213–214
Covalent affinity chromatography, ubiquitin and, 5
CPS1 gene, yeast vacuole and, 116
Cryptic epitopes, antigen catabolism and, 167
C-terminal domain, ubiquitin and, 6, 11–12, 36–37
Cyclic adenosine monophosphate (cAMP)
    autophagy and, 82–83
    hepatic endosomes and, 100
Cyclic nucleotide phosphodiesterase, endoplasmic reticulum retained protein degradation and, 138
Cyclins
    degradation of, 5
    ubiquitin and, 16, 19, 43–44
Cycloheximide
    autophagy and, 65
    endoplasmic reticulum and, 144
    yeast vacuole and, 125, 128
Cysteine
    antigen catabolism and, 167
    ubiquitin and, 5–6, 8, 26, 30, 35, 42
Cysteine protease inhibitors, endoplasmic reticulum retained protein degradation and, 141–142, 144
Cysteine proteases, hepatic endosomes and, 90
Cysteine proteinase inhibitor, autophagy and, 69
Cystic fibrosis transmembrane conductance regulator, endoplasmic reticulum retained protein degradation and, 138
Cytidine, labeled, autophagy and, 66
Cytochalasins B and D, autophagy and, 69
Cytochrome P_{450} (2E1), endoplasmic reticulum retained protein degradation and, 138, 150–151, 153
Cytoplasmic sequestration, autophagy and, 65–73, 77
Cytosol
    antigen processing and, 167–168, 171–172
    insulin and glucagon-degrading activities in, 101–106

13D3 monoclonal antibody, selective proteolysis by lysosomes and, 62
Dansyl cadaverine, hepatic endosomes and, 97–99
DEA1 gene, ubiquitin and, 43
Deamidase, ubiquitin and, 43
Deg genes, ubiquitin and, 44
Degradation signals, ubiquitin and, 41–42
DELTA genes, ubiquitin and, 33, 38

DFP, endoplasmic reticulum retained protein degradation and, 151
Dhr6 gene, ubiquitin and, 29
Diamide, endoplasmic reticulum retained protein degradation and, 141, 144
Dicyclohexylcarbodiimide, hepatic endosomes and, 94
Digitonin, endoplasmic reticulum retained protein degradation and, 144, 150
Dihydrofolate reductase, selective proteolysis by lysosomes and, 55, 61
Dipeptidyl aminopeptidase II exopeptidases, hepatic endosomes and, 90
Dipeptidyl aminopeptidase B, yeast vacuole and, 115, 119
Dipeptidyl peptidase-IV, endoplasmic reticulum retained protein degradation and, 138
Direct protein transport, selective proteolysis by lysosomes and, 56–57
Disulfide reduction, antigen catabolism and, 167–169
Dithiothreitol (DTT)
    endoplasmic reticulum and, 150
    ubiquitin and, 12
DPA2 gene, yeast vacuole and, 116
Drosophila melanogaster
    insulin-degrading enzyme and, 102, 104, 106–107
    ubiquitin and, 27, 29–30, 32, 41
DTT. *See* Dithiothreitol

1E2–25 kd gene, ubiquitin and, 29
E1-E3 enzymes
    ubiquitin and, 4–9, 12–14, 18, 26–31, 35–37
    yeast vacuole and, 127
E1A oncoprotein, ubiquitin and, 14
E2–23 kd gene, ubiquitin and, 29
E64, endoplasmic reticulum retained protein degradation and, 141–142, 151
E64-D, endoplasmic reticulum retained protein degradation and, 142
EDTA
    endoplasmic reticulum and, 150
    hepatic endosomes and, 94, 101, 103
EGF. *See* Epidermal growth factor
Ehrlich ascites tumor cells, autophagy and, 69
Electroporation, antigen processing and, 171
Endocytosis
    autophagy and, 69
    hepatic endosomes and, 89
    lysosomes and, 55, 57, 62
    yeast vacuole and, 122, 129
Endopeptidases
    endoplasmic reticulum and, 151–152
    hepatic endosomes and, 89–90
Endoplasmic reticulum
    autophagy and, 69
    degradation of proteins retained in
        abnormal protein degradation, 140

ApoB and, 145–46
brefeldin A and, 142–43
CD4 molecule and, 148
cell-free assays and, 149–51
characteristics of degradative process, 140–143
destruction of newly synthesized secretory proteins, 139–140
HIV-1 VPU gene product and, 148–149
HMG-CoA reductase and, 143–145
introduction to, 137–139
lysosomal proteolysis inhibitors and, 141
nonlysosomal degradation in yeast secretory pathway, 152–153
permeabilized cells and, 149–151
protease inhibitors and, 141–142
proteolytic activities and, 151–152
resident protein turnover and, 139
subcellular localization of substrates, 140–141
substrate selectivity and, 141
T-cell receptor and, 146–48
UBC6 enzyme and, 153
weak bases and, 141
lysosomes and, 56, 62
yeast vacuole and, 120–121, 123, 128
Endoproteases, yeast vacuole and, 115
Endoproteinases, yeast vacuole and, 115, 123
Endosomes, hepatic
antigen catabolism and, 167–171
glucagon and, 96–106
insulin and, 92–96, 99–108
insulin-degrading enzyme and, 100–108
introduction to, 89
lysosomes and, 89–91
protein degradation in, 91–92
protein disulfide isomerase and, 106–108
Enkephalinase, hepatic endosomes and, 105
Enzymes, viral maturation and, 213–215
Epidermal growth factor (EGF), hepatic endosomes and, 91–92, 94, 96, 103–104, 108
Epinephrine, autophagy and, 82
ER60A-ER60H, endoplasmic reticulum retained protein degradation and, 151–152
Erythrocytes, ubiquitin and, 12
Escherichia coli
HMG-CoA reductase and, 144
insulin-degrading enzyme and, 102, 105–106
lysosomes and, 59
ubiquitin and, 7, 40–41
Exocytosis, selective proteolysis by lysosomes and, 55
Exopeptidases, hepatic endosomes and, 89
Exoproteases, yeast vacuole and, 115

FAO cells, insulin in, 95
Farnesylation, endoplasmic reticulum retained protein degradation and, 145
FAS genes, yeast vacuole and, 123

Fasting, selective proteolysis by lysosomes and, 55–56, 60
Fatty acid synthetase, yeast vacuole and, 123
Fcγ III receptor α-subunit, endoplasmic reticulum retained protein degradation and, 138
Feeding
autophagy and, 77–78, 81–83
lysosomes and, 56
Fetal calf serum, autophagy and, 80
Fibrinogen chains, endoplasmic reticulum retained protein degradation and, 138
Fibroblasts, hepatic endosomes and, 91–92
Flaviviridae, overview of, 200–202
FM3A mouse mammary carcinoma, ubiquitin-activating enzyme in, 26
FMDV. *See* Foot and mouth disease virus
Foot and mouth disease virus (FMDV), protease and, 193–194
Fructose biphosphatase, yeast vacuole and, 124–125, 128–129

GADPH. *See* Glyceraldehyde-3-phosphate dehydrogenase
β-Galactosidase
HMG-CoA reductase and, 144
lysosomes and, 58–59
ubiquitin and, 7–8, 13, 41–42
yeast vacuole and, 123–124
Gap1p protein, yeast vacuole and, 122
Genes
cytosolic antigen processing and, 172
endoplasmic reticulum and, 144–145, 152–153
HMG-CoA reductase, 144–145
ubiquitin, 11–12, 23–45
yeast vacuole and, 115–118, 120–125, 127–129
Glucagon
autophagy and, 82–83
hepatic endosomes and, 91, 96–106
Glucagonase, hepatic endosomes and, 91, 97
Gluconeogenesis, autophagy and, 65
Glucose, yeast vacuole and, 122, 124, 129
Glucose 6-phosphate dehydrogenase, yeast vacuole and, 127
Glutamate, autophagy and, 78
Glutamic acid
lysosomes and, 55–56
ubiquitin and, 7–8, 42
Glutamine
autophagy and, 77–80
lysosomes and, 55–56
ubiquitin and, 42
Glutathione, ubiquitin and, 12
Glutathione-insulin transhydrogenase
hepatic endosomes and, 106
Glyceraldehyde-3-phosphate dehydrogenase (GADPH), selective proteolysis by lysosomes and, 61

Glycine
  ubiquitin and, 4–5, 8–9, 11, 24
  yeast vacuole and, 123
Glycogen, autophagy and, 68–69, 74–75
Glycogenosomes, autophagy and, 83
Glycoproteins, endoplasmic reticulum retained protein degradation and, 140
Golgi apparatus
  autophagy and, 70
  hepatic endosomes and, 90–91, 94
  yeast vacuole and, 120, 122, 125, 128
Golgi fractions, hepatic endosomes and, 93, 95
Gorbalenya model, for 3C proteases, 185–187
GTP. See Guanosine triphosphate
GTP-binding proteins, hepatic endosomes and, 99–100
GTPγS, endoplasmic reticulum retained protein degradation and, 149–150
Guanosine triphosphate (GTP)
  autophagy and, 68
  hepatic endosomes and, 99

H^+-ATPase
  hepatic endosomes and, 94, 97
  yeast vacuole and, 127
Hansenula polymorphis, autophagy in, 127
HCV-encoded serine proteases, flavivirus and, 201–202
Heart
  autophagy and, 67, 83
  lysosomes and, 55, 60
Heat, cellular proteins and, 3
Heat shock cognate protein, of 73 kDa, selective proteolysis by lysosomes and, 55, 60–62
Heat shock 70 kDa proteins (HSP70s)
  lysosomes and, 60–62
  yeast vacuole and, 125
HeLa cells
  hepatic endosomes and, 91, 92
  lysosomes and, 57
Hemoglobin
  lysosomes and, 57
  ubiquitin and, 13
Hepatocytes
  autophagy and, 66–69, 71, 75, 78–80, 82
  endoplasmic reticulum and, 145
  hepatic endosomes and, 91, 99
HepG2 cells, hepatic endosomes and, 91
β-Hexosaminidase
  autophagy and, 71
  endoplasmic reticulum and, 138
HHR6 series genes, ubiquitin and, 29
Histidine
  autophagy and, 77–78
  ubiquitin and, 7–8
Histones, ubiquitin and, 3, 12, 26, 36
HIV-1. See Human immunodeficiency virus type 1

HIV-1 protease, crystal structure of, 205
HMG-CoA reductase
  endoplasmic reticulum and, 138–139, 142–145, 152
  yeast vacuole and, 122
HPV. See Human papilloma virus
HSP70s. See Heat shock 70 kDa proteins
Human immunodeficiency virus type 1 (HIV-1)
  as aspartic protease, 203–204
  Vpu gene product of, 148–149
Human papilloma virus (HPV), ubiquitin and, 14
Hydrolases
  autophagy and, 66–67, 69–70
  hepatic endosomes and, 89
  lysosomes and, 56, 61
  ubiquitin C-terminal, 11–13, 19, 40–41
  yeast vacuole and, 116, 119, 121, 129
l-α-Hydroxyisocaproate, autophagy and, 79
p-Hydroxymercuribenzoic acid, hepatic endosomes and, 101, 103

IDE. See Insulin-degrading enzyme
IFN-γ. See Interferon γ
IgE receptors, ubiquitin and, 16–17
IGF-I. See Insulin-like growth factor I
IGF-II. See Insulin-like growth factor II
IgG, selective proteolysis by lysosomes and, 59–60
IgM, endoplasmic reticulum retained protein degradation and, 139–140
Imidazole, endoplasmic reticulum retained protein degradation and, 151
Immunoglobulins, endoplasmic reticulum retained protein degradation and, 138–139
IMR-90 fibroblasts, selective proteolysis by lysosomes and, 57
Influenza hemagglutinin, endoplasmic reticulum retained protein degradation and, 138
Insulin
  autophagy and, 80–83
  hepatic endosomes and, 91–96, 99, 101–108
  lysosomes and, 57–58
Insulinase. See Insulin-degrading enzyme
Insulin-degrading enzyme (IDE), hepatic endosomes and, 91, 94, 96, 101–108
Insulin-like growth factor I (IGF-I), hepatic endosomes and, 103, 105
Insulin-like growth factor II (IGF-II), hepatic endosomes and, 105
Insulin protease. See Insulin-degrading enzyme
Interferon γ (IFN-γ)
  cytosolic antigen processing and, 172
  macropain and, 17
Iodoacetamide, endoplasmic reticulum retained protein degradation and, 142
^{125}I-Iodotyramine-cellobiitol, autophagy and, 68, 75–76
Ionomycin, endoplasmic reticulum retained protein degradation and, 144

Ionophores, hepatic endosomes and, 94
IOTA gene, ubiquitin and, 33
Isocitrate lyase, yeast vacuole and, 124
Isoleucine
    autophagy and, 80
    ubiquitin and, 8
Isopeptidases, ubiquitin and, 4
Isoproterenol, autophagy and, 83
Isovalerylcarnitine, autophagy and, 79

Jun protein, ubiquitin and, 15

K^+, autophagy and, 77
Karmellae, yeast vacuole and, 122
Kex2 protease, yeast vacuole and, 121–122
^{3}H-KFERA, selective proteolysis by lysosomes and, 59
KFERQ
    lysosomes and, 55–56, 58–62
    yeast vacuole and, 125
Kidneys
    autophagy and, 69, 77
    hepatic endosomes and, 99
    lysosomes and, 55, 60

β-Lactamase-α-globin chimera, endoplasmic reticulum retained protein degradation and, 138
Lactate dehydrogenase, autophagy and, 69
Lamp 1 protein, antigen catabolism and, 167, 169
LAP4 gene, yeast vacuole and, 116, 121
LDLs. *See* Low-density lipoproteins
Leu-βgal, ubiquitin and, 9
Leucine
    autophagy and, 66, 77–81
    lysosomes and, 59, 62
    ubiquitin and, 7–8
    yeast vacuole and, 123
Leucine aminopeptidase, hepatic endosomes and, 91–92
Leucyl-tRNA, autophagy and, 79
Leupeptin
    antigen catabolism and, 167, 170
    autophagy and, 69
    endoplasmic reticulum and, 141, 151
    hepatic endosomes and, 94
    lysosomes and, 61
Ligation system, ubiquitin and, 25–37
Lipoprotein-filled vesicles, hepatic endosomes and, 92–93
Liposomes, antigen catabolism and, 169
Listeria spp., antigen processing and, 171
Liver
    autophagy and, 66–70, 72–74, 76–83
    endoplasmic reticulum and, 150
    endosomes in, 89–108
    lysosomes and, 55–60
LMP genes
    macropain and, 17
    ubiquitin and, 33, 40

LMP proteins, cytosolic antigen processing and, 172
Long-lived protein degradation, autophagy and, 66
Low-density lipoproteins (LDLs), endoplasmic reticulum retained protein degradation and, 138
Low-density vesicles, hepatic endosomes and, 93
Lysine
    lysosomes and, 55–56
    ubiquitin and, 4, 6–9, 11–12, 14, 24–25, 30, 37, 41, 43–44
Lysosomal proteolysis inhibitors, in degradation of endoplasmic reticulum retained proteins, 141
Lysosomal-vacuolar pathway, autophagy and, 67–68, 70–73
Lysosomes
    antigen catabolism and, 167–171
    endosomes and, 89–91
    selective proteolysis by
        activation of lysosomal proteolytic pathways, 56–57
        direct protein transport and, 56
        endocytosis and, 55
        exocytosis and, 55
        future research on, 62
        generality of pathway, 59–60
        heat shock 70 kDa proteins and, 60
        introduction to, 55
        macroautophagy and, 56
        microautophagy and, 55–56
        proteolysis pathways and, 55–57
        ribonuclease A and, 57–59
        selective degradation and, 60–62
        selectivity and, 57
        targeting signal and, 57–59
Lysozyme
    endoplasmic reticulum and, 138
    lysosomes and, 57–58, 60
    ubiquitin and, 7, 10

Macroautophagy
    amino acid control of, 77
    deprivation-induced breakdown of protein and RNA, 70–72
    hormonal regulation and, 80–83
    intralysosomal pools of degradable protein and RNA, 72–73
    selective proteolysis by lysosomes and, 56–57
    vacuole formation and, 67–69
    vacuole maturation and, 69–71
Macropain (MCP), ubiquitin and, 10–11, 17–18
Macrophages, hepatic endosomes and, 91–92
Major histocompatibility complex (MHC)
    endoplasmic reticulum and, 138
    proteasomes and, 40
    protein catabolism and, 163–173
    ubiquitin and, 17–18, 40

Malate dehydrogenase, yeast vacuole and, 124
Mannose-BSA, hepatic endosomes and, 91–92
Mannose 6-phosphate
  antigen catabolism and, 167, 169
  hepatic endosomes and, 91
α-Mannosidase, yeast vacuole and, 115, 119, 121
MAP multiple antigen peptide, autophagy and, 80
MATα2 repressor, ubiquitin and, 15–16, 44
Maturation cleavage, viral proteases and, 192
MC13 gene, ubiquitin and, 33
MCP. *See* Macropain
Mechanism-based inhibitors, viral proteases and, 210
Membrane proteins, yeast vacuole and, 122–123
β-Mercaptoethanol, endoplasmic reticulum retained protein degradation and, 151
Metallo-endopeptidases, hepatic endosomes and, 105
Methionine
  autophagy and, 77–78
  ubiquitin and, 8
Methionine aminopeptidase, ubiquitin and, 9
3-Methyladenine, autophagy and, 68
Methylamine, hepatic endosomes and, 98
3-Methylhistidine, autophagy and, 65
Mevalonate, endoplasmic reticulum retained protein degradation and, 144–145, 150
MgATP, endoplasmic reticulum retained protein degradation and, 151
MHC. *See* Major histocompatibility complex
Michaelis-Menten kinetics, autophagy and, 80
Microautophagy
  hormonal regulation and, 80–83
  mechanism of, 73–75
  lysosomes and, 55–57
Microbial proteases, antigen catabolism and, 166
Microfilament disruption, autophagy and, 69
Mitochondria
  autophagy and, 69
  lysosomes and, 56, 62
  yeast vacuole and, 129
Molecular docks, ubiquitin and, 37
Monensin, hepatic endosomes and, 94, 98
Mosaic model, for endoplasmic reticulum retained protein degradation, 139
Mos protooncogene product, ubiquitin and, 44

Na$^+$, autophagy and, 77
Na$^+$-H$^+$ transporter, autophagy and, 81
Na$^+$,K$^+$-ATPase, hepatic endosomes and, 98, 108
NADP-dedendent glutamate dehydrogenase, yeast vacuole and, 125
N-ethylmaleimide
  endoplasmic reticulum and, 142
  hepatic endosomes and, 94, 97–99, 101
  yeast vacuole and, 127
N-end rule
  ubiquitin and, 7, 9, 36, 42–43
  yeast vacuole and, 123–124

Neurospora crassa, insulin-degrading enzyme in, 102, 105
N-glycan, endoplasmic reticulum retained protein degradation and, 140
NH$_2$-terminal domain, of E2 enzymes, 31, 35–36
Nigericin, hepatic endosomes and, 94
Nitrogen starvation, yeast vacuole and, 125–126, 129
N-myc oncoprotein, ubiquitin and, 14
NS2 protein, togavirus and, 195
NS3-N-terminal domain serine proteases, flavivirus and, 201
N-terminal domain, ubiquitin and, 3, 9, 18
Nucleus, cellular, selective proteolysis by lysosomes and, 56, 62

Oleate, apolipoprotein B and, 146
Oncoproteins, ubiquitin and, 14–15, 19
Oocytes, hepatic endosomes and, 91
Ornithine decarboxylase
  autophagy and, 69
  ubiquitin and, 11, 15
Osmotic shock, antigen processing and, 171
Ouabain, hepatic endosomes and, 98
Ovalbumin
  lysosomes and, 57, 60-61
  ubiquitin and, 18

P34^{cdc2} protein kinases, ubiquitin and, 16
p53 tumor suppressor protein, ubiquitin and, 14–15
Pancreas, autophagy and, 69, 71
Papain, togavirus and, 195
Parathyroid hormone (PTH)
  endoplasmic reticulum and, 139
  hepatic endosomes and, 91–92, 108
Parenchyma, liver, endosomes and, 92–99
PAS genes
  ubiquitin and, 6, 29, 31
  yeast vacuole and, 128
pCMB
  endoplasmic reticulum and, 151
  hepatic endosomes and, 98, 101
PDGF. *See* Platelet-derived growth factor
PEP carboxykinase, yeast vacuole and, 124
PEP4 gene, yeast vacuole and, 116–117, 120–123, 125, 128–129
Pepstatin, hepatic endosomes and, 94
Peptidases, selective proteolysis by lysosomes and, 58
Peptide recognition protein, of 73kDa, 60
Peptidylglutamylpeptide, endoplasmic reticulum retained protein degradation and, 153
Permeabilized cells, endoplasmic reticulum retained protein degradation and, 149–151
Permeases, yeast vacuole and, 122, 128
Peroxisomes
  autophagy and, 68–69

lysosomes and, 56, 62
ubiquitin and, 6
yeast vacuole and, 128–129
PEST sequences, ubiquitin and, 43
PGP 9.5 gene, ubiquitin and, 12, 40
Phagolysosomes, antigen catabolism and, 169
Phagophore, selective proteolysis by lysosomes and, 56
Phagosomes, antigen catabolism and, 169
1,10-Phenanthroline
    endoplasmic reticulum and, 142, 151
    hepatic endosomes and, 94, 98, 101, 103
Phenotypic lag, yeast vacuole and, 120
Phenylalanine
    autophagy and, 66, 77–78
    lysosomes and, 55–56
    ubiquitin and, 7–8
Phenylephrine, autophagy and, 82
Phenylmethanesulfonyl fluoride (PMSF)
    endoplasmic reticulum and, 151
    hepatic endosomes and, 94, 98–99
    yeast vacuole and, 127
PHO9 gene, yeast vacuole and, 116
Phospholipids, acidic, endoplasmic reticulum retained protein degradation and, 151
Phytochrome, ubiquitin and, 13–14
Pichia pastoris, autophagy in, 127
Picornaviridae, overview of, 183–194
Plasma membrane
    antigen catabolism and, 169
    autophagy and, 79–80
    glucagon and, 99–106
    hepatic endosomes and, 90–94, 99, 101
    insulin and, 99, 101–106
    yeast vacuole and, 122
Platelet-derived growth factor (PDGF), ubiquitin and, 16–17
Pma1p protein, yeast vacuole and, 122
PMSF. *See* Phenylmethanesulfonyl fluoride
Pol II-mediated transcription, inhibition of, 192
Pol III-mediated transcription, inhibition of, 192
Poliovirus, replication of, 184
Polyamines, ubiquitin and, 12
Polyphosphate, yeast vacuole and, 115
Potassium chloride, hepatic endosomes and, 98
PRA genes, yeast vacuole and, 116–117, 119, 121–122, 125
PRB genes, yeast vacuole and, 116–117, 119–123, 127
PRC1 gene, yeast vacuole and, 116, 123
PRE genes
    ubiquitin and, 32, 39–40
    yeast vacuole and, 117–118, 124
Pre-Golgi degradation, endoplasmic reticulum and, 138
Prepro-α-factor, endoplasmic reticulum retained protein degradation and, 138

PRG1 gene
    ubiquitin and, 32
    yeast vacuole and, 118
Primary destabilizing residues, ubiquitin and, 42
Procaine, hepatic endosomes and, 99
Proinsulin, hepatic endosomes and, 103
Prolactin, hepatic endosomes and, 94–96
Proline
    autophagy and, 77–78
    ubiquitin and, 8, 35
Pronase, antigen catabolism and, 166
Proofreading function, of hydrolases, 11
PROS genes, ubiquitin and, 32–33
Proteases
    antigen catabolism and, 166–168
    endoplasmic reticulum and, 141–142, 152
    hepatic endosomes and, 91–92, 99, 101–102, 108
    lysosomes and, 58
    ubiquitin and, 6, 9–11, 13, 15, 18–19, 23
    viral, 183–215
    yeast vacuole and, 115–116, 121–122, 129
Proteasomes
    antigen processing and, 169, 172
    endoplasmic reticulum and, 153
    ubiquitin and, 10, 32–34, 38–40
    yeast vacuole and, 115, 117–119, 123–124
Proteinase A, endoplasmic reticulum retained protein degradation and, 152–153
Proteinases
    hepatic endosomes and, 89
    yeast vacuole and, 115, 119–120, 127
Protein catabolism
    antigen processing and
        antigen catabolism and, 166–171
        cytosolic antigen proteolysis and, 172
        introduction to, 163
        MHC molecules and, 164–173
        processed peptides and, 165–166
        subcellular compartments and, 167–171
        transport and, 172–173
Protein disulfide isomerase
    endoplasmic reticulum and, 151
    hepatic endosomes and, 106–107
Protein kinases, ubiquitin and, 16
Protein ligases, ubiquitin and, 4–7, 37
prp73 protein, selective proteolysis by lysosomes and, 60
PRS genes
    ubiquitin and, 32
    yeast vacuole and, 117–118
PTH. *See* Parathyroid hormone
PUP genes
    ubiquitin and, 32
    yeast vacuole and, 118
Pyruvate kinase, selective proteolysis by lysosomes and, 57, 60
Pyruvate/lactate, autophagy and, 78

QSAR studies, viral proteases and, 211–212

RAALGNISN box, cyclins and, 16
RAD6 gene, ubiquitin and, 9, 27–28, 30
^{3}H-Raffinose, selective proteolysis by lysosomes and, 59
RC genes, ubiquitin and, 38–39
Red cell-mediated microinjection, selective proteolysis by lysosomes and, 57
Reductases, hepatic endosomes and, 92
Relaxin, hepatic endosomes and, 103, 105
Resident protein degradation, autophagy and, 66
Reticulocytes, ubiquitin and, 7–10, 12–14, 18, 25, 42
Retinol binding protein, endoplasmic reticulum retained protein degradation and, 138
Retroviridae, overview of, 202–213
rhp6 gene, ubiquitin and, 29
Ribonuclease A, selective proteolysis by lysosomes and, 55, 57–59, 61–62
Ribonuclease S, selective proteolysis by lysosomes and, 55, 57–61
Ribonucleases
  ubiquitin and, 7–8
  yeast vacuole and, 119
Ribophorin I, endoplasmic reticulum retained protein degradation and, 138
Ribosomal extension protein, ubiquitin and, 12
Ribosomes
  autophagy and, 68–69, 72, 74
  ubiquitin and, 11
Ricin A-chain, hepatic endosomes and, 91
RING genes, ubiquitin and, 33–34, 38, 40
RN3 gene, ubiquitin and, 33
RNA
  autophagy and, 66
  deprivation-induced breakdown of, 70–72
Rough endoplasmic reticulum, autophagy and, 68–69
Rous sarcoma virus (RSV), viral proteases and, 203–205
RSV. *See* Rous sarcoma virus

Saccaromyces cerevisiae
  α_1-antitrypsin and, 152
  hepatic endosomes and, 105
  HMG-CoA reductase and, 144
  UBC6 gene and, 153
  ubiquitin and, 9, 11, 15, 23, 27–32, 39, 44
  vacuole in, 115, 126–127
Sby gene, ubiquitin and, 28
Schizosaccharomyces pombe, ubiquitin and, 29
SCL1 gene, yeast vacuole and, 118
scl1+ gene, ubiquitin and, 32
sec mutant, yeast vacuole and, 125
SEC61 gene, endoplasmic reticulum retained protein degradation and, 153
Seminal vesicle cells, autophagy and, 71

Serine, ubiquitin and, 8, 35, 42, 44
Serine protease inhibitors
  endoplasmic reticulum and, 142, 151
  hepatic endosomes and, 94, 98–99
Serine proteases, viral proteases and, 195–202
SFV core protein, togavirus and, 195–200
Signal peptidase, endoplasmic reticulum retained protein degradation and, 151
SIN core protein, togavirus and, 195–200
Skeletal muscle
  autophagy and, 77, 83
  lysosomes and, 60
Smooth endoplasmic reticulum, autophagy and, 74–75
Sodium cholate, endoplasmic reticulum retained protein degradation and, 151
Staphylococcus aureus, α-toxin from, 75–76
Starvation
  autophagy and, 65, 74–78, 81, 83
  cellular proteins and, 3–4
  lysosomes and, 56
  yeast vacuole and, 125–126, 129
Ste2p protein, yeast vacuole and, 128
Ste3p protein, yeast vacuole and, 128
Sterols, endoplasmic reticulum retained protein degradation and, 144
Streptolysin O, endoplasmic reticulum retained protein degradation and, 150
Substrate-based inhibitors, viral proteases and, 208–210
Sulfhydryl endopeptidase, endoplasmic reticulum retained protein degradation and, 152
Synthetic dominant lethality, ubiquitin and, 35

$Ta\alpha$ gene, ubiquitin and, 32
$Ta\beta$ gene, ubiquitin and, 32
Tac-TCR-α chimeric protein, endoplasmic reticulum retained protein degradation and, 150
TAP1-TAP2 transporter, antigen processing and, 172–173
TAS-g64 gene, ubiquitin and, 33
TCA. *See* Trichloroacetic acid
T-cell receptors
  endoplasmic reticulum retained protein degradation and, 138, 141–142, 146–148
  ubiquitin and, 16–17
T-dependent antigens, protein catabolism and, 163
Testes, selective proteolysis by lysosomes and, 60
Tetrahymena pyriformis, autophagy in, 65
TEV. *See* Tobacco etch virus
TGF-α. *See* Transforming growth factor
Thapsigargin, endoplasmic reticulum retained protein degradation and, 144
Thermolysin, hepatic endosomes and, 105
Thermoplasma acidophilum, proteasomes in, 115
Thiolase, yeast vacuole and, 128
Thiol endopeptidase, hepatic endosomes and, 101
Thiol esters, ubiquitin and, 4–6, 12, 27, 35

Thiol metalloendopeptidase, hepatic endosomes and, 101
Thiol proteases
  antigen catabolism and, 166
  hepatic endosomes and, 100
  togavirus and, 195
Thiols, ubiquitin and, 11
Threonine, ubiquitin and, 8, 42
Thyrotrophic hormones, endoplasmic reticulum retained protein degradation and, 141
TLCK. See N-Tosyl-L-lysine chloromethyl ketone
Tobacco etch virus (TEV), 49-kDa protease and, 192–193
Togaviridae, overview of, 194–200
N-Tosyl-L-lysine chloromethyl ketone (TLCK), endoplasmic reticulum retained protein degradation and, 142, 150
N-Tosyl-L-phenylalanine chloromethyl ketone (TPCK), endoplasmic reticulum retained protein degradation and, 142, 150
TPCK. See N-Tosyl-L-phenylalanine chloromethyl ketone
Transferrin receptor, endoplasmic reticulum retained protein degradation and, 138
Transforming growth factor α (TGF-α), hepatic endosomes and, 103–105
Trans-Golgi network, hepatic endosomes and, 89, 91
Trans-targeting, ubiquitin and, 41
Trehalase, yeast vacuole and, 115, 119, 121, 128
Trichloroacetic acid (TCA), hepatic endosomes and, 92, 94–97
Tripeptidyl aminopeptidase, hepatic endosomes and, 99, 101
Trypsin, selective proteolysis by lysosomes and, 61
Trypsin-like protease, hepatic endosomes and, 91–92
Tryptophan
  autophagy and, 77–78
  ubiquitin and, 7–8
ts85 cells, ubiquitin-activating enzyme in, 26
Tumor suppressor proteins, ubiquitin and, 14–15
Tunicamycin, endoplasmic reticulum retained protein degradation and, 140
Tyrosine
  autophagy and, 66, 77–80
  ubiquitin and, 7–8
Tyrosine aminotransferase, autophagy and, 69
Tyrosine kinase, hepatic endosomes and, 108

UBA genes, ubiquitin and, 28
UBC genes, ubiquitin and, 6, 9, 12–13, 15, 27–31, 44
UBC6 enzyme, endoplasmic reticulum retained protein degradation and, 153
UBE genes, ubiquitin and, 28

UBI4 gene, yeast vacuole and, 124
Ubiquitin
  autophagy and, 66, 70
  cytosolic antigen processing and, 172
  endoplasmic reticulum and, 153
  lysosomes and, 60
  molecular genetics of
    cell cycle regulators, 43–44
    conservation and organization of ubiquitin genes, 23–24
    degradation signals, 41–42
    future research, 44–45
    introduction to, 23
    ligation system, 25–37
    MATα2 repressor, 44
    mutants, 24–25
    N-end rule pathway, 42–43
    PEST sequences, 43
    proteasomes, 38–40
    proteasome subunit genes, 32–34
    substrate recognition, 41–44
    RAD6 gene, 27–28, 30
    UBC genes, 27–31
    ubiquitin activating enzyme, 26
    ubiquitin conjugating enzymes, 26–31, 35–37
    ubiquitin C-terminal hydrolases, 40–41
    ubiquitin-like genes, 25
    ubiquitin protein ligases, 37
    UBR1 gene, 37
  proteolytic pathway mediated by
    activation of ubiquitin, 5–6
    α-amino group of protein substrate, 7–9
    cell surface receptors, 16–17
    cellular protein degradation in vivo, 13–18
    cyclins, 16
    general regulation of ubiquitin system, 18
    IgE receptors, 16–17
    introduction to, 3–5
    MATα2 repressor, 15–16
    MHC-restricted class I antigen presentation, 17–18
    non-NH$_2$-terminal signals in ubiquitin-mediated degradation, 9
    oncoproteins, 14–15
    overview of, 3
    phytochrome, 13–14
    platelet-derived growth factor receptor, 16–17
    substrate recognition by ubiquitin-protein ligases, 7–9
    T-cell receptors, 16–17
    ubiquitin-C-terminal hydrolases, 11–13
    ubiquitin-protein conjugate degradation, 9–11
  yeast vacuole and, 123–124, 127
UBP genes, ubiquitin and, 12, 40
UBR1 gene, ubiquitin and, 9, 29, 37, 42

Urea, endoplasmic reticulum retained protein degradation and, 151

Vacuoles
   alkaline phosphatase and, 121
   analog-containing proteins and, 126
   ApI precursor and, 121
   autophagy and, 66, 69, 126–128
   by-product peptides and, 121–122
   carbon source and, 124–125
   carbon starvation and, 125
   carboxypeptidases and, 119, 121
   delivery routes for vacuolar proteolysis, 128
   future research on, 128
   introduction to, 115
   macroautophagy and, 67–71
   membrane protein turnover and, 122–123
   metabolic change and, 124–128
   misfolded peptides and, 121–122
   mutant peptides and, 121–122
   nitrogen starvation and, 125–126
   normal vegetative growth and, 117, 120–124
   peptide metabolism and, 123
   PrA and, 117, 119
   PrB and, 117, 119–121
   precursor processing and, 117, 120–121
   proteasomes and, 115, 117–119
   stationary phase and, 124
   stoichiometry and, 123
   ubiquinated proteins and, 123–124
   vacuolar protease and, 115–116
Valine
   autophagy and, 66, 72–73, 76, 81
   ubiquitin and, 8
Vanadate, hepatic endosomes and, 98
Vasopressin, autophagy and, 82
Vegetative growth, yeast vacuole and, 117, 120–124
Very low-density lipoprotein (VLDLs), endoplasmic reticulum retained protein degradation and, 145–146
Vinblastine, autophagy and, 69
Viral proteases
   2A proteases and, 190–191
   active-site residues and, 205–207
   antipicornavirus therapy and, 194
   Bazan model and, 185–186
   3C proteases and, 185, 188–190, 192–194
   C2-symmetric inhibitors and, 210
   cellular enzymes in viral maturation and, 213–215
   convertases and, 213–214
   early translation and, 194–195
   Flaviviridae and, 200–202
   FMDV protease and, 193–194
   Gorbalenya model and, 185–187
   HIV-1 and, 203, 205
   inhibitor design and, 208–211
   introduction to, 183
   maturation cleavage and, 192
   mechanism-based inhibitors and, 210
   mechanisms and, 205–207
   mutation studies and, 212–213
   NS2 protein and, 195
   papain-like thiol protease and, 195
   Picornaviridae and, 183–194
   Pol II-mediated transcription and, 192
   Pol III-mediated transcription and, 192
   poliovirus relication and, 184
   post-translational effects on host cell protein synthesis, 191–192
   proteolysis and, 214–215
   QSAR studies and, 211–212
   Retroviridae and, 202–213
   Rous sarcoma virus and, 203–205
   serine proteases and, 195–202
   SFV core protein and, 195–200
   SIN core protein and, 195–200
   substrate-based inhibitors and, 208–210
   substrate-specificity studies and, 207–208
   TEV protease and, 192–193
   Togaviridae and, 194–200
   transition state models and, 205–207
Vitellogenin, hepatic endosomes and, 91
VLDLs. See Very low-density lipoproteins
Vpu protein, endoplasmic reticulum retained protein degradation and, 148–149

Weak bases, in degradation of endoplasmic reticulum retained proteins, 141
Western blot analysis, ubiquitin and, 10

$XC3$ gene, ubiquitin and, 33
Xenopus spp., ubiquitin and, 33, 38, 43

$Y7$ gene, ubiquitin and, 32
$Y8$ gene, ubiquitin and, 32
$Y13$ gene, ubiquitin and, 32, 39
Yeast
   endoplasmic reticulum and, 138
   ubiquitin and, 9, 11, 15, 23–31, 32, 39, 42, 44
   vacuoles in, 115–129
Yeast secretory pathway, nonlysosomal degradation in, 152–153
$YUH1$ gene, ubiquitin and, 40

$ZETA$ gene, ubiquitin and, 33
Zn^{2+} metallo-endopeptidase, endoplasmic reticulum retained protein degradation and, 151
ZPCK. See N-Carbenzoxy-L-phenylalanine chloromethyl ketone